Primates

This book is an accessible and comprehensive introduction to primates. It provides both a survey and synthesis of primate history, biology, and behavior. As a survey, it offers a focused review of living and extinct primates in regional and community frameworks. As a synthesis, it applies the community perspective in a unique way to explore primates' adaptive diversity in the context of how evolution works. The book encourages students to study primates as integrated members of regional communities, ecologically, historically, and evolutionarily.

The chapters are organized to emphasize the patterns of primate radiations in the four regions of the world where primates live, and to facilitate comparisons among the radiations. The overviews of communities illustrate how the ecological adaptations of different species and taxonomic or phylogenetic groups enable them to coexist. Illustrations and tools to aid students' learning include case studies, photographs, figures, tables, charts, key concepts, and quizlets to self-test.

This book is an ideal introduction for students studying nonhuman primates, primatology, primate behavior, or primate ecology.

Alfred L. Rosenberger, PhD, is Professor Emeritus at Brooklyn College and the City University of New York Graduate Center, New York, USA, where he taught in the anthropology departments for more than a decade. Dr. Rosenberger is a Fulbright Fellow and an elected Fellow of the American Association for the Advancement of Science. He has authored more than 100 articles and co-edited several volumes on living and fossil primates. Rosenberger's definitive book *New World Monkeys: The Evolutionary Odyssey* was published by Princeton University Press in 2020.

"Rosenberger offers an exciting new take on a survey of nonhuman primates, focusing on continental radiations and the biology and evolutionary history that unites them. In so doing, he provides a rigorous discussion of the anatomical, behavioral, and ecological features that differentiate primate taxa but within the familiar framework of geography rather than taxonomy, which is likely more palatable for undergraduates. A unique and effective approach!"

Larissa Swedell, *Professor and Chair, Department of Anthropology, Queens College*

"This book is a must read for undergraduate and graduate students seeking to understand how evolution, ecology, and adaptation have shaped the behavior and biology of our closest living relatives, the nonhuman primates. Information on all seven major primate radiations is presented in a concise and understandable format, scientific concepts are clearly defined and explained, and examples relating primate form and function serve to illustrate the diverse ways that individual species exploit their social and ecological environments. Each chapter also includes a set of questions for discussion. The volume ends with the sober realization that the majority of primate species are threatened with extinction, along with the optimistic message that if we choose to live sustainably and act now, we can save lemurs, lorises, tarsiers, galagos, monkeys, and apes from extinction."

Paul A. Garber, *Professor Emeritus, Department of Anthropology, University of Illinois*

Primates
An Introduction

Alfred L. Rosenberger

Routledge
Taylor & Francis Group
LONDON AND NEW YORK

Designed cover image: © Clive Horton

First published 2024
by Routledge
4 Park Square, Milton Park, Abingdon, Oxon OX14 4RN

and by Routledge
605 Third Avenue, New York, NY 10158

Routledge is an imprint of the Taylor & Francis Group, an informa business

© 2024 Alfred L. Rosenberger

The right of Alfred L. Rosenberger to be identified as author of this work has been asserted in accordance with sections 77 and 78 of the Copyright, Designs and Patents Act 1988.

All rights reserved. No part of this book may be reprinted or reproduced or utilised in any form or by any electronic, mechanical, or other means, now known or hereafter invented, including photocopying and recording, or in any information storage or retrieval system, without permission in writing from the publishers.

Trademark notice: Product or corporate names may be trademarks or registered trademarks, and are used only for identification and explanation without intent to infringe.

British Library Cataloguing-in-Publication Data
A catalogue record for this book is available from the British Library

ISBN: 9781032189918 (hbk)
ISBN: 9781032189932 (pbk)
ISBN: 9781003257257 (ebk)

DOI: 10.4324/9781003257257

Typeset in Sabon
by codeMantra

This book is dedicated to my grandson, Jude Muskin Geramita, a future steward of our planet.

Contents

List of figures   *x*
List of tables   *xiii*
List of boxes   *xiv*
Acknowledgments   *xv*
Preface   *xvi*

1 What is a primate?   1

2 Arboreal frugivory: the primate adaptive zone   28

3 Madagascar: lemurs   41

4 South America: New World monkeys   69

5 Africa: lorises, galagos, Old World monkeys, and great apes   103

6 Asia: lorises, tarsiers, Old World monkeys, and apes   144

7 Primate communities compared: ecology, morphology, and behavior   167

8 The primate fossil record: highlights   199

9 Primates in crisis   221

Glossary   224
About the author   232
Index   233

Figures

1.1	Geographical distribution of the seven major groups of living primates	3
1.2	Relative sizes of the four main regions and the number of primate genera per region	8
1.3	An indri, Madagascar's largest lemur	14
1.4	A Southeast Asian slow loris	15
1.5	A galago from mainland Africa	16
1.6	A tarsier from Southeast Asia	17
1.7	A saki monkey from South America	18
1.8	A male gelada and his troop from mainland Africa	19
1.9	A mountain gorilla from mainland Africa	20
1.10	Cladogram of the seven major groups of living primates	22
1.11	The 25 most endangered species and subspecies of primates	25
2.1	Open-mouthed baboon showing the standard four tooth groups of the primate dentition	30
2.2	A young baboon riding on the back of its mother	31
2.3	A South American capuchin monkey handling a large fruit	32
2.4	An Asian langur leaping between trees and clutching her infant	33
2.5	A South American howler monkey asleep on a branch	34
2.6	The grasping feet and hands of three primates	36
3.1	Map of Madagascar highlighting the past and remaining rainforests	42
3.2	A fat-tailed, hibernating dwarf lemur	44
3.3	Cladogram and classification of the four living lemur families	45
3.4	Diagram illustrating the position and reflecting function of the tapetum lucidum	46
3.5	Close-up photographs of the rhinarium and toothcomb	47
3.6	Diagram illustrating how streams of dry and wet odor-carrying molecules enter the strepsirhine nose	48
3.7	The smallest living primate, Madame Berthe's mouse lemur	54
3.8	A bark-gouging fork-marked lemur and its specialized cranium, dentition, and hands and feet	57
3.9	A semi-terrestrial ring-tailed lemur	60
3.10	An aye-aye tap-scanning for grubs, its specialized cranium, and long-fingered hand	63
3.11	Threat levels of Endangered and Critically Endangered lemur species in Madagascar	67
4.1	Map of the neotropics highlighting the distribution of rainforests	70

Figures xi

4.2	Facial expressions of capuchin monkeys	73
4.3	The skulls and faces of a lemur and anthropoid are compared to emphasize differences in the snouts, orbits, and eyes	74
4.4	The crania and faces of a platyrrhine and catarrhine are compared to highlight their internal anatomical differences and external nose shapes	75
4.5	Cladogram and classification of the three living platyrrhine families and the six subfamilies	77
4.6	A spider monkey exhibiting the long prehensile tail with its gripping friction pad underneath	78
4.7	The sexually dimorphic fur and face of the white-faced saki monkey	80
4.8	A family group of owl monkeys twining their tails together while resting	89
4.9	The elaborate facial and fur patterns of four callitrichine species	94
4.10	Threat levels of Endangered and Critically Endangered species from three of six subfamilies of New World monkeys	100
5.1	Map of mainland Africa highlighting the distribution of rainforests	105
5.2	Cladogram and classification of the families and subfamilies of galagos and lorises living in Africa and Asia	107
5.3	Cladogram and classification of the families and subfamilies of Old World monkeys and apes living in Africa and Asia	110
5.4	Photographs of Old World monkey molars showing the bilophodont crown pattern shared by animals with different diets	113
5.5	A monkey whose cheek pouches are stuffed with fruit	114
5.6	The lip flip gesture of a female gelada	116
5.7	A comparison of skeletal structure in a terrestrial baboon, semi-terrestrial chimpanzee, and arboreal gibbon	118
5.8	The massive cranium of an adult male gorilla	120
5.9	A quadrumanous potto moving through the small-branch setting in the tree canopy	125
5.10	Photographic portraits of four Tai Forest monkeys	127
5.11	Chart showing how differences in body size, degree of terrestriality, and diet contribute to niche differentiation among seven Tai Forest monkey species	128
5.12	Terrestrial gelada monkeys living in the rugged mountainous terrain of the Ethiopian highlands	131
5.13	Threat levels of Endangered and Critically Endangered species of primates living in Africa and Afroasia	141
6.1	Map of South and Southeast Asia highlighting the distribution of rainforests	145
6.2	Mimicry in the appearance of the Javan slow loris and the spectacled cobra	149
6.3	Photographs of a tarsier in the wild and a tarsier skull	151
6.4	Chart showing how differences in body size, degree of terrestriality, and diet contribute to niche differentiation among Kuala Lompat monkeys, gibbons, and siamang	157
6.5	Sexual dimorphism in proboscis monkeys' nose size and shape	161
6.6	The faces of flanged and unflanged adult male orangutans	162
6.7	Threat levels of Endangered and Critically Endangered species of primates living in Asia and Afroasia	164

xii *Figures*

7.1	Profiles of tropical rainforest structure dominated by African angiosperms and Asian dipterocarps	171
7.2	A young bamboo lemur eating a bamboo stalk	173
7.3	A seed-eating uakari holding a fruit	174
7.4	A male silverback gorilla in knuckle-walking stance	175
7.5	A leaf-eating silvered leaf monkey	177
7.6	A long-legged, vertical-clinging-and-leaping sifaka and a long-armed, brachiating gibbon	185
7.7	Sexual dimorphism of skulls and canine size and shape in hamadryas baboons	190
7.8	Skulls and vocalization mechanisms of a male howler monkey and male spider monkey	192
7.9	Twins being carried by an adult common marmoset	193
7.10	Primates belonging to different regions and taxonomic groups that have similar social organizations	195
8.1	Molar teeth of four fossil primates from a single Eocene site in Libya	201
8.2	The fossil cranium of *Rooneyia*, a proto-anthropoid from the Late Eocene found in Texas, U.S.A.	209
8.3	The cranium of a female fossil *Propliopithecus* (also called *Aegyptopithecus*), the oldest stem catarrhine, from the Oligocene of Egypt	211
8.4	An owl monkey skull compared with the fossil *Tremacebus* from the Miocene of Argentina	212
8.5	Three crania of the subfossil lemur *Archaeolemur* from Madagascar	218
8.6	A subfossil cranium of *Antillothrix* found in a submerged cave in the Dominican Republic	218
9.1	Orphaned juvenile orangutans that live in a reintroduction shelter	223

Tables

1.1	Geographical distribution of the seven major groups of living primates	4
3.1	Ecological profiles of lemur genera in Madagascar	52
4.1	Ecological profiles of platyrrhines at Manu National Park in Peru	83
4.2	Comparison of the body sizes of platyrrhine genera from the Amazon basin and the Atlantic Forest in Brazil	98
5.1	Distinguishing features of living catarrhines and platyrrhines	111
5.2	The main ecological features of Tai Forest primates in Côte d'Ivoire	123
5.3	Traits that distinguish bonobos and common chimpanzees	134
6.1	General features describing the lorises, Old World monkeys, and apes that occur in both Asia and Africa	147
6.2	The main ecological characteristics of Kuala Lompat primates in Malaysia	156
7.1	Examples of convergent evolution in adaptations of primates living in the four geographical regions	179
7.2	Weights of the smallest and largest primate species in five major taxonomic groups	181
7.3	Examples of convergent evolution associated with sociality and mating among species with similar, independently evolved social organizations	196
8.1	Major events of primate history recorded in the fossil record	200
9.1	Factors driving wild primate populations toward extinction and the mitigating strategies	222

Boxes

1.1	Evolutionary principles: natural selection and selective pressure	5
3.1	Locomotion and the intermembral index	49
3.2	Sexual monomorphism and dimorphism	50
3.3	The dental formula	62
4.1	Vocal communication	79
5.1	Bonobos	134
5.2	Jane Goodall	136
5.3	Dian Fossey	139
7.1	Evolutionary principle: coevolution	169
7.2	Primates benefit from being social	188

Acknowledgments

I am particularly grateful to several friends and colleagues who have generously helped me with this textbook project: Anthony B. Rylands, Russell A. Mittermeier, and Stephen D. Nash. Stephen's brilliant artwork enriches this book, and his eagerness to share our enjoyment in the scientific illustration of primates helped me move things along. Anthony, as always, was a rock-solid supporter and font of knowledge shared. Russ helped me start and finish. I am also thankful to have had Eric Delson and Dawn Starin to call whenever I needed a sounding board, facts to be checked, a reference to be located, or an encouraging word.

I am grateful to friends and colleagues who read drafts of material and were helpful in other ways: Sylvia Atsalis, Ann Burrows, Elena Cunningham, Roberto Delgado, Martin Kowalewski, Marilyn Norconk, Leila Porter, Larissa Swedell, and Bernardo Urbani. Jacinta Beehner, Manfred Eberle, Jukka Jernvall, A. P. Levantis, Anna Nekaris, Timothy B. Rowe, and Sara Sorraia (née Clark) allowed me to use their personal photographs or images in their charge. The staff of the A. P. Levantis Ornithological Research Institute was very helpful in this regard as well.

Embedded in this book are lessons I have learned from colleagues with whom I have collaborated on research projects, too many to name. I thank them for sharing their knowledge and collegiality. Also, a salute to a generation (or more) of undergraduate and graduate students who allowed me the rewarding experience of teaching them about primates.

It would not have been possible for me to write this textbook without the wisdom, insight, scholar's pedagogical approach, editorial skills and talent, grace, wit, and support of my wife, Dr. Rivka "Suzie" Rosenberger. Suzie has been my shining inspiration, and I can never thank her enough.

Preface

The objective of this book is to introduce students to the great variety of the world's non-human primates and to highlight the ways in which primates have diversified in each of the four main regions of the world where they live today – Madagascar, the neotropics, mainland Africa, and Asia. Primate species live in social groups that belong to ecological communities that include other primate species. The Big Picture of primate evolution is a recurring pattern of differentiation that has enabled species to coexist by taking advantage of comparable resources in similar ways. This is how primates live in nature, and have for more than 55 million years.

The adaptations found among the hundreds of species of primates, dozens of genera, and many families take on numerous anatomical forms and behavioral expressions. But the adaptive radiations share a common pattern, how the animals differentiate ecological niches based principally on body size, diet, locomotion, and use of space, and how they are organized socially. The adaptations that make their daily lives possible are responses to living among other primates that have overlapping needs, actually or potentially. This recurring dynamic pattern, which is both a cause and consequence of primate diversity, is the pedagogical framework of this book.

This is a book about living primates. For the most part, the fossil evidence of extinct primates is presented as brief case studies of paleocommunities that existed in each of the four major regions where extant primates live. Additionally, one chapter is devoted to highlights of the fossil record, focusing primarily on remains that pertain to the origins of the seven major taxonomic groups – Madagascar's lemurs, neotropical New World monkeys, Afroasian lorises, African galagos, Afroasian Old World monkeys and apes, and Asian tarsiers.

Students will learn that phylogeny shapes adaptation, so having a grasp of history helps us make sense of what we see. The history that is evident in the diversity of primates is carried in their genes and is a product of their genealogical relationships. Closely related animals are destined to have comparable lifestyles, so that becomes a starting point for examining their similarities and differences.

This book also examines existential threats that primates all around the world are facing as their habitats are being decimated at an alarming rate, and looks at some of the efforts to save these animals.

1 What is a primate?

Chapter Contents

The science of primatology	1
Adaptive radiation of living primates	2
Taxonomy	5
Primate species coexist as communities in nature	6
Ecological opportunities	7
The biological diversity of primates	8
Ten factors that describe a primate profile	9
The seven major groups of primates at a glance	13
How the seven major groups of primates are related and taxonomically organized	21
The conservation status of primates today	24
Key concepts	26
Quizlet	26
Bibliography	27

The science of primatology

Primatology is the scientific study of primates and their evolution, meaning the genetically based changes of a population over time, often as adjustments or adaptations to their environment. It is an organized body of knowledge based on collections of verifiable facts, data, and the explanations of those facts, hypotheses, which are grounded in science and in theories about how the world works – so-called natural laws. Primatologists, the scientists who study primates, are concerned with all matters relating to documenting and understanding the nonhuman primates, which in this book we simply call primates. Primatologists look at primate biology, genetics, adaptations, ecology, behavior, reproductive patterns, methods of communication, fossils, and more. Data are gathered in the wild, in nature reserves, species-specific sanctuaries, zoos, museums, fossil sites, and in laboratories. This knowledge is also vitally important to conservationists aiming to preserve the lives of endangered primates and their threatened natural habitats, now and into the future.

Like all scientific endeavor, primatology is an ongoing enterprise, always developing and subject to change based on new information and innovative tools. Hypotheses are tested over and over again – that is the nature of the scientific method – challenged by experiments or observations, by devising new ways to look at an issue, or by discovering new facts. The results of testing may diminish a hypothesis or add confidence to its

DOI: 10.4324/9781003257257-1

2   *What is a primate?*

viability and robustness. The ability to test an idea or explanation using scientific methods is what distinguishes a hypothesis from an opinion or belief, no matter how logical it may seem.

It is important to note that in science we do more than try to explain what is behind observations or episodes in absolute terms. We search for patterns and trends, knowing that there are always variations and exceptions in species and biological systems. For example, while monkeys are typically **diurnal**, active during the day, decades-long studies of South American owl monkeys had shown them to be **nocturnal**, active at night. Yet, only recently has one population been discovered to be different from the other owl monkeys. They are **cathemeral**, meaning they are active both at night and during the day.

Primates are a highly varied group, yet there is a pattern in the ways in which they have diversified in the world. Their lifestyles have become adjusted to the environment in multiple ways via evolution, in a pattern that is repeated everywhere that primates live, in extant (living) primate communities: in Mexico, Central and South America (collectively called the neotropics), Madagascar, mainland Africa, and Asia. The pattern of primate diversification in size, diet, locomotion, social organization, and more far exceeds the varieties in appearance, behavior, and lifestyles exhibited by the few animals with which most people are familiar, such as chimpanzees and baboons.

The approach taken in this textbook's survey of living primates and the underlying evolutionary pattern that shapes their diversity is called **ecophylogenetic**. This means it is a synthesis of information about ecology (the relationships between organisms and their environment), phylogeny (the genetic or genealogical connections between species or other classified groups of organisms), and adaptation (genetically based adjustments to specific environmental conditions) – terms and concepts discussed in more detail later in this chapter and throughout the book. We will look at how various **species** of primates live together in communities in nature.

Primates consist of a very large number of species. There are more than 500 extant species belonging to 81 **genera** (plural for **genus**) according to the International Union for the Conservation of Nature (IUCN) Species Survival Commission (SSC) Primate Specialist Group, an international network of scientists devoted to cataloging the diversity of primates and working to promote their conservation.

In this book we follow a more conservative estimate of primate genera than the number recognized by the IUCN SSC Primate Specialist Group – 68, not 81 – although we follow their determinations regarding species in discussions of primate conservation.

Adaptive radiation of living primates

The majority of primates live in trees in tropical rainforests. **Arboreality**, tree-dwelling, is the central factor of a primate's existence in nature. Only some are **terrestrial**, ground-dwellers who spend a lot of time on the forest floor or in open country.

In this book we will take a close look at the seven major groups of the **adaptive radiation** of living primates in the world today, in four geographical regions, as well as some of their extinct relatives whose fossilized remains have been discovered (Figure 1.1). Adaptive radiation means the evolutionary development and proliferation, over time, of related species and genera into a variety of **ecological niches**. This is made possible by the evolution of various new adaptations as adjustments to specific environmental conditions. As adaptations accumulate, the genetic structure of a population changes and a

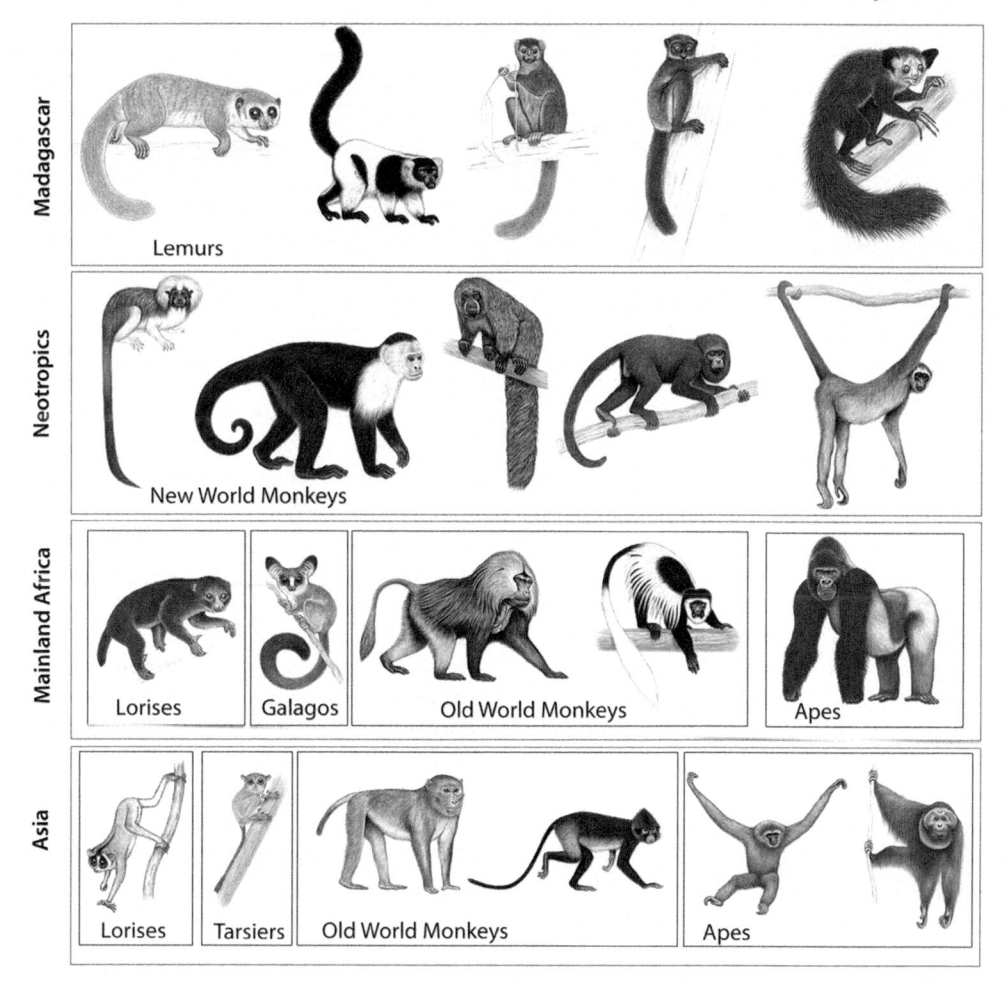

Figure 1.1 Geographical distribution of the seven major groups of living primates. Artwork by Stephen D. Nash, courtesy of Stephen D. Nash and the IUCN SSC Primate Specialist Group.

new species may be formed. This is the underlying evolutionary process that results in the origin of a new species, and often a radiation of new species. The process is called **speciation**.

Each of the seven primate groups that we will study represents a distinctive adaptive radiation. They are:

- Lemurs
- Lorises
- Galagos
- Tarsiers
- New World monkeys
- Old World monkeys
- Apes

4 *What is a primate?*

Table 1.1 Geographical distribution of the seven major groups of living **primates**

	Madagascar	Neotropics	Africa	Asia
Lemurs	√			
Lorises			√	√
Galagos			√	
Tarsiers				√
New World monkeys		√		
Old World monkeys			√	√
Apes			√	√

These seven groups are not all present in each of the four regions where primates live (Table 1.1). Lemurs are found only in Madagascar – and only lemurs and no other primates exist there. The same is true of New World monkeys, called platyrrhines. They are found only in the neotropics (in Mexico and South and Central America) – where no other primates exist. Three groups, lorises, Old World monkeys, and apes live both in mainland Africa and in Asia. The galagos are found only in mainland Africa. Tarsiers are found only in Southeast Asia, a region where other primates also live. As we shall see in later chapters, the geographical distributions of some groups have changed over time in significant ways as the world's geography and climate have changed.

A species (this word is both singular and plural) is a biological unit defined by the ability of its members to reproduce with one another in nature, exclusively, and not with other species. To ensure that the adult members of each species recognize one another, they exhibit a variety of traits that are species-specific. These traits generally result in a distinctive appearance which may also include a unique subset of features distinguishing males from females, or a certain pattern of vocalizations, characteristic odors, and more.

This definition of species, the one we use in this book, is known as the **biological species concept**. It emphasizes the process by which species arise, for example, if a subgroup of a population is exposed to new circumstances that isolate them and become a barrier to reproducing with other subgroups. The biological species concept is associated with geographical, ecological, morphological, behavioral, and genetic conditions or mechanisms that can influence species formation. These properties are useful tools for the purpose of identifying a species, living or extinct. It should be noted that scientists may have other, narrower definitions of the species concept that do not use the criterion of reproduction as a defining feature. (This sometimes leads to different classifications and varying estimates of diversity.) As individuals of a species mate, produce offspring, and pass on their genes, either evolutionary change occurs or evolutionary stability is maintained, from one generation to the next.

A genus is a scientific construct used to classify a collection of species that are uniquely related to one another and share a distinctive lifestyle because of their shared characteristics. While species are biological, dynamic evolutionary units, genera, in contrast, are categories used to convey information. Each genus is defined by its own pattern of adaptations reflecting the combination of traits among its included species. In some cases, a single species may be classified in its own genus if it is particularly distinctive. Primatologists may have different views on the sorting of primate populations into species and genera.

For example, the common chimpanzee and the bonobo are two species comprising the genus called *Pan*. These two primates resemble one another and have many traits in common. But they do not reproduce with one another. This makes them separate species. It is their similarities and close genetic relationship that leads them to be classified in one genus. The aye-aye of Madagascar presents a different example. It consists of a single species that is markedly different anatomically, behaviorally, and adaptively from all the other lemurs on Madagascar, so it is classified in its own genus.

Taxonomy

The practice of classification is called **taxonomy**, from the Greek word (*taxis*) meaning arrangement, and every indexed entity is called a **taxon** (plural: **taxa**). The genus is one of the main taxonomic categories (organizational units) used in **biological classification**. Biological classifications provide the formal scientific names of animals and other organisms specified and arranged in a special format of tiered, or hierarchical, categories. The species, followed by the genus, are the lowest tiers of the hierarchy.

The basis of all biological classifications is phylogenetic relatedness, or **phylogeny**, the genetic or genealogical connections between populations and taxonomic groups. The classification of primates considers multiple forms of phylogenetic information, including the DNA of living species and anatomical evidence based on fossil discoveries. Information drawn from ecology, behavior, and studies of adaptation involving many biological systems are also pertinent.

Phylogeny and adaptation are the two main kinds of evolutionary information embodied in classifications. They are inextricably tied and are evident at all levels of the classification hierarchy. That is because closely related groups share many unique genes that define them and determine adaptations in morphology, behavior, physiology, and more.

Phylogeny is both the context describing how primates are related to one another, and it is also the process by which they came to be related. Returning to the prior example of the common chimpanzee and the bonobo, two species of the genus *Pan*, we describe them as being more closely related to each other than either is to another species of African ape, the gorilla, genus *Gorilla*. That is their phylogenetic context. The process, phylogenetically, by which the first two species of the same genus became uniquely related was their descent from one common ancestor, while the gorilla descended from a different ancestor.

Similarly, adaptation refers to both a condition and a process. As a condition, adaptation is an inherited change of a trait that works best in relation to the environment in which the organism lives. As a process, it occurs over generational and geological time by affecting individuals' lifetime reproductive success, and ultimately the species, through evolution (Box 1.1).

Box 1.1 Evolutionary principles: natural selection and selective pressure

- Natural selection – the process by which adaptations evolve
- Selective pressure – ever-present environmental factors that drive natural selection

Natural selection is one of the most important concepts in evolutionary theory. It is the process by which adaptations evolve. This process works by way of differential reproduction, which means the increased number of offspring born to individuals with beneficial adaptive traits. The concept of natural selection was one of the central ideas developed by Charles Darwin in his book, *The Origin of Species*, published in 1859.

A related, critical evolutionary concept is selective pressure. Selective pressure refers to the ever-present environmental factors that drive natural selection. For example, the South American capuchin monkeys rely on a staple diet of soft fruits, but when these are not available in the dry season the monkeys have to expend more energy to eat a substitute of harder fruit. The shortage of the dietary staple imposes the selective pressure. Natural selection will favor those monkeys that possess genetically based traits that enable them to access substitute foods most efficiently. Such traits may include stronger jaws and teeth to bite into very hard-rind fruits. As a consequence, the animals with these traits are able to eat tougher fruit and will have a higher likelihood of surviving and reproducing. The increased number of their offspring – differential reproductive output – will be more likely to inherit the beneficial traits and promote their spread in following generations.

All of the world's primates share a single phylogenetic history and a core set of ecological adaptations. They occupy an exclusive **adaptive zone**, which may be thought of as an amplified ecological niche, made possible by a discrete set of anatomical and behavioral traits. **Arboreal frugivory** is the primate adaptive zone, which means they are adapted to living in trees and eating fruit. Arboreal means tree-dwelling. Frugivory means fruit-eating.

Primate species coexist as communities in nature

Each species occupies its own ecological niche and has a particular geographical distribution. The ecological niche refers to the position or role of a species in nature, how it relates to environmental conditions, local resources, and other coexisting organisms. All species belonging to a genus share similar ecological niches and lifestyles supported by a distinctive pattern of anatomy and behavior. These shared characteristics are derived from the exclusive sets of genes the species have in common because they are closely related. This is what sets each genus apart from others ecologically.

Genera also have a distinctive geography based on the spatial distributions of the species included in them. Because the geographical distribution of a multi-species genus is usually larger than any one of its species, a genus that is spread across a landscape often encounters differences in some aspects of the environment to which the species must adjust. As a result, there is more variation among the different species of one genus than there are differences among the individuals of any one species.

Primates do not live as isolates. All primate individuals live in social groups comprised of individuals that belong to the same species. Typically, social groups of several primate species live in **sympatry**, meaning more than one species occupies the same habitat at the same time because of their overlapping ranges. It is extremely rare for any primate species to be the only one that lives in a particular habitat. The coexistence of various species forms a **primate community** that occupies an area. In a community, each of the co-located species may be represented by more than one social group. Together, all the species in a primate community form a distinctive ecological system with its own resource needs and internal relationships.

There can be many primate communities in each of the major geographical regions where primates exist that are composed of different species. Within each region, the species composition of a community may vary in connection with habitat. For example, the wet forests of eastern Madagascar are occupied by a different composition of species than the dry areas in the west.

The majority of primate species live in habitats in the richest land-based **ecosystems** on the planet – the tropical and subtropical rainforests. The rainforests are home to a tremendous number and variety of plants and animals, making for complex interactions among them. Within these ecosystems, habitats and community compositions vary depending on abiotic and biotic components, non-biological elements including climate and physical features like mountains, plains, rivers and lakes, and soil conditions, as well as biological elements like local plants, animals, and fungi.

All primates share a distinctive way of life, an adaptive uniformity, based on arboreal frugivory. Extant primate communities as well as primates that lived in **paleocommunities**, communities some of which are millions of years old, show similar adaptive patterns, as we shall see in Chapters 3–6.

Ecological opportunities

The tropical rainforests where most primates live today are concentrated in a broad band north and south of the Equator, largely between the northern and southern latitudes called the Tropics of Cancer and Capricorn. The sun there is more directly overhead for more of the year than in any other rainforests on Earth, causing it to be generally warm. It is also wet because the perpetually warm air carries moisture into the atmosphere that eventually turns into rainclouds. The rainclouds, coupled with atmospheric winds produced by the Earth's rotation, send more cloud-forming moisture overhead that evaporates from the warm oceans. These climatic conditions are excellent for supporting a profusion of life, and they are ideal for primates.

The biological richness and complexity of the rainforest and its trees make it possible for primate social groups to coexist by providing ecological opportunities for adaptations to arise via natural selection that minimize inter-specific conflicts. The different ecological opportunities stem from the use of natural resources that differ in their abundance, dispersion, and quality over space and time. Along with predation, these elements are the immediate drivers that determine many adaptations, including those involving the social and reproductive systems of primates.

The ecological diversity of primates (and other mammals) becomes most apparent at the genus level, which is why the survey of living primates at the heart of this book is organized around the genus. Each of the four major regions where extant primates live has roughly the same number of genera (Figure 1.2). This is remarkable given the four regions' range of differences in location, area, landscape configuration, and the amount of available rainforest habitat.

Madagascar is tiny in size compared to the other three regions, and the primate habitats there cover only a fraction of the island. South and Southeast Asia, where much of the landmass is broken up into many islands on which primates live, is enormous by comparison but has the same number of genera as Madagascar. Its geography is very different from the massive, continuous landscapes of South America and mainland Africa. Mainland Africa has the largest number of genera because it is the only region where

8   *What is a primate?*

Distribution of the 68 Genera of Living Primates

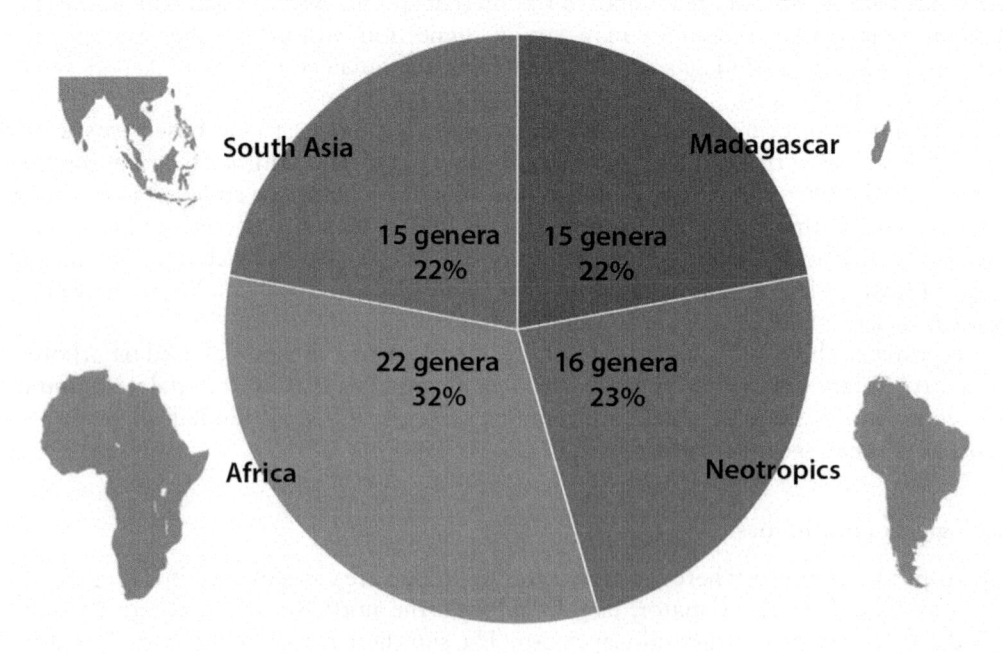

South Asia — 15 genera 22%
Madagascar — 15 genera 22%
Africa — 22 genera 32%
Neotropics — 16 genera 23%

Figure 1.2 Scaled maps depicting sizes and shapes of the four main regions where primate habitats are found, and a pie chart illustrating the number of primate genera in each region according to the taxonomy favored in this book. Note that despite differences in the sizes of the regions, the number of living genera in each of the areas is similar, except for Africa where there also is a major adaptive radiation of terrestrial primates that exist nowhere else. Figure by A. L. Rosenberger.

more than half-a-dozen terrestrial and semi-terrestrial genera have adaptively radiated into habitats outside of the rainforest.

The biological diversity of primates

The 68 genera of primates have evolved adaptations to divide resources in different ways based on variations in body size, dietary preferences, locomotor patterns, use of space, and more.

In every region where primates are found there is an overall pattern of biological diversity made up of:

- Arboreal primate species and genera that range in size from small to large
- Primates that eat leaves or tree gum in addition to fruit
- Primates that eat insects and small vertebrates (including mammals, birds, and reptiles)
- Primates that locomote mostly by walking and running quadrupedally, on all fours, or by leaping using powerful hindlimbs
- Primates that range longer distances each day to get food and those that do not go far
- Primate social groups that include adult male-female pairs and their offspring, and larger groups comprised of many adults and their young

What enables social groups belonging to multiple primate species to co-occur locally as a community are numerous niche differences. The differences are based on distinctive biological profiles that set each species of a community apart from the others. The result is that sympatric species do not need to compete for the same resources.

In regions where resources – basic foods primarily – are abundant, the communities may be made up of species belonging to more than a dozen primate genera. Where resources are sparse yet serviceable, and where ecological opportunity is more limited, there may be fewer than five coexisting genera.

A few species are also found living far from the rainforest, in outlying habitats that cannot sustain a primate community because trees that produce adequate edible foods are rare or lacking. For example, in northern Africa the semi-desert habitats at the edge of the Sahara and the mountainous grasslands of Ethiopia are each inhabited by only one monkey species. Another isolated primate species surviving in a divergent climate where rainforest foods are not available to them are the Japanese macaques, also called snow monkeys. These animals endure several months of winter with heavy snowfalls.

In Uganda, Gabon, and the Democratic Republic of the Congo there are areas where semi-terrestrial chimpanzees and gorillas, species belonging to two different genera, live in the same habitat, eat some of the same foods, and may rest in the same individual trees. But the rich rainforest allows them to live different lifestyles. Chimpanzees rely on fruit, while gorillas prefer leaves, stems, terrestrial herbs, and new plant growth. Chimpanzees spend considerably more time in the trees and use about three times as much space in their daily activities in search of food. But no two species of chimpanzees (i.e., the common chimpanzee and bonobo) and no two species of gorillas (lowland and mountain) live in this same habitat, or anywhere else in Africa, because they are too much alike, ecologically, to sustain coexistence.

Ten factors that describe a primate profile

To describe the basics of primate life we must first identify and organize pertinent information. This can be done by examining the factors introduced here and discussed in greater depth later. Some factors influencing primate life are extrinsic, relating to environment, such as the composition of the habitat, climate, rainfall, and seasonality, which affect the variety and abundance of food throughout the year. Other factors are intrinsic to the animals themselves, their anatomical, behavioral, and genetic makeup. The following factors will be further discussed in detail throughout the book.

There are ten primary factors used to distinguish the seven major groups of living primates:

- Geography and habitat
- Phylogeny
- Body weight (size)
- Activity rhythm and sensory modality
- Diet
- Locomotion
- Social system and mating
- Communication
- Reproductive pattern
- Parenting style

All the intrinsic factors are functionally associated with one another. A powerful logical correlation, for example, is that between diet and locomotion, and these are interconnected with body mass or weight. A small-sized arboreal species that eats animal matter may search for food by scampering about quadrupedally. Large species that eat fruit will climb and clamber through the forest canopy. In other words, feeding and foraging, eating and the process of obtaining food, are intimately connected and correlated.

Another example of functional correlation of factors is activity pattern and communication. Diurnal species rely on vocal signals as well as visual displays, including facial gestures that express emotion or patches of skin that change color to indicate arousal, which can be seen in daylight. Nocturnal species communicate by relying more on olfaction, depositing scent by rubbing a gland against a branch.

All ten factors are important in examining extant primate species and communities.

1 **Geography and habitat.** Where do they live?

- Madagascar
- South and Central America (or the neotropics, which includes parts of Mexico)
- Mainland Africa
- East, South, and Southeast Asia

   Why is this important? Knowing where in the world primates live and in what type of habitat determines the external, environmental factors that influence their lives and evolution. These include climate, weather, available food, coexisting plants and animals, and the physical surroundings, such as forest structure and the density or scarcity of tree cover.

2 **Phylogeny.** What are their genetic or genealogical interrelationships?

- Where they belong on the primate Tree of Life.

   Why is this important? Every evolutionary lineage, or branch, of the primate Tree of Life is a unique line of descent, with its own genetic makeup, that determines the outlines of its ecological and behavioral adaptations. Closely related lineages and taxonomic groups tend to have similar resource needs and lifestyles.

3 **Body weight.** What size are they? The range of sizes includes:

- Small (1 oz. – 1 lb.; 28 g –.45 kg)
- Medium (1.5 lbs. – 10 lbs.; .7 kg – 4.5 kg)
- Large (10 lbs. – 20 lbs.; 4.5 kg – 9 kg)
- Extra-large (20 lbs. – 60 lbs.; 9 kg – 27 kg)
- Super-large (60 lbs. – 400 lbs.; 27 kg – 181 kg)

   Why is this important? Size defines basic biological requirements and prospects, such as how much food a species needs and how vulnerable it is to certain types of predators. Size difference is the most fundamental way in which the ecological positions of coexisting primate species are sorted out evolutionarily, as it directly relates to diet, locomotion, reproductive rate, and other factors.

4 **Activity rhythm and sensory modality.** When are they active during the 24-hour day/night cycle, and what sensory systems are emphasized?

- Diurnal: vision dominates

- Nocturnal: olfaction dominates
- Cathemeral (active at various times of day and night): olfaction and vision

  Why is this important? Primate evolution unfolded as two major phylogenetic and ecological divisions, one nocturnal and/or cathemeral and the other originally diurnal. This means they live dramatically different lifestyles, perceiving and experiencing the world in different ways. Either scent or sight predominates.

5 **Diet.** What do they eat and what is their diet called?

- Fruit (frugivory; frugivorous)
- Leaves (folivory; folivorous; includes herbivory, the use of ground-based, non-woody plants)
- Seeds (seed-eating; seedeater)
- Tree gum (gumivory; gumivorous)
- Animal matter (the terms faunivory and insectivory are used almost interchangeably, and include eating insects and small vertebrates)
- Flesh (carnivory; occasional hunting and eating mammals including other primate species)

  Why is this important? Food, nutrition, is fundamental to life. The critical ways in which species coexisting in the same community are ecologically separated are based on the variations in their feeding patterns. These dietary terms are operational categories, not exclusive classes. Almost all primates eat from all of these categories in different proportions, but for most species one food type, or a combination such as fruits and leaves or fruits and insects, tends to be most influential. Animal byproducts are also occasionally eaten, such as eggs and insect secretions.

6 **Locomotion and posture.** How do they get from one place to another and position themselves to feed?

- Quadrupedalism (walking and running on all fours, and leaping horizontally)
- Climbing and clambering (pulling and/or pushing the body upwards and through the tree branches using arms and legs)
- Vertical-clinging-and-leaping (holding on to a tree trunk and then jumping)
- Brachiation (swinging by the arms below the tree branches)

  Why is this important? Different styles of locomotion and posture are required for species to forage for food, depending on their body size, and the methods used to extract, handle, and consume various dietary items. These locomotor styles serve to separate species ecologically. Within the same community, for example, large-bodied, ripe-fruit eaters like spider monkeys may cover long distances to feed, traveling quadrupedally and relying on climbing and clambering to reach end-of-branch fruits. Small-sized gum-eaters like marmosets, travel low in the forest canopy or understory, leaping from one tree trunk to another, then hanging on using their claws as grappling hooks while lapping up semi-liquid tree gum. They remain spatially separated by using different parts of the forest.

7 **Social systems.** How do individuals live together in social units, whether large or small, and secure mating opportunities? There are many types of primate social systems, including:

12   *What is a primate?*

- Semi-solitary: an adult male and an adult female with her dependent offspring occupy the same area. Adults may not interact while foraging and feeding, but they monitor each other's presence and may share sleeping sites. Male ranging patterns can overlap with the ranges of multiple females, and a male may mate with several females.
- Pair-bonded: one male and one female live together as a unit over the course of several breeding seasons, sometimes for much or all of their adult lives, along with their offspring until they mature. Males may contribute to offspring care.
- Unimale-multifemale groups: one adult male monopolizes the breeding of several females and plays an important role in protecting females and offspring, especially when there is potential for hostile group takeovers by an unaffiliated bachelor male or a roving group of unattached males seeking to establish their own group and mating opportunities.
- Multimale-multifemale groups: several adult males and females live in a large social group and both sexes have multiple breeding partners. Dominance hierarchies of either males or females or both play an important role in regulating social interactions and mating.
- Fission-fusion groups: a variation on the multimale-multifemale social and breeding systems in which the core social group regularly splits into smaller, temporary foraging parties of varying sizes and compositions.
- Multi-level societies: a highly complex, rare social system that usually occurs in severe habitats. It involves hundreds of individuals foraging herd-like in open country or sharing access to sleeping sites. It may involve a massive aggregate of three or more different types of social subgroups, such as many small, unimale-multifemale reproductive units and many bachelor male subgroups.

Why is this important? Primates are inherently social animals. Different forms of social groupings have evolved in connection with varied environments, patterns of resource usage, types of predation pressure, paradigms of interpersonal behaviors, mating systems, reproductive output, and more.

8 **Communication.** Which special senses are most important to exchange information?

- Smell (olfaction; scent)
- Sight (vision; gestural displays)
- Sound (hearing; vocalizations)
- Touch (tactile; physical contact)

Why is this important? Social behaviors are regulated by the senses of smell, sight, sound, and touch in order to communicate. Sociality depends on the exchange of information between and among individuals. All primates live year-round in social groups, which means individuals interact, often intensely and with high frequency. For example, sitting side-by-side in bodily contact is a tactile gesture expressing amiability, while baring large, tusk-like canine teeth is typically a visual sign of aggression. Communicating by sound and scent is advantageous for nocturnal primates and even for diurnal primates in an arboreal environment where visual signals can be obstructed by foliage. Activity rhythm is closely correlated with particular senses.

9 **Reproductive pattern.** What is the frequency and timing of breeding cycles and what is the litter size?

- Annual or multi-annual breeding periods
- Seasonal or non-seasonal breeders: breeding limited to particular times of the year or breeding year-round
- Singletons or twins, and triplets: females in almost all species give birth to one infant at a time. In some species the females always give birth to twins. Triplets are rare.

Why is this important? Pregnancy, gestation, birth, lactation, and infant care require investments of energy by females that must be met with good nutrition, and more so in the genera that always have twins.

10 **Parenting style.** Who helps to take care of the young?

- Maternal care: mothers provide primary infant care, feeding, and carrying
- Other caregivers: fathers, siblings, or others provide critical secondary care, carrying infants and assisting with feeding as they are weaned
- Indirect care (infant parking): infants are left alone on a branch, in a nest or in a sleeping-tree hole while mothers forage

Why is this important? Infant primates are usually carried by their mother. Living in social groups allows for the possibility that levels of support can be provided by other troop members. Different parenting styles are correlated with the number of newborns per litter, their growth and development, among other factors.

The seven major groups of primates at a glance

The following is a snapshot description of each of the seven major primate groups using the ten distinguishing factors that describe a primate profile. Each group exhibits a distinctive combination of traits, with the characteristics overlapping among the groups. Only the most important variations are noted. These descriptions are generalizations, and not comprehensive. Additional discussion is provided in the following chapters as we examine more closely the communities of primates living in each of the four regions on the planet where they are found today.

Added to each profile is an additional factor that is extremely important, the group's conservation status as of December 2022. The data provided here are based on the taxonomy and analysis of the SSC Primate Specialist Group under the auspices of the IUCN, the premier global organization devoted to conserving species and promoting the sustainable use of natural resources. The IUCN categorizes several levels of risk associated with each living species. Here we note the two most severe categories:

- Critically Endangered species are those with an *extremely high risk* of extinction in the wild.
- Endangered species are those with a *very high risk* of extinction in the wild.

The two major causes of extinction are loss of habitat and reduction in genetic viability. To determine the threat level to a species, conservation biologists assess whether its habitat is shrinking and how fast, and if the size of the population is declining at a rate that jeopardizes its future. Other types of threats faced by primates are discussed throughout this book, as well as in Chapter 9, along with the various efforts to mitigate them.

14 What is a primate?

LEMURS

Geography: Madagascar
Phylogeny: Most closely related to lorises and galagos
Body weight: small, medium, large
Activity rhythm and sensory modality: Nocturnal, diurnal, cathemeral; olfaction
Diet: Frugivory, folivory, insectivory, gumivory
Locomotion: Quadrupedalism, vertical-clinging-and-leaping
Social system: Semi-solitary pairs, pair-bonded, multimale-multifemale groups
Communication: Olfaction, tactile
Reproductive pattern: Seasonal breeders; singletons
Parenting styles: Maternal care, other caregivers, infant parking
CONSERVATION STATUS: 77 Endangered or Critically Endangered species (71%)

Figure 1.3 An indri, Madagascar's largest lemur.
Photo credit: Indri Head, by Zigomar (CC-BY-SA-3.0,2.5,2.0,1.0).

LORISES (includes pottos and angwantibos)

Geography: Mainland Africa, South, East, and Southeast Asia
Phylogeny: Most closely related to galagos
Body weight: Small, medium
Activity rhythm and sensory modality: Nocturnal; olfaction
Diet: Frugivory, insectivory, faunivory, gumivory

Locomotion: Quadrupedalism
Social system: Semi-solitary, unimale-multifemale groups
Communication: Olfaction, tactile
Reproductive pattern: Non-seasonal breeders; singletons, twins
Parenting styles: Maternal care, infant parking
CONSERVATION STATUS: 6 Endangered or Critically Endangered species (37%)

Figure 1.4 A juvenile Southeast Asian slow loris.
Photo credit: *Nycticebus bancanus* by M. E. Gunay (CC-BY-SA).

GALAGOS

Geography: Mainland Africa
Phylogeny: Most closely related to lorises
Body weight: Small, medium
Activity rhythm and sensory modality: Nocturnal; olfaction
Diet: Frugivory, insectivory, gumivory
Locomotion: Vertical-clinging-and-leaping, quadrupedalism
Social system: Semi-solitary, unimale-multifemale groups, multimale-multifemale groups
Reproductive pattern: Seasonal and non-seasonal breeders; singletons, twins
Communication: Olfaction, tactile
Parenting styles: Maternal care, infant parking
CONSERVATION STATUS: 1 Endangered species (6%)

16 *What is a primate?*

Figure 1.5 A galago, the Mohol bushbaby, from mainland Africa.
Photo credit: Mohol bushbaby (*Galago moholi*) at Thalamakane River, Botswana, by A. Giljov (CC-BY-SA).

TARSIERS

Geography: Southeast Asia
Phylogeny: Most closely related to monkeys and apes
Body weight: Small
Activity rhythm and sensory modality: Nocturnal; vision, hearing
Diet: Faunivorous (eats no vegetation)
Locomotion: Vertical-clinging-and-leaping
Social system: Semi-solitary, pair-bonded, unimale-multifemale groups, multimale-multifemale
Reproductive pattern: Seasonal breeders; singletons
Communication: Vocalization
Parenting styles: Maternal care, infant parking
CONSERVATION STATUS: 5 Endangered or Critically Endangered species (36%)

Figure 1.6 A Bohol tarsier from the Philippines, Southeast Asia.
Photo credit: Tarsier Sanctuary, Corella, Bohol, by yeowatzup (CC-BY).

NEW WORLD MONKEYS

Geography: South and Central America and Mexico
Phylogeny: Most closely related to Old World monkeys and apes
Body weight: Small, medium, large
Activity rhythm and sensory modality: Diurnal, nocturnal (and cathemeral in one population); vision
Diet: Frugivory, folivory, insectivory, seed-eating, gumivory
Locomotion: Quadrupedalism, vertical-clinging-and-leaping
Social system: Pair-bonded, unimale-multifemale groups, multimale-multifemale groups, fission-fusion groups
Communication: Vision, olfaction, vocalization, tactile
Reproductive pattern: Seasonal and non-seasonal breeders; singletons, twins
Parenting styles: Maternal care, other care givers
CONSERVATION STATUS: 39 Endangered or Critically Endangered species (21%)

18 *What is a primate?*

Figure 1.7 Gray's bald-faced saki monkey, a New World monkey from South America.
Photo credit: Rio Tapajós Saki (*Pithecia irrorata*) in Brazil, by Ana Cotta (CC-BY).

OLD WORLD MONKEYS

Geography: Mainland Africa, South and Southeast Asia
Phylogeny: Most closely related to apes
Body weight: Medium, large, extra-large
Activity rhythm and sensory modality: Diurnal; vision
Diet: Frugivory, folivory, seed-eating
Locomotion: Quadrupedalism
Social system: Unimale-multifemale groups, multimale-multifemale groups, multi-level societies
Reproductive pattern: Seasonal and non-seasonal breeders; singletons
Communication: Vision, vocalization, tactile
Parenting styles: Maternal care, other caregivers
CONSERVATION STATUS: 72 Endangered and Critically Endangered species (44%)

Figure 1.8 A male gelada, an Old World monkey from mainland Africa with members of his troop.
Photo credit: Gelada Baboons, Simien Mtns, by Rod Waddington (CC-BY-SA).

APES

Geography: Mainland Africa, Southeast Asia
Phylogeny: Most closely related to Old World monkeys
Body weight: Extra-large, super-large
Activity rhythm and sensory modality: Diurnal; vision
Diet: Frugivory, folivory, and occasional carnivory
Locomotion: Quadrupedalism (knuckle-walking), brachiation
Social system: Semi-solitary, pair-bonded, unimale-multi-female groups, multimale-multi-female groups, fission-fusion groups
Reproductive pattern: Non-seasonal breeders; singletons
Communication: Vision, vocalization, tactile
Parenting styles: Maternal care, other care-givers
CONSERVATION STATUS: 26 Endangered or Critically Endangered species (96%)

20 *What is a primate?*

Figure 1.9 A male mountain gorilla, an ape from mainland Africa.
Photo credit: Mountain gorilla (6461821349).jpg, by Richard Ruggiero/USFWS (CC-BY).

In each of these seven primate groups, the factors we use to describe them are interconnected. Diet, locomotion, social system, reproductive patterns, and the other intrinsic components as well as some extrinsic components, all interrelate to define a unique ecological niche. All these interconnections exist within each of the seven primate groups and they reflect similar selective pressures, such as the need to find food during periods of scarcity. Tropical rainforests have distinct wet and dry seasons, and the dry seasons produce fewer fruits.

The two factors of diet and locomotion are strongly correlated as elements of a **feeding-and-foraging** system. The important connection between diet and locomotion determines a primate's daily energy budget, how much effort is spent in looking for food, acquiring, ingesting and processing food, converting it to energy, and what is gained in the process. This is the case in every taxonomic category of primates and every regional assemblage.

The contrast between frugivores and folivores reflects two different kinds of energy budgets, for example. Frugivores travel long distances and expend a lot of energy to find their preferred, widely dispersed, nutritious fruits. In contrast, edible leaves are found everywhere and do not require much travel, but they are a poor source of nutrients. Folivores gain little energy by feeding on leaves but they do not have to expend much in order to find them.

How the seven major groups of primates are related and taxonomically organized

Phylogenetic relatedness of species and other taxonomic groups is defined by their shared ancestry, which means they have a common origin. All primates are interrelated but some are more closely related to one another. Apes, Old World monkeys, and New World monkeys, for example, are taxonomically categorized as anthropoids because they all share a single remote common ancestor. But apes and Old World monkeys are more closely related to one another than either is to New World monkeys. That is because apes and Old World monkeys share a more recent common ancestor that New World monkeys do not share (Figure 1.10).

A simplified phylogenetic tree describes how the seven major groups of living primates are related to one another (Figure 1.10). This type of stylized diagram is called a **cladogram**, from the Greek word *clados*, meaning branch. The method used to develop it is called **cladistic analysis**. All forms of gene-based evidence, like morphology, DNA, and species-specific behaviors such as vocalization patterns, can be analyzed cladistically in order to develop a cladogram. The aim is twofold: to place individual taxa on the Tree of Life based on common ancestry, and to provide a framework for tracing the sequence by which characteristics, adaptations, and ecological context have changed during the course of primate evolution. In a cladogram, common ancestors are indicated by the intersections of the lines representing each lineage. This is the point at which descendant groups diverge from a common ancestor.

The seven major groups of primates are sorted into lineages that belong to two major phylogenetic divisions, the **strepsirhines** (also spelled strepsirrhines) and **haplorhines** (also spelled haplorrhines). Strepsirhines include lemurs, lorises, and galagos. Among strepsirhines, lorises and galagos are each other's closest relatives. Another way of saying this is: lorises and galagos are **sister-groups**, or sister-taxa, and together they are the sister-group of lemurs.

The strepsirhines form a **monophyletic** group, meaning they can be traced back phylogenetically to a single common ancestor (S). The haplorhines, including tarsiers, New World monkeys, Old Word monkeys and apes, are the other major monophyletic group of primates. They are all descendants of a common ancestor that is unique to them (H). Further down in the tree, strepsirhines and haplorhines share a unique ancestor (P). It represents the origins of all living primates.

The branching patterns of the cladogram correspond to primate taxonomy. Producing a cladogram is an important step in developing a classification because the foundation of all biological classification is phylogenetic relatedness. Here, for example, lorisoids is used as a term for lorises plus galagos; catarrhines encompass Old World monkeys and apes; anthropoids include platyrrhines and catarrhines.

Several taxonomic terms used in this cladogram reflect the various shapes of the primate nose: strepsirhines, haplorhines, platyrrhines, and catarrhines (Figure 1.10). These terms are discussed below, as is the distinction between "wet-nosed" for strepsirhines and "dry-nosed" for haplorhines.

The classification used in this book includes formal and informal taxonomic names. There are also informal common names for the taxonomic groups in widespread usage, except for the three families of New World monkeys, as listed below.

22 What is a primate?

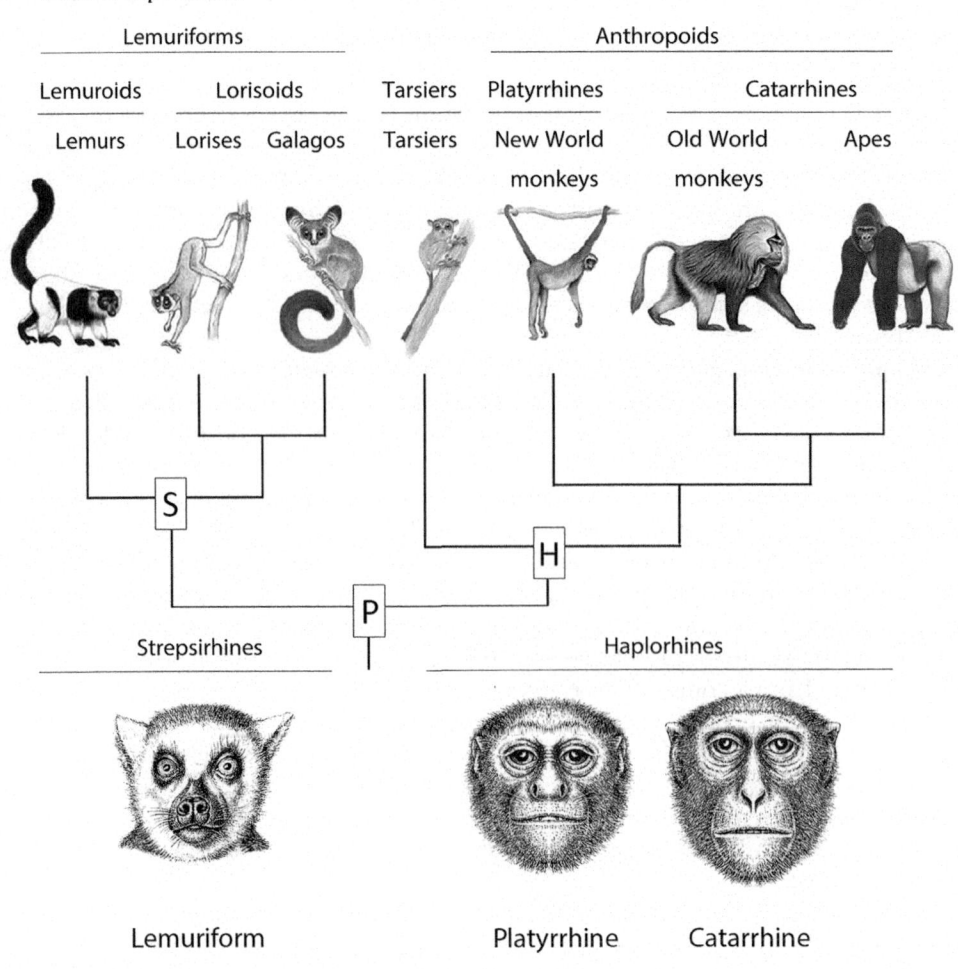

Figure 1.10 Cladogram showing the genealogical relationships of the seven major groups of living primates, with the faces of a lemur, platyrrhine, and catarrhine below to highlight their typical, differently shaped noses. Capital letters in rectangles represent the common ancestors of taxonomic groups: S, strepsirhines; H, haplorhines; P, primates.

Upper illustrations by Stephen D. Nash, courtesy of Stephen D. Nash and the IUCN SSC Primate Specialist Group. Lower images adapted from Schultz, A. H. (1969). *The Life of Primates*. New York, Universe Books.

Strepsirhini (strepsirhines) – wet-nosed primates

　Superfamily Lemuroidea (lemuroids) – Madagascan lemurs

　　Family Lemuridae (lemurids) – lemurs
　　Family Cheirogaleidae (cheirogaleids) – dwarf and mouse lemurs
　　Family Indriidae (indriids) – leaf-eating lemurs
　　Family Daubentoniidae (daubentoniids) – aye-ayes

Superfamily Lorisoidea (lorisoids) – lorises and galagos

>Family Lorisidae (lorisids) – lorises
>Family Galagidae (galagids) – galagos

Haplorhini (haplorhines) – dry-nosed primates

>Superfamily Tarsioidea – tarsiers

>>Family Tarsiidae (tarsiids) – tarsiers

Anthropoidea (anthropoids) – platyrrhines and catarrhines
Platyrrhini (platyrrhines) – New World monkeys

>Superfamily Ateloidea – New World monkeys

>>Family Cebidae (cebids) – predaceous frugivorous platyrrhines
>>Family Pitheciidae (pitheciids) – platyrrhine fruit-huskers and seedeaters
>>Family Atelidae (atelids) – prehensile-tailed platyrrhines

Catarrhini (catarrhines) – Old World monkeys and apes

>Superfamily Cercopithecoidea (cercopithecoids – Old World monkeys

>>Family Cercopithecidae (cercopithecids) – Old World monkeys

>>>Subfamily Cercopithecinae (cercopithecines) – cheek-pouched monkeys
>>>Subfamily Colobinae (colobines) – leaf monkeys

>Superfamily Hominoidea (hominoids) – apes

>>Family Hominidae (hominids) – great apes
>>Family Hylobatidae (hylobatids) – lesser apes

The terminology used in all biological classifications aims to standardize the names of species and genera and reflect phylogenetic relationships. The species and genus are the lowest taxonomic categories. Technically, species names are compound terms that combine that name with a genus name in what is called a binomen, a two-name set, written this way, for example, for the chimpanzee – *Pan troglodytes* – with the genus capitalized, the species in lower case, and both terms italicized.

As we survey the neotropical New World monkeys and the Old World monkeys, it is important to note that the term "monkey" has no scientific or phylogenetic significance. In fact, Old Word monkeys are more closely related to apes than to New World monkeys.

The categories and terms of classification above the genus level that are used frequently in this book are three family-level terms: subfamily, family, and superfamily. They are organized as a hierarchy: superfamily encompasses the family, which in turn encompasses the subfamily. Just as a genus usually contains several species, there may be multiple subfamilies within a family and multiple families within a superfamily.

The three family-level terms for primates are compounds. Each name has a prefix, or root-word, and a suffix. The prefix refers to a genus, living or extinct, that is included in each of the groups in the set. A prefix is combined with one of three suffixes to identify the rank in the hierarchy: -oidea, -idae, and -inae, as in Superfamily Ateloidea, Family

Atelidae, and Subfamily Atelinae. Informal names for these groups are simplified this way: ateloid(s), atelid(s), and ateline(s).

The words strepsirhine and haplorhine are also compound terms. They share a common root-word, *rhine*, from the Latin, meaning snout or nose, as in rhinoplasty (plastic surgery of the nose). The term strepsirhine means "twisted nose." The term haplorhine, means "simple" nose. "Twist" refers to the slit-like or comma-shaped lateral extension of the nostrils of the nose of lemurs, lorises, and galagos. "Simple" refers to the more open, round, or oval nostrils of the noses of tarsiers, New World monkeys, Old World monkeys, and apes (Figure 1.10).

Among the anthropoids, there are two types of simple noses depending on the placement and orientation of the nostrils. Platyrrhines, New World monkeys, have widely separated, laterally facing nostrils. Catarrhines, Old World monkeys and apes, have closely spaced, downward or forward-facing nostrils (Figure 1.10).

It is not only the shapes of their noses that divide these two primary branches of the primates, strepsirhines and haplorhines. A vital functional distinction is how the anatomy of the external nose serves to collect scent. The skin of a strepsirhine's nose is textured and perpetually moist, which makes it efficient at gathering odors. Haplorhine noses are smooth and dry and much less efficient at trapping scent. These features reflect dramatic differences in how strepsirhines and haplorhines perceive and experience the world, what sensory systems are most important for inputting information from the surrounding environment, and from one another. Lemurs, being strepsirhines, are nocturnal and cathemeral for the most part. They emphasize olfaction over vision. Haplorhines, almost all of which are diurnal, rely on vision rather than olfaction. This phenomenon is called the **olfaction-vision tradeoff**. It represents a major adaptive shift in the course of primate evolution, a transition from a more or less nocturnal lifestyle to a diurnal lifestyle, as further discussed in the following chapters.

Phylogeny and adaptation are inextricably connected. There is a pattern to the distribution of certain adaptive variations among the major phylogenetic and taxonomic groups. Some adaptations and combinations of adaptations and lifestyles are shared by specific taxonomic groups because they have a common phylogenetic history. This is true for the seven major groups of living primates and for each of the subgroups they encompass, down to the species level.

The various adaptations among the seven groups of primates are products of evolution meant to ensure their survival in the primate adaptive zone – arboreal frugivory. Unfortunately, primates today face increasingly accelerated threats to their survival due largely to the destruction of their forest habitats which provide the food and shelter they need to exist.

The conservation status of primates today

Primates are the most endangered large order of mammals today and face threats to their very existence. This is especially shocking given the fact that they have lived on the planet for tens of millions of years, as will be discussed further in Chapter 8.

A report identifying the world's 25 most endangered primate species has been published every two years since 2000 by the Primate Specialist Group of the IUCN Species Survival Commission. The most recent version of the top 25 most critically endangered primates as of this writing is presented graphically in Figure 1.11. The threats that primates are experiencing in all of the regions where they now live are escalating rapidly. There are a number of reasons for this.

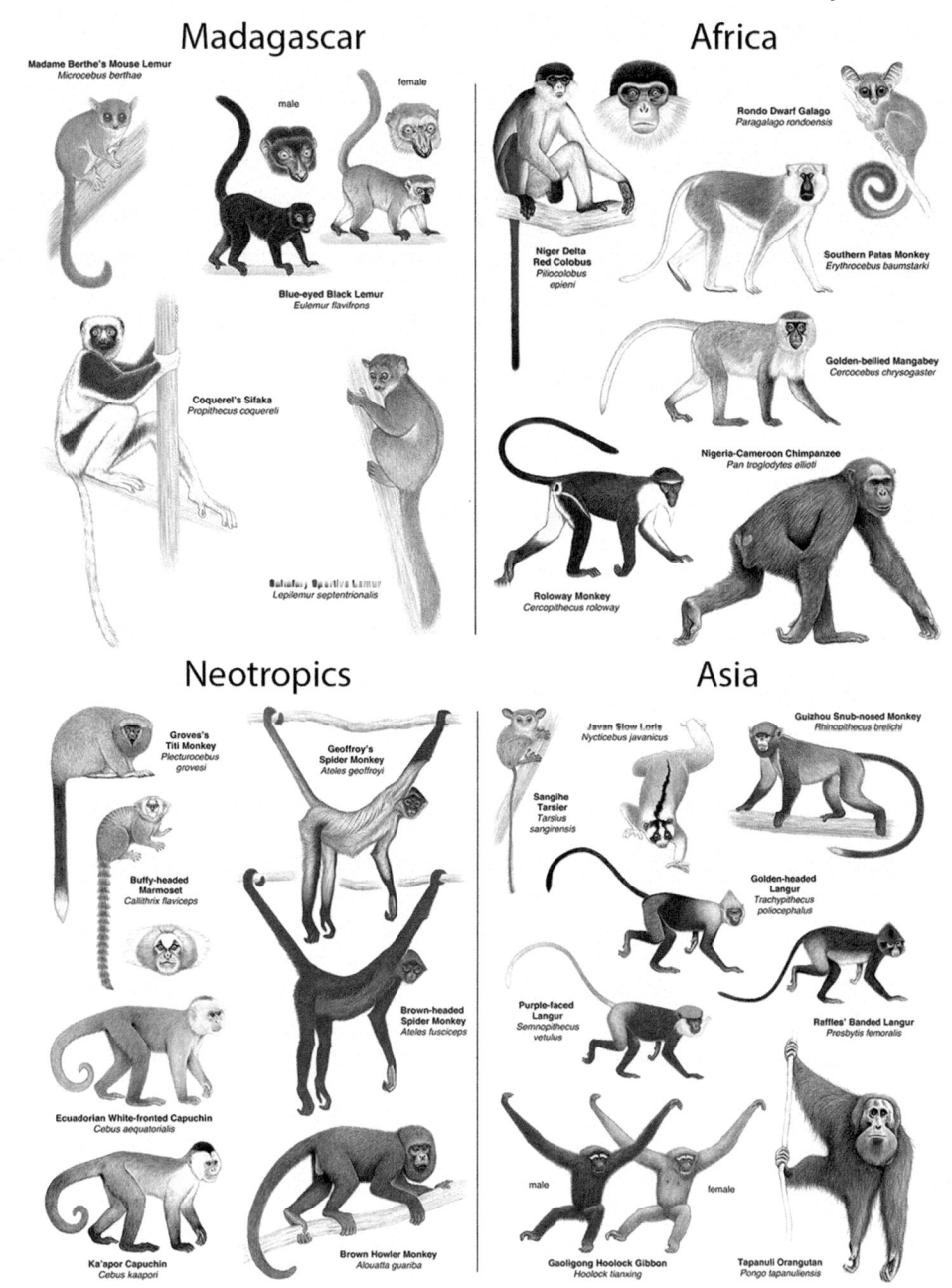

Figure 1.11 The 25 most endangered species and subspecies of primates.

Artwork by Stephen D. Nash, courtesy of the IUCN SSC Primate Specialist Group, reprinted with permission from *PRIMATES IN PERIL: The world's 25 most endangered primates 2022–2024*. Courtesy of Russell A. Mittermeier.

26 What is a primate?

The specific threats that primates are facing in each of the four regions, and efforts at conservation, will be examined in the following regionally organized chapters, and in Chapter 9. The major threats are deforestation and loss of habitat due to a variety of reasons.

What almost all of the threats to primate survival have in common is their **anthropogenic** cause. Anthropogenic means "… relating to or resulting from the influence of human beings on nature," according to the Merriam-Webster dictionary. While this has been going on for a long time, it has only recently begun to accelerate at an exceptionally rapid rate.

Awareness of the existential threats to primate existence have led to organized efforts to mitigate them. The conservation strategies include:

- Establishing preserves, protected areas, parks, reserves, and sanctuaries of many different kinds to safeguard habitats and species, while supporting the needs of Indigenous People and local communities.
- Supporting field research stations where important studies are conducted that are critical to the work of conservation.
- Advocating for legal protections, such as the United States Endangered Species Act that became a federal law in 1973 and the Convention on International Trade in Endangered Species of Wild Fauna and Flora (CITES) also of 1973.
- Developing programs at zoos and sanctuaries that fund and promote primate conservation efforts in the wild and maintain "reserve populations" that can be reintroduced into their native habitats to replenish and genetically diversify decreasing populations.
- Organizing local, national, and international groups that sponsor conservation awareness and education programs to highlight the plight of primates.

In the chapters covering each of the four main regions where primates are found we look more closely at their conservation status. A global perspective is offered in Chapter 9, which summarizes an appeal by 31 of the world's leading experts on primate conservation, who assure us that despite the challenges, primate conservation "… is not yet a lost cause and … the environmental and anthropogenic pressures leading to population declines can still be reversed."

Key concepts

Evolution
Selective pressure
Sympatry
Primate communities
Niche differentiation
Phylogeny

Quizlet

1 How are primates studied?
2 What is adaptive radiation?
3 What is cladistic analysis?

4 What can we tell from a primate's nose?
5 What features are associated with diurnal, nocturnal, and cathemeral primates?
6 Describe two different locomotor styles.
7 Name two of the seven major groups of primates. Where are they found?

Bibliography

Anonymous. (2023). *The IUCN Red List of Threatened Species*. IUCN Red List of Threatened Species. https://www.iucnredlist.org/en

Barton, R. A. (2004). Binocularity and brain evolution in primates. *Proceedings of the National Academy of Sciences of the United States of America*, *101*(27), 10113–10115. https://doi.org/10.1073/pnas.0401955101

Estrada, A., Garber, P. A., Gouveia, S., Fernández-Llamazares, Á., Ascensão, F., Fuentes, A., Garnett, S. T., Shaffer, C., Bicca-Marques, J., Fa, J. E., Hockings, K., Shanee, S., Johnson, S., Shepard, G. H., Shanee, N., Golden, C. D., Cárdenas-Navarrete, A., Levey, D. R., Boonratana, R., ... Volampeno, S. (2022). Global importance of Indigenous Peoples, their lands, and knowledge systems for saving the world's primates from extinction. *Science Advances*, *8*(31), eabn2927. https://doi.org/10.1126/sciadv.abn2927

Fleagle, J. G. (2013). *Primate Adaptation and Evolution* (3rd edition). New York, Academic Press (A principal source for data on body mass and limb proportions used in this book).

Galán-Acedo, C., Arroyo-Rodríguez, V., Andresen, E., & Arasa-Gisbert, R. (2019). Ecological traits of the world's primates. *Scientific Data*, *6*(1), Article 1. https://doi.org/10.1038/s41597-019-0059-9

Gregory, T. R. (2008). Understanding evolutionary trees. *Evolution: Education and Outreach*, *1*(2), Article 2. https://doi.org/10.1007/s12052-008-0035-x

Lambert, J. E. (2012). Primates in communities: The ecology of competitive, predatory, parasitic, and mutualistic interactions between primates and other species|Learn Science at Scitable. https://www.nature.com/scitable/knowledge/library/primates-in-communities-the-ecology-of-competitive-59119961/

Larson, S. G. (2018). Nonhuman primate locomotion. *American Journal of Physical Anthropology*, *165*(4), 705–725. https://doi.org/10.1002/ajpa.23368

Mittermeier, R. A., Reuter, K. E., Rylands, A. B., Jerusalinsky, L., Schwitzer, C., Strier, K. B., Ratsimbazafy, J., & Humle, T. (Eds.). (2022). Primates in Peril: The World's 25 Most Endangered Primates 2022–2023. IUCN SSC Primate Specialist Group, International Primatological Society, Re:wild, Washington, DC. https://cdn.www.gob.pe/uploads/document/file/3574458/Primates_in_Peril_2022_2023.pdf.pdf

Rosenberger, A. L. (2014). Species: Beasts of burden. *Evolutionary Anthropology: Issues, News, and Reviews*, *23*(1), 27–29. https://doi.org/10.1002/evan.21392

Sussman, R. W., Tab Rasmussen, D., & Raven, P. H. (2013). Rethinking primate origins again. *American Journal of Primatology*, *75*(2), 95–106. https://doi.org/10.1002/ajp.22096

Zachos, F. E. (2016). An annotated list of species concepts. In F. E. Zachos (Ed.), *Species Concepts in Biology: Historical Development, Theoretical Foundations and Practical Relevance* (pp. 77–96). New York, Springer International Publishing.

2 Arboreal frugivory
The primate adaptive zone

Chapter Contents

Defining the adaptive zone	28
Six key functional-adaptive patterns and attributes	29
The uniqueness of primate arboreality	33
What makes primates mammals?	34
What distinguishes primates from other mammals?	35
The evolution of primate traits: phylogeny reconstruction	37
The primate fossil record	39
Paleocommunities	39
Key concepts	40
Quizlet	40
Bibliography	40

Defining the adaptive zone

The world's population of primate species constitutes one of the 21 orders of living mammals. The **order** is a classification category that, in the case of mammals, usually includes a large number of genera. All primates belong to the group of mammals formally called Order Primates. Each mammalian order occupies a unique adaptive zone, which may be thought of as an enlarged ecological niche, made possible by a discrete set of anatomical and behavioral traits. Those traits relate to adaptive breakthroughs that opened the gateway to a potentially new way of life.

The most important elements defining an adaptive zone are:

- The environment in which the adaptive zone is situated
- The specific food resources available there
- The requirements for accessing and processing the food

Arboreal frugivory is the adaptive zone of the primate order. With few exceptions, primates live all of their lives in trees, mostly in tropical or subtropical rainforest ecosystems. This is where they find food and shelter, reproduce, and care for their young. Arboreality and all that it entails is a primate's primary characteristic. It is the essence of the order's ecological success. Africa's terrestrial baboons are one of the few important exceptions to arboreality among primates. Many of them live in open country where trees are sparse.

Exceptional cases are not surprising. As we see repeatedly in studying the primates, nature produces patterns, but they are not necessarily universal. In the case of baboons, it is notable that despite their terrestrial lifestyle, and the details of their anatomy that are designed for ground-dwelling, they can be arborealists at times and can go into a tree for food or shelter. Their arboreal skills are a deep-seated sign that baboons evolved from tree-dwelling ancestors.

Primates eat fruit no matter what else they may consume. It is important to note that the wild-growing fruits eaten by primates in the rainforest are different from the cultivated, commercially available fruits with which we are familiar. Fleshy fruits are most abundant and diversified in the treetops and understories of the world's tropical rainforests. They are an excellent source of nutrition. Fruits are easy for the primates' body to digest rapidly and are a good source of sugar, as well as vitamins and minerals. Primates do not require specialized teeth for chewing most fruits or a particular type of gut for processing. Fruit is the most nutritious type of vegetation of all the plants, plant parts, and plant products in the rainforest that all primates eat – leaves, nuts, seeds, bark, pith, flowers, nectar, pollen, gum, sap, roots, tubers, grasses, and more. There are only two exceptions to this diet of vegetation, the tarsiers and a species of loris. They are the only exclusively predaceous primates. Tarsiers eat insects and a variety of vertebrates, and they do not eat vegetation of any kind.

Primates themselves play a vital ecological role in maintaining their own adaptive zone. As a major consumer of fruits and seeds, which are typically eaten, swallowed, carried off, and excreted by hundreds of frugivorous primate species in all four regions of the world where they exist, both the animals and the plants benefit. The primates obtain nutritious food, and the plants gain an advantage in reproduction by having their seeds, potential offspring, moved to locations where they can grow and develop into another generation of trees. This is called seed dispersal.

Six key functional-adaptive patterns and attributes

The primate body is what makes arboreal frugivory in the tropical rainforest possible. No other mammals have evolved a complement of arboreal adaptations comparable to the extensive integrated systems found among primates. The primate body is designed for versatility and flexibility, for positioning and moving among the branches within a tree's crown, and for traveling between them. A primary structural feature of the rainforest is that the crowns of its trees interlace to form a closed canopy above the ground, effectively becoming an interconnected network of travel paths for a primate.

The *combination* of six key functional-adaptive patterns or attributes enables primates to flourish as arboreal frugivores. Some features sustain arboreality, some are specifically food-related, and still others have overlapping importance for living in trees and eating a diet centered on fruit. These attributes distinguish primates from other mammals as occupants of a distinct adaptive zone. These patterns, presented first as an overview, will be examined in more detail as we proceed in this and later chapters.

1 **Diet and dentition.** Primates have an omnivorous, generalized dentition that enables them to eat foods with highly varied physical properties – soft mushy fruits, thin pliant leaves, and brittle or squishy insects. Fleshy fruits are nutritious but cannot provide a fully balanced diet because they lack a high protein content. So, primates combine eating fruit with eating a protein-rich complement, either leaves or insects. But, as will be discussed in detail later, in all taxonomic groups, there are primates that have dentitions that are particularly suited, or even specialized, to access and chew certain types of foods.

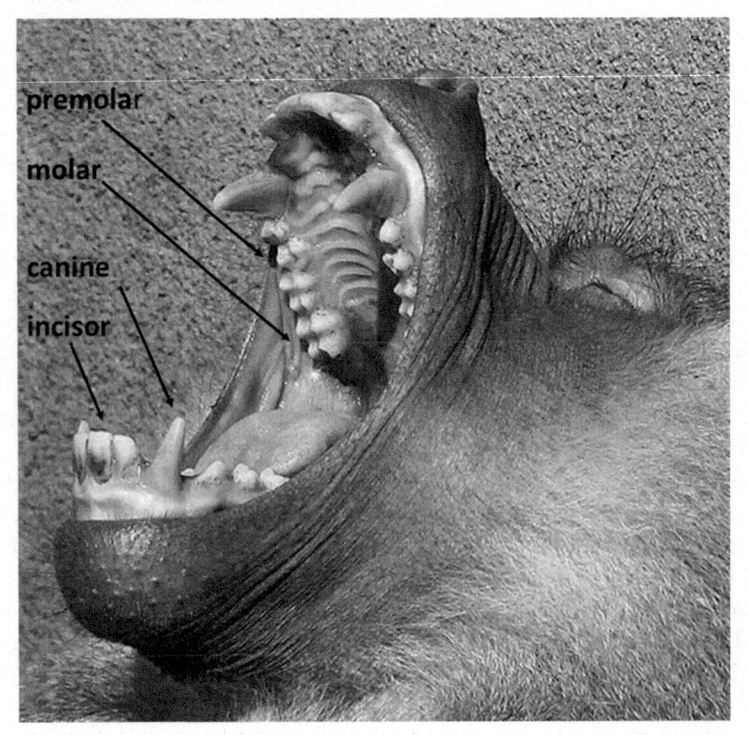

30 *Arboreal frugivory: the primate adaptive zone*

Figure 2.1 Open-mouthed subadult male baboon showing the standard four tooth groups of the primate dentition.

Photo credit: Ha, ha, ha...., by R. D. Brecher (CC-BY-SA).

Primates typically eat a wide range of food items daily, from various types of vegetation to the insects and small vertebrates that are also abundant in the rainforest. Living in a super-rich ecosystem that grows vegetation year-round, with an all-purpose dentition composed of four functionally differentiated tooth types, primates are able to adjust the focus of their diet according to seasonal fluctuations in the food supply (Figure 2.1). Their flexibility also allows for species-specific feeding patterns that enable species living together in each community to focus on different foods, thereby limiting competition for the same resources.

2 **Locomotion, posture, and food-handling.** Primates have evolved grasping feet, and clasping hands, and gripping tails in some species to provide stability and maintain balance in the trees. Grasping requires an opposable **hallux** (large, innermost toe) or thumb; clasping does not. It involves closing the hand or foot around an object by flexing the digits. Highly mobile shoulders and hips allow for highly varied angles of reach when using the arms and legs. The primate anatomical package, more than in any other mammal, provides arboreal mobility, steadiness, and maneuverability when traveling, along with postural flexibility while stationary. Dexterous hands and feet are also an asset benefitting offspring care (Figure 2.2). The ability of most primate infants to clasp and grasp their mother's fur soon after birth makes it possible for mothers to forage for food while the infant holds on to her fur.

Arboreal steadiness gives primates the opportunity to reach for food in a unique way. They are hand-feeders, not strictly face-feeders like virtually all other mammals

Figure 2.2 A young baboon riding on the back of its mother.

Photo credit: Olive baboon (*Papio anubis*) with juvenile.jpg, Charles J. Sharp (CC-BY-SA 4.0).

(Figure 2.3). This means primates generally use their hands to bring the food to their mouths, rather than putting their faces into the food, like dogs and cats, for example. Some face-feeding mammals, like squirrels, can use both "hands" to firmly hold food while bringing their mouths to it, but because primate hands are **prehensile**, able to grasp an object between the digits and palm, they can hold food in one hand. Hand-feeding in primates places a premium on selection for good hand-eye coordination. This trait is ultimately crucial to the development of tool-use as a feeding strategy in a New World monkey and in apes. It creates opportunities to eat fruits that are relatively large and cannot be ingested with a single bite. It also enables primates to efficiently manipulate fruits with thick husks or hard shells in order to get at the desired material inside, which may be a fleshy inner layer, seed coat, or the seeds themselves.

3 **Sensory systems**. The senses of vision, balance, and touch are well developed in primates. Their eyes are relatively large and forward-facing, an orientation that is conducive to seeing distant objects and depth perception. This enables primates to accurately monitor and assess distances, discriminate between foreground and background, and distinguish one overlapping layer of branches from another. The skin on the soles of primate feet, and on the palms of their hands, and on the tips of their toes and fingers, are highly tactile, sensitive to touch and directional pressure, and covered with tiny

Figure 2.3 A South American capuchin monkey handling a large fruit.

Photo credit: A White-Headed Capuchin...Panama, by Holtocw (CC-BY-SA).

ridges, dermatoglyphs, that help resist slippage. So, primates are efficient at securing handholds and footholds while moving about, precisely monitoring limb position to maintain balance. Locating food also requires a well-balanced, efficient sensory apparatus because fruits and leaves change size and tend to change color as they grow and ripen; mature fruits may have a distinct smell; and, the movements of insects and small vertebrates can be detected by sight, hearing, and particular odors.

4 **The brain.** Primate brains are large relative to body weight and other standards used to measure brain size in mammals. Primates have larger brains than other mammals of similar body size. The large primate brain allows for advanced sensory integration and cognitive abilities in connection with large volumes of visual input from a highly complicated environment. Regions in the brain that relate to locomotion, including balance, vision, eye movement, and head control are well developed, as are areas relating to hand-eye coordination and tactile sensitivity.

5 **Social systems and group size.** Primates tend to be more social than most other animals, and they exhibit a greater variety of social systems. Social living is a cooperative lifestyle that benefits individuals residing in the arboreal environment. It provides the ability to share the locations of many types of fruit, where sleeping sites can be found, and the presence of predators. A primate social group may contain males and females, and individuals of different ages, social ranks, and kinship. Many primate social systems are quite complex in their structure, dynamics, and degrees of cooperation by individuals. Group size can range from a few to several hundred individuals.

Figure 2.4 An Asian langur leaping between trees and clutching her infant.
Photo credit: Dusky Langur infant holding on while its mother leaps.jpg, by Roughdiamond21 (CC-BY-SA-4.0).

6 **Life-history strategies**. Primates typically give birth to one offspring at a time, although several species always give birth to twins. The infant is able almost immediately to clutch the fur of its mother with its hands and feet and hang onto her so she can move with it on her back, and soon on her front (Figure 2.4). This is important to frugivorous arboreal animals who need to have their hands free to locomote and access food. Primate infants and young develop slowly; it may take years before they can live independently, learning what to eat, how to find food, and how to behave socially in a group setting. Primates characteristically have long life stages and a long lifespan. They have long interbirth intervals, a long gestation, a long infancy, a lengthy pre-adolescent juvenile period, and a prolonged adolescence before reaching sexual maturity and reproductive age. Many species live for several decades in the wild.

The uniqueness of primate arboreality

The Order Primates is offset from all other mammals by an array of lifestyles made possible by anatomical and behavioral characteristics that reflect life above the ground and in the trees. These characteristics diverge from other mammalian patterns because the architecture of trees, first and foremost the most dominant physical facet of a primate's environment, presents a fundamental challenge to a mammal. Trees are a highly complex, variably stable, three-dimensional framework of woody branches and surfaces oriented at all angles, in which it is difficult to maneuver, and where there are always gaps. The critical foods that primates rely on are fruits and leaves that are most plentiful at the ends of slender branches or, as in many palm trees, on long drooping tendrils. These structural

34 *Arboreal frugivory: the primate adaptive zone*

features of the forest are a major source of strong selective pressures, environmental conditions that impact survivorship and reproduction, which influence the locomotor skeleton and sensory systems relating to posture and locomotion (Figure 2.4).

The tree canopy is also an environment where sight is impeded every day of the year at all hours by an endless visual barrier – foliage – that also absorbs sounds. The dense surrounding of leaves adds another layer to the challenge of negotiating the treetops, to determine where food is located. It also impacts how primate troop members behave, and how they maintain contact with each other when they are dispersed while foraging.

Primates have a special ability to locomote *within* trees, in contrast to the only other two orders of mammals that are exclusively arboreal, bats that roost in trees and fly between them, and colugos that live in trees and glide between them. Primates have flexible locomotor patterns that make them versatile quadrupeds, good climbers, and good leapers able to travel well on branches. In some cases, they are able to hang on to branches by their tails.

The range of primates' body sizes exceeds that of all other mammals. It ranges from the one-ounce (28 g) Madame Berthe's mouse lemur to a 400-plus pound gorilla (>181 kg). Due to their exceptional locomotor versatility, primates of all sizes are able to exploit trees because they can maintain their balance on all manner of branches, including bendable twigs, and can navigate around or across gaps in the tree canopy or its understory. They can locate and access foods in the smallest tangles or in the wide-open perimeters of tall trees that stand above the closed canopy. With a superior sense of balance, they can securely sleep in trees as well (Figure 2.5).

What makes primates mammals?

Living mammals are divided into three groups based on their reproductive systems and other attributes:

Figure 2.5 A South American howler monkey asleep on a branch.
Photo credit: Mono aullador rojo...by Petruss (CC-BY-SA 4.0).

- Monotremes – a small group that lays eggs and is confined to the Australian region. The platypus is an example.
- Marsupials – a diverse, widespread group that gives birth to young in a highly immature, almost embryonic state that continues development outside the mother's body in an external pouch. The kangaroo is an example.
- Placentals – give birth to relatively well-developed live young and are the most taxonomically abundant and ecologically diversified mammals. Primates are examples.

Primates and all other mammals share four notable soft-anatomy traits:

- Mammary glands with which mothers nurse offspring
- A hairy skin-covering providing thermal insulation
- An advanced, relatively large brain to process information, with an enhanced olfactory system that boosts the sense of smell
- Whiskers around the nose and mouth that are tactile

Six key hard-anatomy features shared by primates and other mammals are:

- A structurally simple skeleton with relatively few separate bones, and limbs placed under the trunk to provide efficient support for forward locomotion.
- A species-specific, target body size, with finite growth patterns and articulating joints that fuse to the ends of bone shafts when adulthood is reached.
- A solidly built skull with a large braincase housing the brain, specialized sense organs, and the feeding apparatus (the dentition and attachment surfaces for jaw muscles).
- Three middle-ear ossicles (tiny bones) for hearing that are located in a compartment within the cranium.
- Teeth that are functionally differentiated into four types: incisors, canines, premolars, and molars.
- Teeth that erupt into the mouth in two finite waves, with deciduous (baby) teeth replaced by permanent teeth.

What distinguishes primates from other mammals?

The combination of seven characteristics and three functional-adaptive complexes distinguishes primates from other mammals. Several are specifically arboreal adaptations that allow them to interface directly with the structurally complex environment in which they live.

1 A grasping foot with a specialized innermost toe, the hallux. In nearly all primates the hallux is quite large, set off to the side, and capable of producing a strong muscular grasp around a branch via a rotation about the toe's axis that places it opposite the rest of the foot. With a divergent big toe, primates' feet look and act like hands. Their hands have a differentiated thumb that is often offset and, in some species, can produce an opposable grip against the fingers and/or palm (Figure 2.6). The foot, somewhat like the hand, can grasp and rotate to adjust position.
2 Long digits (bones of the palm, sole, and visible fingers and toes) of the hands and feet, proportioned to fold around a support or apply stabilizing pressure to the fingertips when the digits are arched.

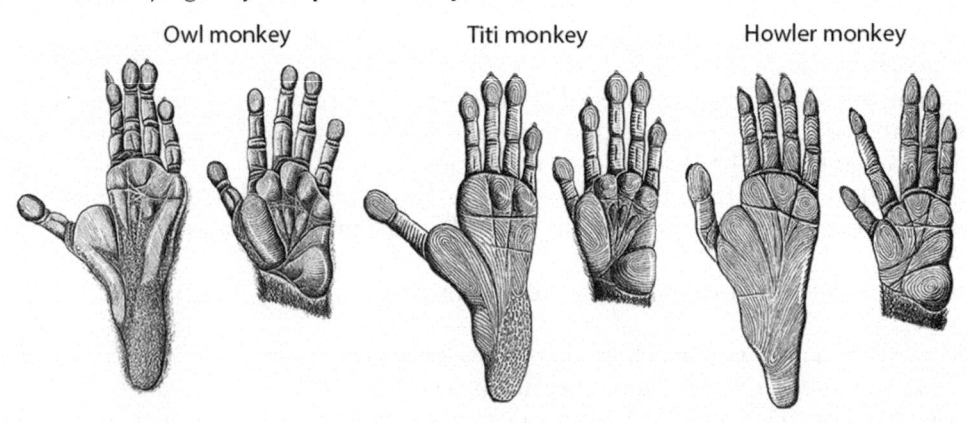

Figure 2.6 The grasping feet and hands of three primates.
Adapted from Rosenberger, A.L. (2020). *New World Monkeys: The Evolutionary Odyssey*. Princeton, Princeton University Press.

3 Soft, fleshy touch pads on the tips of digits and relatively flat nails (with a few exceptions as will be discussed) to enhance fingertip sensitivity to touch and sense directional pressure.
4 Relatively large eyes, facing forward to enhance the quality of vision, including the ability to accurately judge distance.
5 A large brain relative to body size, for information storage and processing.
6 Well-developed incisor teeth suitable for biting fruits that may be fleshy, thick-skinned, or relatively large.
7 A bony petrosal **auditory bulla,** which is a shell-like cap on the underside of the cranium that seals off the space containing the middle ear and its ossicles, and the inner ear, structures dedicated to hearing and balance. Only in primates is the auditory bulla formed by a bone called the **petrosal**.

These seven characteristics distinguish primates from other mammals, contributing to three system-wide adaptations relating to arboreal frugivory: locomotion, special senses, and feeding. All three systems are coordinated by the large primate brain, which stores and integrates high volumes of sensory and locational information, and is crucial to the primates' superior proprioceptive ability (awareness of body and limb positions).

- **Locomotion.** Primate morphology promotes two unique locomotor abilities:
  - Leaping. Relatively long legs and a long hindfoot (mainly the two ankle bones) combine to make primates strong leapers.
  - Grasping. Both the foot and hand are able to grasp. The foot has an opposable hallux. The hand often has a divergent thumb. Feet and hands have long digits and sensitive fingertips, making both the forelimb and hindlimb effective grasping organs.
  - The combination of these locomotor functions defines a style of arboreal locomotion that is unique to primates, **grasp-leaping**, springing away from a support while holding onto it until the final moment so muscular thrust is maximized by allowing full extension of the hindlimb.

- **Special senses.** The visually oriented sensory complex enabling primates to travel, maneuver, and live in the trees is centered around relatively large and forward-facing eyes. In the cranium, this is made possible structurally by a shorter, narrower and lower snout than in other mammals and a reduced olfactory apparatus, parts of which are situated in the space between the orbits as discussed in Chapter 4. Enhanced tactile sensitivity of the hands and feet aid in proprioception, and areas of the brain responsible for balance are also enhanced.
- **Feeding.** Hand-feeding augments the functional potential of primate teeth, in contrast to long-snouted, face-feeding non-primate mammals. Primates' low-crowned cheek teeth with blunt cusps are biomechanically suited to crush and grind fruits. The ability to prepare a fruit by peeling a husk or rind with the teeth, or by biting off a piece so it can be chewed, is a critical process that relies mostly on well-developed incisor teeth and/or the manual ability to position an item between upper and lower incisors or elsewhere in the mouth where it can be efficiently bitten.

The evolution of primate traits: phylogeny reconstruction

These seven characteristics serve to describe the primates' unique features and systems in broad ecological and behavioral terms. Yet, similarities can arise in various species in two different ways, as jointly inherited traits or independently evolved traits. This means that we need to look at these seven distinguishing features in a different way, phylogenetically, as part of a cladistic analysis. Cladistics is the method of phylogeny reconstruction described in Chapter 1 in the discussion of how the seven major groups of primates are interrelated. Cladistic analysis is based on biology's comparative method, which studies the evolution of diversity by comparing and contrasting attributes in a phylogenetic context.

Phylogenetically, there are two fundamentally different kinds of similarities shared by taxa:

- **Homologous** characteristics (homologies) are present in different species because they were inherited from their last common ancestor.
- **Analogous** characteristics (analogies) evolved separately in different species and were not inherited from the last common ancestor that they shared.

Homologous characteristics are phylogenetically significant because they provide information about species genetic relatedness. For example, the absence of a tail and the development of a coccyx, the tail's vestigial bones that form the base of the spinal column is a trait found in chimpanzees, bonobos, gorillas, orangutans, gibbons, and siamang. It is a homologous characteristic inherited from their last common ancestor, a species of ape from which all the lines of living apes ultimately descended.

Analogous characteristics are important because they contain information about adaptation and natural selection. How do two different species evolve similar traits when they do not share a common ancestor from which the similar traits could have been inherited? The process by which analogies evolve in species is called **convergent evolution**. More about it is discussed in Chapter 7. An example is the enormous eyes of tarsiers, an Asian haplorhine, and galagos, African strepsirhines, two relatively distantly related primates whose last common ancestor did not have huge eyes. In both tarsiers and galagos massively large eyeballs evolved independently as adaptations to hunting animal prey at night.

There is a further distinction that needs to be made about homologous features:

- **Primitive** (or **ancestral**) **homologous features** are holdovers from the remote past that are widely shared by different species. A primitive feature that is shared by two or more taxa is called a **symplesiomorphy**.
- **Derived homologous features** are *modifications* of primitive features that are shared exclusively by a certain cluster of species. A shared derived feature is called a **synapomorphy**.

Features are referred to as primitive because they are unchanged in a particular phylogenetic context, not because they are crude functionally or adaptively. For example, almost all primates have five fingers on their hands; that is a primitive feature, the ancestral primate condition. But there are two different neotropical primate species that have only four fingers on their hands; they are missing the thumb. This is a shared derived feature that appears in spider monkeys and muriqui. Their thumbless hands are considered a synapomorphy, a shared derived feature, not independently evolved. It is part of a wide-ranging adaptive complex involving many other parts of their skeletons that are also morphologically unique and related to their exceptional locomotor skills.

What makes unique shared derived features important as phylogenetic indicators is that they signify a separate evolutionary history. Phylogenetic information can only be obtained by identifying homologous, shared derived features. Within a cladogram, specific synapomorphies are found only in certain branches or subsets of branches, which means the derived features can be traced back to a specific point of origin – the point where one parent branch splits and two lineages begin evolving unique anatomical, behavioral, and genetic configurations.

Distinguishing between primitive and derived traits is a crucial step in a cladistic analysis. One objective of cladistic analysis is to determine the direction of evolutionary change in a characteristic that is exhibited in two (or more) versions, or states, among species. For example, if some primates have a long tail and others have a short tail – two different states – did the tails of these species evolve from short-to-long or from long-to-short? Cladistic analysis also uses derived states to cluster species into monophyletic sets and place them in a phylogenetic framework that shows their broader genealogical relationships. This is depicted in a cladogram and is expressed in taxonomic classifications (Figure 1.10).

The seven characteristics describing the functional-adaptive complexes unique to primates are derived homologous characteristics, synapomorphies shared by living primates and their fossil relatives. They demonstrate that primates are a separate monophyletic group of mammals, which justifies placing primates in their own order. Together, they show that primates diverged from other mammals by entering a new adaptive zone defined by arboreal frugivory.

All living species, and all fossils, present a mixture of primitive and derived features. The mixture is called a mosaic, after the concept of **mosaic evolution,** which is based on the principle that different traits and different functional complexes within a species can evolve at different rates. There are several ways to determine within a taxonomic group or clade if a trait is primitive or derived. In many cases the evidence comes from fossils related to the species in question, which often exhibit the primitive state of a primitive-to-derived transformation. One reason for the mosaic evolution of individual traits is that selective pressure is not exerted evenly on all systems simultaneously. For example, as demonstrated by fossils, in the evolution of apes, dietary adaptation – modern-looking teeth – preceded full development of the locomotor system – loss of the external tail. More about this in Chapter 5.

The primate fossil record

Fossils are the petrified (mineralized), rock-hard remains of species from an earlier geological era. Fossils can tell us when important primate characteristics or behaviors existed, in what sequence the traits of a functional complex were assembled over time, and when ecological shifts into new niches or **microhabitats** occurred. It is important to note that a fossil can tell us when an attribute or species *existed*, but it cannot tell us when it *originated*. The primate fossil record, which is discussed in Chapter 8, is a unique source of information about primate life in the distant past, what kinds of species existed at a given time, what species coexisted with them, what ecological niches were represented, and how species and lineages may have changed or not.

Only a small fraction of the biodiversity of life is discovered preserved in the fossil record. It is exceedingly rare for any remains of an individual of any species to become subject to the environmental conditions leading to fossilization, and the chance that any remains will be discovered by a paleontologist is exceptionally small. This is why ancestral-descendant hypotheses at the species level, which are explicitly *phylogenetic* hypotheses, are rarely presented. Rather, we focus on a *taxonomic construct* for the ancestor, called the **ancestral stock**.

Primates have existed, living in the arboreal frugivorous adaptive zone, from tens of millions of years ago to the present day. This is documented and dated by the remains discovered in the fossil record.

- Most fossil specimens of primates, and other mammals, consist of isolated teeth or several teeth set in a fragment of a jaw. The next type of most frequently found fossils includes sturdy, dense bones and skeletal pieces, such as the ankle bones and the ends of limb bones that form the joints. A fossil cranium (plural: crania), meaning the upper portion of the head including the face and braincase, is rarely found intact because most of its bones are delicate and prone to postmortem damage. Today's steadily improving collection methods and digital, three-dimensional visualization tools are now able to reveal a wealth of information even from crushed skulls or fragile parts that are embedded in rock.
- Primate **subfossils** are specimens that have not yet mineralized and come from more recent periods, about 12,000 years ago to the present. The bones and teeth of subfossils are frequently found undamaged and, even when individually scattered, often can be reconstructed into relatively complete skeletons. Many primate subfossils have been found in caves, having been washed in by heavy rain or floods, or in the shallow mud of marshes.
- Primate trace fossils include petrified dung (coprolites), like those found associated with subfossil lemur specimens in Madagascar.

Paleocommunities

The fossil and subfossil records contain not only evidence of individual primate species and lineages. They also provide evidence of primates as they existed in paleocommunities. By paleocommunities we mean an assemblage including the remains of species that no longer exist, that occupied the same habitat together for a period of time. Paleocommunities of primates have been found in all four regions where primates live today.

By studying paleocommunities in the four major geographical regions where primates have adaptively radiated, we are able to examine patterns that unfolded in the course of primate evolution. We can see how communities and regions were either stable or subject to change over time. In northern South America, for example, we find a 12–14 million-year-old community of platyrrhines that is essentially ecologically and phylogenetically continuous with modern communities of the Amazon Basin.

In north Africa we find 30 million-year-old paleocommunities that included many anthropoid and strepsirhine primates that were more distantly related to the living genera. They represent another stage of primate evolution in Africa. Yet, the African paleocommunities were ecologically differentiated along the same lines, such as body size, activity rhythm, diet, and locomotion, that separate the ecological niches that enable today's species to coexist sympatrically.

In the following four chapters we will look more closely at each of the regions on Earth where primates and communities of primates are found today and had existed in the past.

Key concepts

Adaptive zone
Primate arboreality
Homology and analogy
Primitive and derived features
Convergent evolution

Quizlet

1 What is arboreal frugivory?
2 What are the four tooth groups that are standard in the primate dentition?
3 Describe three of the features that make primates mammals.
4 Describe two of the characteristics that make primates primates.
5 In what ways do primates' feet resemble their hands?
6 What is meant by the primate fossil record?
7 What is a paleocommunity?

Bibliography

Gebo, D. L. (2013). Primate Locomotion|Learn Science at Scitable. https://www.nature.com/scitable/knowledge/library/primate-locomotion-105284696/

Livingston, S. (2016). How did primate brains get so big? https://news.ufl.edu/articles/2016/08/how-did-primate-brains-get-so-big-.html

Rosenberger, A. L. (2013). Fallback foods, preferred foods, adaptive zones, and primate origins. *American Journal of Primatology*, 75(9), 883–890. https://doi.org/10.1002/ajp.22162

Silcox, M. T., & López-Torres, S. (2017). Major questions in the study of primate origins. *Annual Review of Earth and Planetary Sciences*, 45(1), 113–137. https://doi.org/10.1146/annurev-earth-063016-015637

Toussaint, S., Llamosi, A., Morino, L., & Youlatos, D. (2020). The central role of small vertical substrates for the origin of grasping in early primates. *Current Biology*, 30(9), 1600–1613.e3. https://doi.org/10.1016/j.cub.2020.02.012

Vasey, N., Burney, D. A., & Godfrey, L. R. (2013). Coprolites associated with *Archaeolemur* remains in North-Western Madagascar suggest dietary diversity and cave use in a subfossil prosimian. In J. Masters, M. Gamba, & F. Génin (Eds.), *Leaping Ahead: Advances in Prosimian Biology* (pp. 149–156). Springer. https://doi.org/10.1007/978-1-4614-4511-1_171

3 Madagascar
Lemurs

Chapter Contents

Madagascar's geography and climate	41
Lemurs: the most diverse group of living primates	44
Lemurs are strepsirhine primates	45
The lemuroid profile	48
Lemurs living in communities: diversity, ecology, and adaptation	51
Cheirogaleids – dwarf and mouse lemurs	53
Phaner, the gumivore	56
Lemurids – cathemeral lemurs	58
Lemurids living elsewhere	59
Indriids – leaf-eating specialists	61
Daubentoniids – aye-ayes	62
Paleocommunities	65
The conservation status of Madagascar's lemurs	66
Key concepts	68
Quizlet	68
Bibliography	68

Superfamily Lemuroidea (lemuroids) – Madagascar lemurs
    Family Lemuridae (lemurids) – lemurs
    Family Cheirogaleidae (cheirogaleids) – dwarf and mouse lemurs
    Family Indriidae (indriids) – leaf-eating lemurs
    Family Daubentoniidae (daubentoniids) – aye-ayes

Madagascar's geography and climate

In this chapter we focus on the lemurs, the only primates that live on the African island of Madagascar. Lemurs are **endemic** to Madagascar and live nowhere else in the world. Madagascar is a very large island by world standards – the fourth largest – and it is situated within the tropical belt that is fundamental to primate ecology. It is a long, narrow island, 1,000 miles (1,609 km) long and 250 miles (402 km) wide, off the southeast coast of mainland Africa in the Indian Ocean (Figure 3.1).

DOI: 10.4324/9781003257257-3

Figure 3.1 Geographical location of major primate habitats in Madagascar. Dotted, dark gray areas are the remaining rainforests. Inset shows the size of Madagascar in relation to mainland Africa and its rainforest habitats, which are shaded.

Image credit: Ecoregions and forest types in Madagascar.jpg, by Ghislain Vieilledent (CC-BY-SA-4.0). Inset from Corlett, R., & Primack, R. (2011). *Tropical Rain Forests: An Ecological and Biogeographical Comparison* (2nd edition). New York, Wiley-Blackwell.

The climate is hypervariable: extremely seasonal and unpredictable from year to year. It is strongly influenced by the surrounding waters, by local topography that splits the island into eastern and western biological domains, and regional global factors, such as the powerful tropical cyclones (equivalent to hurricanes and typhoons) that develop over the Indian Ocean and hit eastern Madagascar about seven times each year. The cyclones generate heavy rain, strong winds, flooding, and landslides, and can be very destructive to vegetation, defoliating trees or snapping off their canopies.

Madagascar is a region with a very large assemblage of animals and plants that are endemic, meaning they are found nowhere else on Earth. Only a few native mammal groups – primates, rodents, bats, carnivorans, and tenrecs (hedgehogs and related shrews) – exist there. The island has been geographically isolated for about 88 million years. This has made it very difficult for plants and non-flying vertebrates to get there from the African mainland. As a consequence, more than 90% of its plants and more than 98% of its mammals, reptiles, and amphibians are geographically unique. This makes Madagascar one of the world's most crucial **biodiversity hotspots**, recognized for its endemic diversity, limited remaining natural vegetation, and high level of threat to its flora and fauna. The lemurs are a preeminent group, comprising 40%–50% of the island's endemic mammal species.

Much of the terrain on Madagascar is dominated by ecosystems that may appear inhospitable to rainforest-loving primates. Yet, because of their unique adaptations, lemurs are a significant presence there, too. Nearly as many lemur species live in the hot, dry forests and desert conditions of the west and south as in wet evergreen forests of the east. The fat-tailed dwarf lemur, *Cheirogaleus medius*, is an example (Figure 3.2). It is able to survive in the resource-poor environment without eating for more than half a year by gaining weight when food is available and storing fat in its tail, that is converted to energy while the animal hibernates. **Hibernation** is an inactive physiological state similar to sleep.

The lush forests are concentrated in a long, narrow, north-south lowland strip where it rains year-round, up to 150 inches (381 cm) annually. The amount of rain is highly unpredictable from month to month and the vegetation is sustained via rainclouds carried by the equatorial trade winds that stream east to west off the Indian Ocean. These are perennial global air currents that flow around the equator in roughly the same direction under the influence of the Earth's rotation, and they also affect other tropical regions where primates live. In Madagascar, the trade winds deliver a distinct six-month season of warm, rainy weather along the east coast.

The western region of the island is characterized by a long dry season and drought-resistant, vegetation-poor habitats. These include dry deciduous tropical forests, where leaves are shed seasonally, and dry spiny forests, shrubland, and desert. It is far drier here because of a long mountain chain that runs nearly the length of the island, and a large, elevated plateau that accounts for about 40% of Madagascar's surface. The mountain chain impedes the flow of trade-wind clouds and moisture. The little rain that the western region receives forms over a much smaller body of water, the Mozambique Channel that separates Madagascar from the African mainland. One city, Toliara, in the southwest near a site where primates have been studied, averages about 16 inches (42 cm) of rainfall annually. This is only about 13% of the amount of rain that falls in the east, and there is no more than one inch (2.5 cm) of rain per month for five months straight. These conditions support different types of lemur species.

Figure 3.2 The fat-tailed dwarf lemur, *Cheirogaleus medius*, also known as the hibernating lemur. Photo credit: David Haring, Duke Lemur Center.

Lemurs: the most diverse group of living primates

Lemurs are one of the most remarkable, and the most diverse adaptive radiation of living primates. The different climates and habitats in which they live on Madagascar have resulted in many varied responses to selective pressure. The radiation has been strongly influenced by Madagascar's novel biota, the animal and plant life which establish the broader ecological context.

Contrasting habitats have significant consequences for the lemur communities found in each of these ecosystems. Because arboreal frugivory defines the primate adaptive zone, the quality of available fruit is always important. In the east, richer, more diverse forests offer more ecological opportunities. In the east, there are more nutritious fleshy fruits that are often soft and juicy when ripe. So, there are more species of lemurs in those communities that are frugivores and folivores. In the more arid western deciduous forests, the fruits are adapted to the parched conditions and tend to be drier and more fibrous. At the same time, tree gum may be more available year-round in the west as a dietary option, even a specialization. As a consequence of the dry conditions in the west, there are fewer frugivores and folivores and more omnivores and gumivores.

Lemurs are an exceptionally varied radiation of primates in terms of anatomy and behavior, with 15 anatomically distinct living genera classified in four family-level clades (Figure 3.3). This level of diversity is remarkable considering Madagascar's size compared

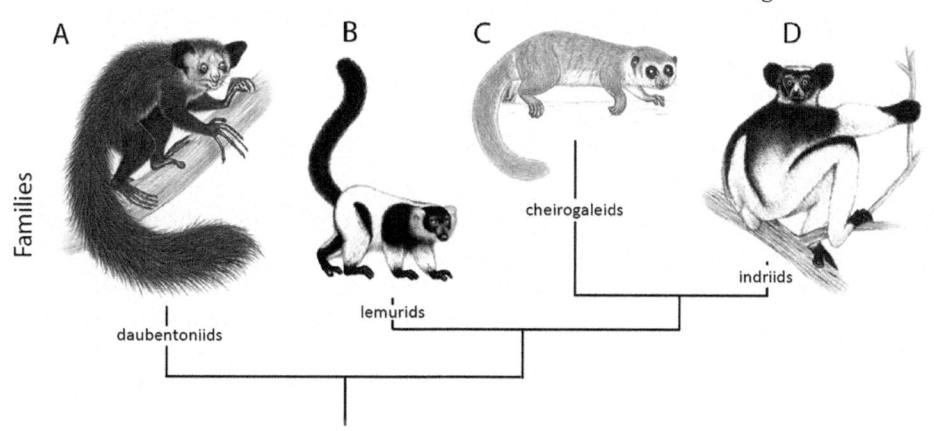

Figure 3.3 Cladistic relationships of the four living lemur families represented by (a) the aye-aye, *Daubentonia*; (b) black-and-white ruffed lemur, *Varecia*; (c) fat-tailed dwarf lemur, *Cheirogaleus*; (d) indri, *Indri*. Not drawn to scale.

Artwork by Stephen D. Nash, Courtesy of Stephen D. Nash and the IUCN SSC Primate Specialist Group.

with the other regions where primates are found. Madagascar is only 2% the size of mainland Africa and 3% the size of South America, yet it supports a roughly similar number of families and genera as these much larger continents, as discussed in Chapter 1 (Figure 1.2).

The lemurs include nearly every pattern described previously in the section in Chapter 1 that provided a synopsis all of the primates at a glance, ecologically, morphologically, and behaviorally. There are also several outliers among the lemurs that have unique characteristics. In addition to the fat-tailed dwarf lemur that hibernates (Figure 3.2) there is, for example, the aye-aye, *Daubentonia*, that has evolved echolocation, a sonar-like hearing system, and the bamboo lemur, *Hapalemur*, that ingests large amounts of cyanide from the leaves and shoots that it eats every day. These outliers will be discussed further later in this chapter.

Lemurs are strepsirhine primates

Lemurs embody all the main features of strepsirhine primates, which are related to nocturnality and a reliance on olfaction to gather information about the environment, and for communication. This contrasts with the vision-oriented haplorhines, as described below and in Chapter 1. Also included in the strepsirhine category are the lorises and galagos of mainland Africa, and the lorises of Asia (Figure 1.10).

Strepsirhines and haplorhines are the two main divisions of living primates (Figure 1.10). They represent two branches of primate evolution that are based on major differences in anatomy, behavior, and ways of life – how they perceive the world, what sensory systems are most dominant: the olfaction-vision tradeoff.

The major features of living strepsirhines include:

- Moist rhinarium
- Well-developed internal (bony) nose

46 Madagascar: lemurs

- Well-developed vomeronasal organ
- Large olfactory bulb
- Tapetum lucidum
- Toothcomb

Lemurs and other strepsirhines are largely nocturnal, active when ambient light is low and the potential for accurately sensing the details of objects by sight is limited. Their eyes are attuned to darkness but vision can only go so far without adequate light to illuminate surfaces. That is why having a very good sense of smell is highly advantageous for species whose visual system can rely only on moonlight. Olfaction is a powerful way to sense the world – by the smells of flowers, fruits, insects, vertebrates, and predators. A keen sense of smell also provides a means to collect specific sorts of detailed information about conspecifics, individuals of the same species, which is transmitted in the form of odors that are natural bodily byproducts. All strepsirhines have eyes that are functionally designed for night vision, while several lemurs are cathemeral, active at various times of the day and night, and a few are diurnal.

How do lemurs and other strepsirhines see in the dark? They have a structural layer at the back of the eye called the **tapetum lucidum** (Figure 3.4) which amplifies light. It is situated behind the retina, a transparent tissue that supports light-sensing photoreceptor cells. The tapetum consists of a mirror-like sheet of reflective cells, found in many vertebrates. When light passes through the retina, the tapetum bounces it back to instantaneously stimulate the photoreceptor cells a second time, from behind, thus enhancing the visual value of limited light. That reflection is a familiar glowing eyeshine, also seen in other mammals adapted to living in darkness.

The ability to produce, detect, and utilize odors is deeply ingrained in strepsirhine behavior. It is featured in every aspect of their lives. They find food in the dark by smelling it from a distance, exchange personal odors to identify one another, send

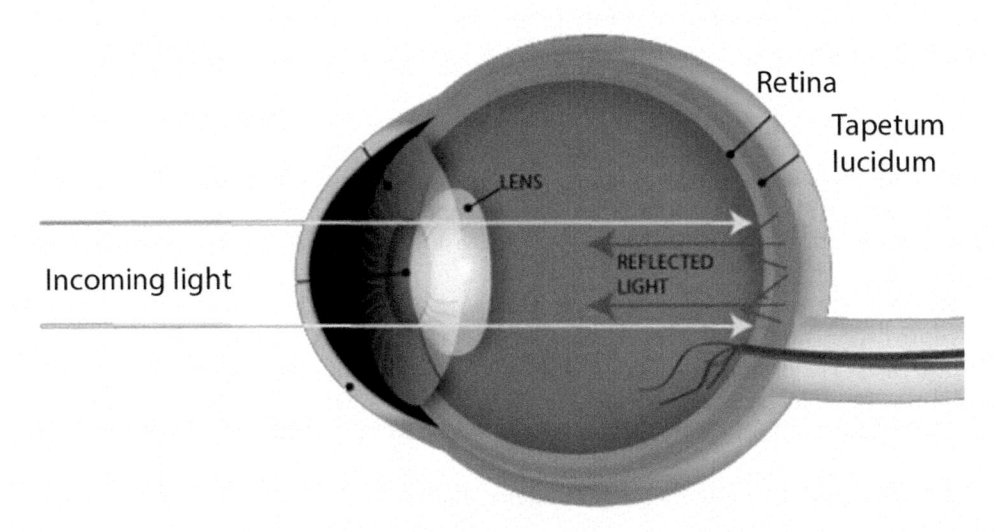

Figure 3.4 The strepsirhine eye amplifies light by stimulating retinal cells twice, once by incoming light and a second time by light reflecting off the mirror-like tapetum lucidum.

Adapted from Quora.com. https://www.quora.com/Do-human-eyes-reflect-light-at-night.

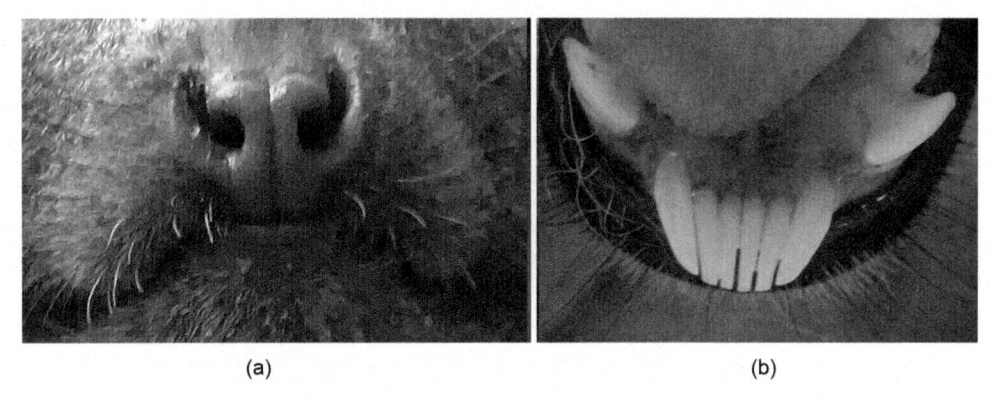

Figure 3.5 Close-up photographs of the rhinarium and toothcomb. (a) Rhinarium of a bamboo lemur, showing the slit-shaped nostrils. (b) Toothcomb in lower jaw of a ring-tailed lemur, composed of four incisors in the middle and two flanking canines. The tall pointed teeth behind the toothcomb are canine-like premolars.

Photo credit: a, adapted from Eastern Lesser Bamboo Lemur at Lemurs' Park, by J. Surrey (CC-BY-SA-4.0). b, *Lemur catta* toothcomb.jpg, by Alex Dunkel (Maky) (CC-BY-3.0).

social signals, and determine when it's time to mate, when females are fertile and sexually receptive to males. In one lemur species, during the breeding season, males engage in stink-fights as they compete for access to females, rubbing their own scent onto their tail and waving it in the air to disperse the smell in the presence of rivals. Two features that are critical to the prominent role of olfaction are the external nose, the **rhinarium,** and the **toothcomb,** part of the dentition that is a less obvious olfactory element (Figure 3.5).

As in most mammals, at the forefront of the strepsirhine's highly sophisticated olfactory system is the dog-like external nose, the rhinarium. The rhinarium is made of textured skin that is furless and perpetually moist. It is continuous with the upper lip and attached to the inside of the mouth through a gap between the upper incisor teeth, splitting the upper lip in two. The rhinarium is efficient at capturing smells from the air, and shunting them to scent-sensitive tissues inside the internal bony nose and mouth. The rhinarial skin and the whiskers surrounding it are also sensitive to touch.

The rhinarium is designed to route two types of smells – one carried by dry air and the other type by moist air (Figure 3.6). Each type moves along a different pathway toward two different parts of the **olfactory bulb,** a knob-like part of the forebrain that is responsible for relaying information about odor. Smells carried by air-borne molecules pass through the nostrils to stimulate tissues covering the delicate bones of the internal nose, which are then converted into signals that are sent through a bundle of nerves to the **primary olfactory bulb.** This is part of the olfactory system that is responsive to a wide range of environmental smells.

Odors carried in moisture are channeled into the mouth by way of rhinarial skin through the gap between the upper incisors, where they stimulate the **vomeronasal organ** (also called Jacobson's organ). This structure consists of a specialized patch of cells located just behind the incisors at the front end of the palate. The vomeronasal organ relays impulses directly to the brain through its own separate nerve, to the accessory olfactory bulb that is located behind the main bulb. This part of the system specializes in

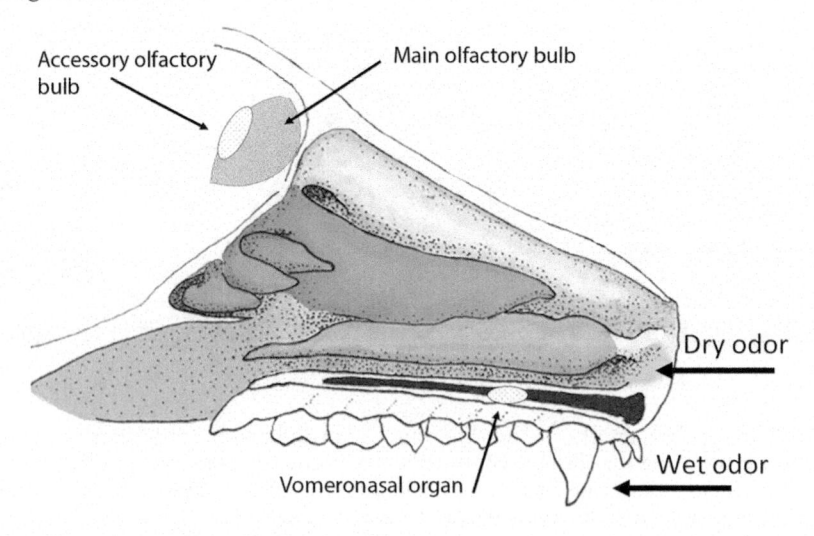

Figure 3.6 Dry odors enter the strepsirhine nose and stimulate nerves going to the main olfactory bulb. Wet odors enter through the mouth and stimulate nerves going to the accessory olfactory bulb.

Adapted from Smith, T. D., Laitman, J. T., & Bhatnagar, K. P. (2014). The shrinking anthropoid nose, the human vomeronasal organ, and the language of anatomical reduction. *The Anatomical Record*, 297(11), 2196–2204. https://doi.org/10.1002/ar.23035.

conveying information relating to an individual lemur's identity, and is especially important in regulating sexual activity in adults of the same species. It is also stimulated by the scent of predators.

The toothcomb is an important structure relating to social behavior and self-maintenance that is also associated with olfaction (Figure 3.5). The lower front teeth of lemurs and all living strepsirhines, with the exception of the aye-aye, are structurally adapted to comb and stroke the fur. The incisors and canines are arranged as a set of narrow tines separated by thin slots that can be raked through the fur. The animals groom themselves with it and also groom one another, cleaning the fur and removing parasites. Using the wet tongue during toothcomb grooming may also help stimulate the vomeronasal organ with smells carried in moisture. As a result, when a lemur nuzzles a conspecific while grooming, the process is not only hygienic, it is a social activity. Personal scent, carried in molecules of saliva, becomes part of the exchange. Lemurs also have a specialized upturned **grooming claw** on the second digit of the foot that they use to self-groom by scratching.

The universality of the strepsirhine nose among nocturnal lemurs, lorises, and galagos indicates that this feature was present in their last common ancestor. The reliance on olfaction is the primitive primate condition.

The lemuroid profile

All of Madagascar's primates are colloquially called lemurs, a simplification of lemuroid, based on Superfamily Lemuroidea.

The lemuroid profile, and the other taxonomic profiles provided in later chapters, is a summary description of ecological and biological aspects that characterize the group. It is based on the ten primary distinguishing factors developed in Chapter 1, in the section describing primates at a glance. The profile also includes other aspects that are diagnostic criteria used for classifying taxonomic groups, interpreting their phylogeny, and describing their ecology and adaptations, such as:

- Ecological domain where the animals are most active – arboreal, terrestrial, or semi-terrestrial
- General appearance – body shape, pelage (fur)
- Secondary sexual characteristics – dimorphism or monomorphism
- Craniodental morphology

The lemur profile is largely a reflection of their phylogenetic status as strepsirhines. Their features are tightly correlated and work together, and are separated as distinct parts for descriptive purposes only.

- **Phylogeny.** The lemurs' nearest relatives are lorises and galagos.
- **Body mass.** Living lemurs typically fall within the small (<1 lb., <0.4 g) and medium (1.5–10 lbs., 0.7–4.5 kg) body mass classes of living primates. One or two species exceed the medium range and fall near the lower end of the large-size class (10–20 lbs., 4.5–9 kg).
- **Activity rhythm.** Lemurs are the only primate radiation with species that have evolved three alternative activity rhythms – nocturnal, diurnal, and cathemeral.
- **Ecological domain.** All the genera are arboreal except for one. The semi-terrestrial ring-tailed lemur, a genus consisting of one species, *Lemur catta*, is the only lemur that uses the ground habitually, for up to 30% of its daily routine.
- **Diet.** Species of lemurs exhibit a seasonally varied, diversified plant and animal diet. As smaller-sized primates, lemurs have various gum- and insect-eating strategies, in addition to eating a broad spectrum of foods like stems, buds, flowers, nectar, insect secretions, and small vertebrates. While lemur diets tend to include fewer fruits than leaves as staples, the group's fundamental pattern is one of dietary breadth, flexibility, and ability to switch food sources seasonally.
- **Locomotion.** Hindlimb-dominated locomotion is the typical lemur pattern. Leg-powered vertical-clinging-and-leaping locomotion evolved in several lemur genera. Even species classified as quadrupeds are adept leapers that have evolved long legs and feet. Vertical clinging is a typical posture during sitting, resting, and sleeping, as well as the take-off position that sets up a jump in specialized vertical-clingers-and-leapers (Box 3.1).

Box 3.1   Locomotion and the intermembral index

The hindlimbs of leapers are relatively longer – sometimes much longer – than the forelimbs. One way to quantify this is a measure called the intermembral index. The intermembral index is a measurement that compares the length of the forelimb to the length of the hindlimb. It takes the ratio of forelimb-to-hindlimb length (humerus+radius length/femur+tibia length) and converts the percentage to a whole number by multiplying the value by 100. Vertical-clinging-and-leaping primates like lemurs have among the lowest indices, meaning the shortest forelimbs and longest hindlimbs.

Where do the lemurs fall within the range of indices? The index for cheirogaleids, dwarf and mouse lemurs, ranges from 68 to 72, i.e., the forelimb is 68%–72% the length of the hindlimb. The range of another vertical-clinging-and-leaping Madagascar genus, the sifaka, is 64. With their relatively long hindlimbs, the lemurs are at the lower end of the intermembral index range for primates. Leapers generally have indices below 65.

At the other end of the range of intermembral indices are primates with very long arms and a different locomotor specialization, the brachiating lesser apes, gibbons and siamang. Their indices range roughly between 125 and 150 (Figure 7.6).

- **Craniodental morphology.** The lemurs' cranium is structured around a long, large snout and a comparatively small braincase. The relatively large snout partly reflects the importance of olfaction – the extensive surfaces inside the bony nose that collect scent – and the prominence of the olfactory bulb at the front end of the braincase that is accommodated by a wide space between the orbits.

  The toothcomb is a hallmark of lemurs and other living strepsirhines, except for aye-ayes. While it is universally important in grooming, in some species the toothcomb plays a prominent role in feeding. It is used to scrape off tree gum. It is also used to scrape bark off trees to be eaten when other resources are scarce. Some species use the toothcomb to gouge trees and then mark the spot by rubbing scent glands against it to deposit secretions, as a means of communication.
- **Pelage.** Nocturnal lemur coats and markings are beige, black, and grey – good colors for camouflage. But some cathemeral and diurnal forms are more visible, with reddish, white, or two-color fur.
- **Sexual monomorphism/dimorphism.** Lemurs are monomorphic. There is very little or no difference in the morphology of male and female lemurs. Lemurs are not dimorphic (Box 3.2).

Box 3.2 Sexual monomorphism and dimorphism

The phenomena of sexual monomorphism and sexual dimorphism refers to gender-related similarities and/or differences exhibited by adults of the same species, other than the anatomical and physiological differences relating specifically to male and female reproductive systems, the primary sexual characteristics. All living strepsirhine primates, lemurs, lorises, and galagos, are monomorphic – males and females of each species are the same size, have the same craniodental morphology, and almost always have the same pelage. Examples of how this phenomenon plays out morphologically, socially, and ecologically will be addressed in later chapters. The patterns vary in other primates but, in general, features relating to the monomorphism/dimorphism dichotomy are exhibited in body mass and skeletal dimensions relating to body size, such as cranial size, where males are larger. The colors of skin patches may differ and the sexes also may have pelage differences, in color and patterns, like capes of fur in baboons that develop around the shoulders. Sexually dimorphic males also have larger canine teeth and may have a distinctive pelage color and pattern. Sexual dimorphism can be expressed in one or more of these features in a species. Behaviors relating to social roles and sexual activity also obviously differ. In monomorphic lemurs, males are not dominant over females.

- **Communication.** Scent is the essential medium of lemur communication. Vocalizations are also important. Lemurs have few facial expressions because the strepsirhine rhinarium makes it impossible to encircle the upper lip with musculature, which limits lip mobility.

  Characteristics of the face and brain, like the rhinarium and vomeronasal organ and relatively large olfactory bulb, are indicative of lemurs' dependence on olfactory input in communication. There are also features and elaborate behaviors associated with olfactory output, with odors the animals produce and spread. The lemurs have scent glands distributed on various parts of the body, including the neck, throat, chest, wrist, and genital area. The glands exude odoriferous secretions that contain **pheromones**, chemical compounds that carry species-specific personal information about gender, social group affiliation, reproductive status, physical condition, and more.
- **Social systems and mating.** Mutual toothcomb-grooming plays a very important role in lemur social life. Social groups revolve around a single or small number of breeding females, their offspring, and one or more attendant males, so the groups are consistently small. Mating seasons differ among species of lemurs.
- **Reproductive pattern.** Lemurs usually give birth to one offspring at a time, though some species give birth to two, and very rarely to three.
- **Parenting styles.** The degree of male involvement in infant caretaking varies. It is negligible in some lemur species but males can be quite helpful in others, for example by caring for newborns for long periods while the mother forages widely for food. Allowing mothers to eat first and get the choicest fruits while lactating is also an indirect form of caring for offspring that may be exhibited by males.
- **Low basal metabolic rate.** The amount of energy lemurs expend to maintain bodily functions at rest is less than in other primates, as an energy-saving adaptation.

Lemurs living in communities: diversity, ecology, and adaptation

The coexistence of multiple primate species living as a community is made possible by niche partitioning, or **niche differentiation**, the evolution of variety in the anatomical and behavioral traits among sympatric species that allows them to utilize different resources – to basically stay out of each other's way. Each species has a different way of exploiting and relating to the habitat in which it lives. Each one has carved out, by the process of natural selection, its own position, or ecological niche, within the ecosystem. This is often associated with a species-specific microhabitat, a particular place in the forest environment where essential foraging and feeding activities take place, and where they find shelter to safely rest and sleep.

How niches are differentiated in three family-level cades of lemurs (not including the aye-aye family) is detailed in studies of sympatric genera living in Madagascar's eastern rainforest at Andasibe-Mantadia National Park. The studies focus is on seven of the ten species living there and in Ankarafantsika National Park at Ampijoroa, in the western dry deciduous forest where six of the same genera reside. The primates at these sites, collectively referred to in the text as Andasibe, represent the major forms of lemur ecological diversity in the rainforests and elsewhere on the island.

There is a close correspondence between phylogeny and adaptation in the ecological characteristics of these lemurs (Table 3.1). Each of the families is characterized by a distinctive adaptive pattern linking body mass, activity rhythm, locomotor style, and diet

Table 3.1 Ecological characteristics of lemurs at Andasibe and elsewhere

Family	Common name	Genus and species	Size	Activity cycle	Locomotion	Diet
Cheirogaleidae	Greater dwarf lemur	*Cheirogaleus major*	Small	Nocturnal	Quadrupedal	Omnivorous (fr, i, n, g)
Cheirogaleidae	Rufous mouse lemur	*Microcebus rufus*	Small	Nocturnal	Quadrupedal	Omnivorous (fr, fl, i, l, g)
Cheirogaleidae	Weasel sportive lemur	*Lepilemur mustelinus*	Medium	Nocturnal	VCL	Folivorous (fr, fl, s)
Cheirogaleidae	Hairy-eared dwarf lemur	*Allocebus trichotus*	Small	Nocturnal	Quadrupedal	Frugivorous (n)
Lemuridae	Brown lemur	*Eulemur fulvus*	Medium	Cathemeral	Quadrupedal	Frugivorous (fl, l, i, n)
Lemuridae	Gray bamboo lemur	*Hapalemur griseus*	Medium	Cathemeral	VCL	Bamboo (gr)
Lemuridae	Greater bamboo lemur	*Prolemur simus*	Medium	Cathemeral	VCL	Bamboo
Lemuridae	Black-and-white ruffed lemur	*Varecia variegata*	Medium	Cathemeral	Quadrupedal	Frugivorous (l, n, fl)
Indriidae	Eastern woolly lemur	*Avahi laniger*	Medium	Nocturnal	VCL	Folivorous
Indriidae	Indri	*Indri indri*	Large	Diurnal	VCL	Folivorous (fr, fl, s)
Indriidae	Diademed sifaka	*Propithecus diadema*	Large	Diurnal	VCL	Folivorous (fr, fl, s, b)
Daubentoniidae	Aye-aye	*Daubentonia madagscariensis*	Medium	Nocturnal	Quadrupedal	Insectivorous (fr, s, fu, n)
Lemur genera elsewhere						
Cheirogaleidae	Giant mouse lemur	*Mirza*	Small	Nocturnal	Quadrupedal	Omnivorous (fr, fl, g, i)
Cheirogaleidae	Fork-maked lemur	*Phaner*	Small	Nocturnal	Quadrupedal	Gumivorous (fl, i)
Lemuridae	Ring-tailed lemur	*Lemur*	Medium	Cathemeral	Quadrupedal	Omnivorous (fr, fl, l, i)

Dietary classifications include additional food types in parentheses in order of their importance. Abbreviations: b, bark; i, insects and other animal matter (i.e., faunivory); fr, fruit; fl, flowers; fu, fungus; l, leaves; g, gum and other plant or insect secretions; gr, grasses; n, nectar, s, seeds; VCL, vertical-clinging-and-leaping.

that helps maintain their ecological separation. For example, the smallest forms, cheirogaleids, are all nocturnal quadrupeds, omnivores that tend to eat fruit, flowers, gums, and insects unless they have become dietarily specialized. The largest forms, the indriids, are diurnal or nocturnal folivores with a locomotor specialization, vertical-clinging-and-leaping. The middle-sized lemurids are cathemeral or diurnal, and are either omnivorous, frugivorous or specialized bamboo-eaters, which is a variation of folivory.

The table shows that variations within the families are centered around diet. This is to be expected for several reasons. Food resources offer the most varied types of ecological prospects because of the sheer variety of plant life. Maintaining a nutritionally balanced diet almost always requires that each species eats a variety of foods. That is why food is a pivotal point of potential competition as well as opportunity among species living sympatrically. However, even species that do not live in the Andasibe community, and live in contrasting habitats, exhibit the same strong connection between phylogeny and adaptation within families. Each clade exhibits a distinctive adaptive strategy.

Another detail that is shown in the table is the importance of fruit in the diet of all taxa. Even those species that are operationally defined as omnivores, folivores, and insectivores also eat fruit. In addition to the three species characterized as fully frugivorous, eight others consume fruit as a secondary resource. This is consistent with the concept of arboreal frugivory as the primates' adaptive zone, despite the fact that in comparison with the other regions where primates have adaptively radiated Madagascar's forests do not produce many edible fleshy fruits.

Cheirogaleids – dwarf and mouse lemurs

The cheirogaleid family includes the smallest lemurs of the Andasibe primate community, the mouse and dwarf lemurs. *Microcebus*, the mouse lemur, and *Cheirogaleus*, the dwarf lemur, occupy the small-body-size niche. They are the only primates at Andasibe that are both small and omnivorous. There is one cheirogaleid genus that is marginally larger, *Lepilemur*, the medium-sized sportive lemur, that is a folivore.

- **Activity rhythm**

    The cheirogaleids are all nocturnal, quadrupedal, mostly omnivorous, and geographically widespread, with many species living in the west of the island as well. The family includes the smallest living primate, Madame Berthe's mouse lemur from western Madagascar, *Microcebus berthae*, which weighs only one ounce (28 g) (Figure 3.7). Cheirogaleids are small-headed and large-eyed, with a pointy snout, nondescript body shape, an unspecialized dentition, and unremarkable limb proportions, features in keeping with their nocturnal habits and quadrupedal style of locomotion.

There is another, longer-term cycle of activity exhibited by some lemurs. The cheirogaleid mouse and dwarf lemurs are exceptional in having a feature called **adaptive hypothermia**, which is rare among primates. It is the ability to lower their body temperature as a way to minimize energy expenditure when food is scarce. It is exhibited by cheirogaleids living in both wet and dry climatic zones, and is manifest in two ways, **torpor** and **hibernation**. In torpor, the animals enter a lethargic state that may last for hours or several days as they "power down" and become inactive, resting in tree holes. Doing so enables some species to save as much as 40% of their normal daily energy expenditure. Torpor is a short-term response to a temporary scarcity of food, or to excessive heat or cold.

Figure 3.7 The smallest living primate, Madame Berthe's mouse lemur.
Courtesy of Manfred Eberle, www.phocus.org.

Hibernation in cheirogaleids is a long-term seasonal phenomenon that extends the process of torpor and lasts for months. This enables the animals to endure a prolonged, hot dry-season drought and food shortage. Normal activities like foraging and feeding stop. The animals survive by using an internal source of food energy, converting stored fat as the nutritional fuel needed to stay alive. The fat-tailed dwarf lemur, *Cheirogaleus medius*, prepares for fasting hibernation by gaining body weight beforehand, storing most of the extra mass in its tail (Figure 3.2). Some individuals may come close to doubling their weight to prepare for hibernation. They may spend eight months in a dormant state, holed up in the hollow of a tree, huddling with a small group consisting of an adult pair and several offspring.

- **Diet**

The thin-branch setting, low in the tree canopy, is where many of these small, omnivorous species live, searching for the insects they eat, as well as fruits and flowers. They can also be found using all levels of the forest in some locations, ecologically separated from other lemurs by the elevation at which they forage and the narrower diameter of branches that they use. Tree gum is also an important food source, especially in the drier parts of the island where gum is available year-round even when choice fruits are scarce. The gums eaten by cheirogaleids are naturally exuded by trees where the bark has been penetrated and wounded by insects. The gum at first has a syrupy consistency, but it later hardens into a crust. Nutritionally, the plant secretions eaten by some cheirogaleids may be comparable to fruit, but they are more difficult to digest. Another secretion eaten by cheirogaleids is produced by insect larvae.

Cheirogaleids are omnivorous primates, and fruits are also important to them. In a well-studied population of brown mouse lemurs, more than 60 kinds of fruits were eaten. The animals ate a larger amount of fruit from a greater variety of species as they became available seasonally. Over the course of a year, the most important foods were the berry-like fruits of mistletoe trees (generally inedible to humans) and beetles.

- **Locomotion**

Locomotion in cheirogaleids combines quadrupedalism with leaping. Quadrupedalism, walking and running on all fours, is an efficient method for a small primate to negotiate the countless small twigs it encounters, while leaping is essential to get around the inevitable gaps in the tree canopy. That is why nearly all arboreal primates are able to leap though each genus is not anatomically specialized to do so. The degree to which a species relies on a general form of quadrupedalism as opposed to leaping is reflected in its limb proportions. The hindlimbs of arboreal quadrupeds are somewhat longer than the forelimbs but in leapers they are considerably longer.

A characteristic postural behavior of cheirogaleids is extending the torso horizontally away from a tree while holding on to a branch by their feet, in order to stretch and reach for a food. This tactic requires an acute sense of balance, and considerable leg and foot strength. It can be done rapidly, lunging to grab at prey while still anchored to a support by the grasping feet.

The small-sized, omnivorous, quadrupedal Andasibe genera, *Cheirogaleus* and *Microcebus*, are typically adapted nocturnal cheirogaleids. Occupying a different niche is a third member of the clade, the sportive lemur, *Lepilemur*. It is separated from its sympatric cheirogaleid relatives by a set of three traits: larger body mass, which places the genus in the medium body size class, a folivorous diet, and a tendency to use vertical-clinging-and-leaping locomotion. The sportive lemur is the smallest of all folivorous primates. These traits also distinguish the ecological niche of *Lepilemur* from other co-located genera that belong to the folivorous indriid clade. Both of the indriids are larger than the sportive lemur and two of them are diurnal.

The sportive lemur's dentition and digestive tract have folivorous adaptations resembling other primate leaf-eaters. Compared to other cheirogaleids, its cheek teeth are larger and more crested, to produce more cutting and shearing capacity. This species has no upper incisors, which has been interpreted as a cropping feature to facilitate harvesting leaves.

Folivores need to break down the structural components in leaves, like cellulose, and toxic ingredients that may be present. *Lepilemur* has evolved several such digestive adaptations. It has a very low **basal metabolic rate**, the amount of energy needed to keep the body functioning at rest. It moves leafy material through the digestive tract slowly. Its gut hosts a colony of micro-organisms, **microbiota**, that ferment cellulose. These features give the micro-organisms time to neutralize ingested toxins, and it minimizes the amount of energy expended while the animal is otherwise inactive.

Such traits are advantageous to a folivore, given the limited nutritional value of leaves and the large amounts that they must consume. Like other leaf-eating primates, sportive lemurs spend a lot of time, roughly 50% of their daily time budget, resting and digesting. Another 31%–35% of their awake time is spent feeding on leaves, also a low-energy activity. Little time and energy are given to traveling. On a daily basis, as much as 85% of their active life is devoted to eating one type of food, leaves.

- **Social systems and mating**

Cheirogaleids tend to forage alone quietly, so as not to disturb potential prey or alert predators. They rely on eating one thing at a time, like an insect or a lick of tree gum, once every so often. For species that feed on naturally produced tree gums, it is advantageous when individuals feed independently. Gums (and sometimes insects) are **spot foods**. They are widespread but highly localized to a small spot in the environment, and difficult to harvest. Gums become exhausted or depleted with every feeding bout, so they tend to provide food for only one individual at a time. In contrast to spot foods, desirable fruits and leaves can be found hanging from many twigs at a canopy's edge and can potentially feed several individuals simultaneously. The same species of fruit trees are also sometimes found growing together in forest patches, giving a social group multiple access points to a single resource that is able to provide food for several individuals.

Cheirogaleids live in a variety of arrangements that may be called semi-solitary, but this does not mean the animals are not social. In daily life, the male and female adults do not consistently use a shared space in the habitat. Males and females plus offspring occupy different, but overlapping, territories. A **territory** is an area that an individual or a social group actively defends from neighboring conspecifics or troops. It is typically smaller than a **home range**, which encompasses all of the space used by an animal during its adult life. Though cheirogaleids rarely interact at night while foraging, the adults do monitor each other's movements through vocalizations and olfaction, and they sometimes bed down together during the day. Several such related families may sleep together in a nest, in a tangle of vegetation, or in a tree hole, and unpaired males living in the same neighborhood may join them. Up to 15 individual gray mouse lemurs have been found nesting together.

Being small and having small social groups is also an advantage in finding shelter. In golden-brown mouse lemurs, lactating females build nests out of twigs. The female collects small, leafy twigs and carries them back to a nest site. These nests may be used by a variety of related individuals.

Cheirogaleids give birth to one, two, or three offspring at a time. They are an exception to the typical primate pattern of having one birth per litter. The only other groups that have multiple births per litter are South American marmosets and tamarins; they normally give birth to twins.

Phaner, the gumivore

Phaner, the fork-marked lemur, is a cheirogaleid that lives outside of Andasibe, mostly in the west and north of Madagascar. *Phaner* is a small, nocturnal, specialized gumivore, with a set of correlating dental, digestive, postural, and social system adaptations. Eighty-five percent of the animals' feeding time is devoted to eating gum and sap, another exudate that comes from a deeper layer of a tree below the bark. Their adaptations make it possible for *Phaner* to subsist on a gumivorous diet throughout the year. This differentiates the genus from other sympatric cheirogaleids that also eat gums but shift to a more insectivorous diet during the dry season.

The fork-marked lemur's anterior teeth are specialized to scrape or gouge tree bark (Figure 3.8).

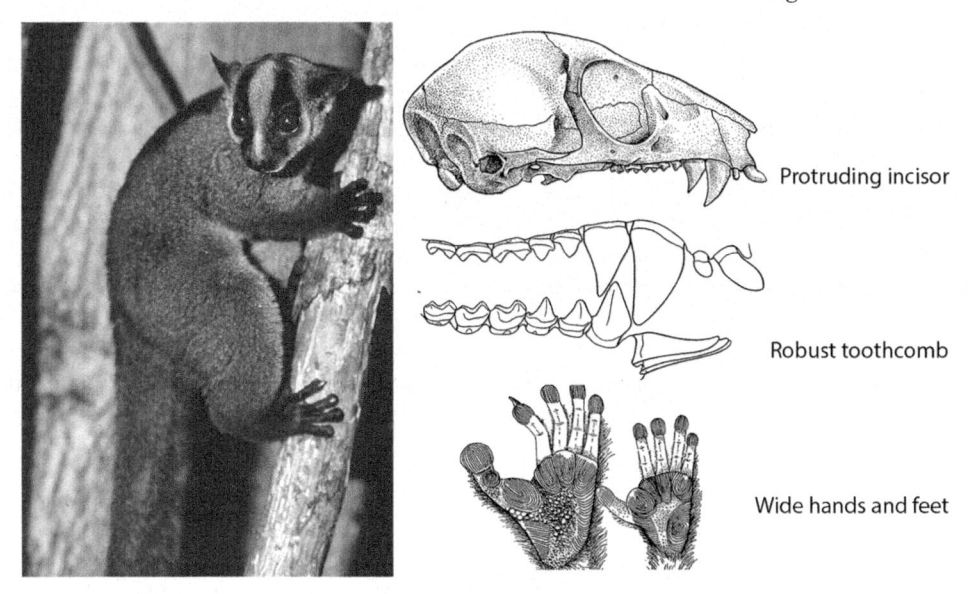

Figure 3.8 The fork-marked lemur (left). Adaptations to bark-gouging (right, from top): skull, dentition, and hand and foot.

Image credit: **Left,** *Phaner pallescens* 1985. JPG, by R. A. Mittermeier (CC-BY-SA-4.0). Right panel figures (top to bottom) adapted from Mahé, J. (1976). *Craniométrie des Lémuriens: Analysis Multivariable-Phylogénie*. Mémoires du Muséum National D'Histoire Naturelle. Series C, Tome XXXII; Maier, W. (1980). Konstruktions-morphologische Untersuchungen am Gebis der rezenten Prosimiae (Primates). Abhandlungen der Senckenbergischen Naturforschenden Gesellschaft 538:1–158; Biegert, J. (1961). *Volarhaut der Hände und Füsse*. Primatologia 2: 1–326.

- The entire *Phaner* toothcomb is elongated and reinforced on the sides by unusually robust lower canines.
- The upper central incisors are large, robust, and horizontally protruding.
- The hands and feet are broad, with large tactile pads at the base of the digits to minimize slippage while gripping.

Tilting its head to the side, the animal uses its large upper incisor, the upper canine, and the tall upper and lower anterior premolars behind it, to bite into tree bark to stimulate production of gum. The *Phaner* tongue is very long, to facilitate lapping up oozing exudates.

The caecum in the large intestine is well developed in *Phaner* to enable the digestion of gum which, like digesting leaves, is chemically challenging. Microbiota facilitate the conversion of gum into energy.

Postural adaptations help these animals secure an advantageous position on medium-diameter vertical tree trunks where tree gum is often located several meters above the ground. These include pointy nails on digits other than the flat-nailed hallux. Their nails are reinforced against bending by a ridge that runs longitudinally down the center.

The spot-food diet of *Phaner* is consistent with a semi-solitary lifestyle comprised of a female and her offspring and one male. The gums eaten by the fork-marked lemur in the dry western forests of Madagascar come predominantly from a small number of rare trees belonging to one tree species in the ecosystem. They are spaced well apart in the habitat. The production of gum varies from tree to tree and the supply at each site is drained after feeding, but eventually becomes available again. This means it is advantageous for the animals to make regular rounds of their territories to identify when a site is replenished. Small social group size minimizes intra-group competition for access as well as inter-group competition from other conspecifics in the vicinity.

Studies that employed radio-telemetry tracking over several years have shown how *Phaner* individuals, and as many as eight social groups occupying the same area, are organized. Adult male-female pairs live in a commonly shared, defended territory. Territorial defense may involve vocalizations, chasing, and fighting. The behavior is effective. There is no spatial overlap among neighboring social groups. Male-female pairs may sleep together during the day in tree holes or nests made of twigs, and use the same sparsely distributed trees as feeding sites, but they do not forage or feed together. The pair maintains a distance of about 100 m (328 ft.) during their nightly foraging rounds and avoid meeting at the same tree.

Lemurids – cathemeral lemurs

Lemurids are a medium-sized primate family that is widespread on Madagascar. They are the only family of primates whose activity cycle is cathemeral, active during blocks of hours throughout the 24-hour day-night cycle. The timing of active periods relates to seasonality and habitat. One of the ecological consequences of this pattern is that it allows for niche differentiation in the presence of sympatric species that are diurnal or nocturnal.

Lemurids have long muzzles and widely spaced eyes. They are quadrupeds with distinctly long hindlimbs and, like many of Madagascar's primates, exhibit a wide range of locomotor and postural behaviors – leaping, vertical-clinging-and-leaping, foot-hanging, suspension by all four limbs, and others. There are five living genera that sort into two groups according to their dietary adaptations. The more omnivorous forms are the ring-tailed lemur, *Lemur*, brown lemur, *Eulemur*, and ruffed lemur, *Varecia*. They rely on fruits and leaves. The other lemurids, the bamboo lemur, *Hapalemur,* and greater bamboo lemur, *Prolemur*, rely on bamboo. The bamboo-eaters have a variety of cranial and dental adaptations enabling their highly specialized diet. As a clade, lemurids are the most sociable of Madagascar's primates. They tend to live in the largest, most interactive social groups.

- **Diet and locomotion**

Two cathemeral lemurids at Andasibe, the brown lemur and the bamboo lemur, each inhabit distinct ecological niches, from each other and from the other primates in the community, such as the medium-sized, nocturnal sportive lemurs and the woolly lemurs, both folivores. *Eulemur*, the brown lemur, is frugivorous. *Hapalemur*, the bamboo lemur, eats bamboo, a highly abrasive diet that heavily wears down the cheek teeth. The genera are also distinguished by locomotor pattern. *Eulemur* is a quadruped and *Hapalemur* is a vertical-clinger-and-leaper.

The strong niche differentiation of bamboo lemurs from other primates is made possible by their divergent morphology, including the ability to tolerate significant levels of **secondary compounds,** chemicals in foods that tend to be unpalatable, are difficult to digest or are toxic. *Eulemur* ingests these substances when eating mature leaves and unripe fruit. The bamboo diet of *Hapalemur*, which contains large amounts of cyanide, is even more extreme and further distinguishes its ecological niche. Some of the material that *Hapalemur* eats from inside the hard trunk of the plant is also exceptionally difficult to access. On a daily basis, the amount of food *Hapalemur* eats relative to body mass, is estimated to contain 12 times more than a human lethal dose of cyanide. This includes a quantity of bamboo parts, young shoots (the stems where leaves form), leaves, and pith (the softer core material of a branch or trunk, for example).

A third cathemeral lemurid at Andasibe is the quadrupedal ruffed lemur, *Varecia*. *Varecia* is the largest lemurid genus, and that alone distinguishes it ecologically from the others. At 7–10 lbs. (3.2–4.5 kg), all three lemurids fall into the medium body size class, but ruffed lemurs are larger. *Varecia* is highly frugivorous and frequents large trees, where it spends most of its time foraging in the canopy well above the ground. Seventy-five to ninety-five percent of its diet consists of fruit.

- **Social systems and mating**

Eulemur and *Hapalemur* species exhibit a variety of social organizations. Some live in long-lasting, **pair-bonded** monogamous groups, essentially exclusive social and reproductive pairings of a single male and female plus their immature offspring, while others live in groups of about a dozen individuals. The larger groups may temporarily split up into subgroups that reunite, a behavioral pattern termed **fission-fusion**. Both of these systems, which are rare among primates, will be discussed further in later chapters. Monogamous and fission-fusion societies may be advantageous for several reasons relating to dietary ecology, mating systems, group dynamics, and life-history strategies.

Several of these factors come into play in *Varecia*, a genus that depends on the ripe fruit of a few tree species of widely dispersed large trees. They live in troops comprised of multiple adult males and females that break up into subgroups of two to six individuals. These units collaborate in rearing offspring. Foraging female *Varecia* "park" their infants with other individuals for safeguarding in a nest that was built before giving birth. The helpers may be genetically affiliated mothers or other males and females of the troop that may not be such close kin. Breeding and non-breeding males may engage in guarding nests that belong to several females, giving them the opportunity to forage widely. The males also assist in rearing juveniles by spending time interacting and traveling with the young, and helping them learn to feed.

Lemurids living elsewhere

Members of the lemurid family found elsewhere on Madagascar have comparable dietary and locomotor adaptations, split between bamboo-eating and fruit-eating. But one genus is uniquely semi-terrestrial, the ring-tailed lemur, which consists of one species, *Lemur catta* (Figure 3.9). It spends 30% of its time on the ground, the most of any lemur, often in relatively open habitats. It is highly social, living in large multimale-multifemale groups that may include more than 20 individuals, the largest social groups of all Madagascar lemurs. Ring-tails are common in the drier southern parts of the island where they are highly omnivorous, eating fruit, leaves, flowers, insects, and more. Their dietary flexibility and use of semi-terrestrial microhabitats are assets in that sparse, resource-poor environment. These are the features that enable *Lemur catta* to inhabit their own ecological niche, separate from other lemurs in the primate community.

Figure 3.9 The semi-terrestrial ring-tailed lemur.

Photo credit: Ring-tailed Lemur (*Lemur catta*) at Berenty, by Alex Dunkel (Maky) (CC-BY-3.0).

Like most of Madagascar's lemurs, females in ring-tail groups enjoy social dominance over males. For example, females have priority access to feeding and resting sites, and they have been observed to push undesirable males away with a slap of the hand or a brief chase.

The social status of individual females and males is determined by gender-specific **dominance hierarchies,** a linear ordering of the society's power structure. Hierarchies tend to be stable among females but males regularly vie among themselves for authority. Higher-ranking males advertise their status by carrying aloft and waving their long, black-and-white striped tail that carries their own personal scent.

Dominance rank determines access to resources. It is evident during terrestrial travel also, when females and their close kin may coordinate troop movements with higher-ranking males while lower-ranking males follow behind, with tail and head lowered in a sign of submissiveness. Social status strongly influences mating patterns in both sexes. While females will copulate with more than one male during the brief mating period, higher-ranking males tend to mate first. But females have latitude in selecting mating partners, and higher-ranking females have more options.

The male dominance hierarchy dissolves during the breeding season when they battle one another intensely and frequently for access to females. Many males may suffer serious bite wounds at that time. The intensity is related to several factors. The ring-tail's breeding season is brief, lasting about two weeks each year. During that time the adult females enter **estrous,** the phase of the reproductive cycle when they are fertile and display a willingness to mate. But that period for each female only lasts for several hours, and not more than a day, and different females are in estrous on different days. As a consequence, the more dominant males are continuously engaged in efforts to be selected as a female's mate.

In ring-tailed lemurs social organization is regulated to an important degree pheromonally, via olfaction. Odors are often deposited by **scent-marking**, such as rubbing a throat gland on a branch or rubbing the rump against a tree trunk, a behavior seen in both females and males. During the breeding season, males not only "stink fight" to fend off the competition, they also get ready for "stink-flirting" to attract females by pulling their tail across a patch of glandular skin in the forearm above the wrist. They follow up by waving the aromatized tail in the air to advertise their social and sexual status pheromonally. Males with stronger odors may have better success. Lab experiments demonstrate that there is greater odor strength in breeding males than non-breeding males. In another ring-tail behavior that combines sound and smell, the wrist gland, which has a horny spur embedded in it, is snapped against a tree to deposit scent with a rub of the gland.

Indriids – leaf-eating specialists

Indriids are a family of medium- and large-bodied lemurs. They are an adaptively uniform clade set apart ecologically from all other lemurs by a combination of primary characteristics: size, diet, and locomotion. They are all folivorous and the most adept vertical-clingers-and-leapers on Madagascar, with very long hindlimbs. Their premolar and molar teeth are relatively large and have prominent sharp crests for shearing leaves.

The nocturnal woolly lemur, *Avahi*, is camouflaged with a dull-colored brown coat. Most of the diurnal indriid species, the sifakas, *Propithecus*, and indris, *Indri*, usually have a dense pelage that is visible during the day, white, white and black, or white and brown, and a contrasting black-skinned face topped by a different fur color on their heads, like a cap. They have relatively short, broad faces and wide-set eyes. Their hands and feet are large. The hallux and thumb are proportionately very large and strong, and widely offset to facilitate grasping. The indri is distinguished from all other lemurs by its rudimentary tail, which is barely visible. They also have a unique vocalization, an early morning **dawn call** that carries for significant distances through the forest. A variety of primates in other taxonomic groups also call loudly each morning.

Two of the three indriid genera are diurnal and large, *Indri* and *Propithecus*, essentially setting apart their ecological niches from all other sympatric lemurs. They are also distinguished by folivory, with a dentition that is morphologically specialized. The toothcomb is robust but consists of only four teeth – instead of six – and there are only two premolars per quadrant, for a **dental formula** of 2.1.2.3/1.1.2.3 (Box 3.3). Toothcomb robusticity may relate to the habit of tree-gouging as seen in the sifaka, *Propithecus*. Male sifaka use the toothcomb to gouge preferred feeding and sleeping trees before scent-marking the spots, often near territorial boundaries. The reinforced toothcomb enables indriids to peel and eat tree bark when other foods are scarce.

There are two indriids at Andasibe, the small woolly lemur, *Avahi*, and the large indri, *Indri*. While both are folivores and dedicated vertical-clingers-and-leapers, the woolly lemur is nocturnal and the indri is diurnal. These differences in activity patterns allow them to inhabit separate ecological niches. A third indriid lives elsewhere on Madagascar, *Propithecus*, the sifaka. It is adaptively similar to *Indri*: a large, diurnal, folivorous, vertical-clinger-and-leaper.

Indriids present an important functional morphology that distinguishes them from other specialized leapers, such as tarsiers and galagos. Indriids are thigh-powered leapers while the others are foot-powered leapers. Vertical-clinging-and-leaping is the indriids'

> **Box 3.3 The dental formula**
>
> The dental formula is a commonly used descriptive and diagnostic feature of all mammalian taxa. It is a short-hand method that specifies the organization of the teeth in the upper and lower jaws, how many and of what type. Because the right and left sides of upper and lower jaws are symmetrical, the dental formula only records the numbers and types of teeth for one of the four quadrants, front to back, beginning at the midline. For example, the cheirogaleid formula is 2.1.3.3, which represents two incisors, one canine, three premolars, and three molars in each quadrant. That is considered the basic formula of all lemurs. There are several exceptions. One occurs in the indriids. They have only one lower incisor that complements the two upper incisors. To express that, we add a second set of counts offset by a slash (/): 2.1.2.3/1.1.2.3. In *Lepilemur*, where there are no upper incisors, the dental formula is 0.1.3.3/2.1.3.3.

main locomotor style, but they also are adept at clambering and climbing in the canopy, hanging beneath branches while foraging and feeding, and moving about upside-down, using their arms and large hands and feet. When they come to the ground, indriids use an unusual style of bipedal locomotion. They hop sideways.

Socially, indris and woolly lemurs are organized in monogamous groups while the sifaka tends to live in multimale-multifemale groups that may include more than a dozen individuals. The dawn calls of indris are a trait associated with a pair-bonded social system and territorial behavior. An indri group defends its territory, announcing its presence by vocalizing, which discourages potential intruders. The calls may be performed by one of the adults, by both together in duettes as they coordinate notes and rhythms, and by adults and offspring. The combination of dawn calling, duetting, social monogamy, and small territory size is a pattern also seen among other primates, as will be discussed in later chapters. Territoriality may play a role in securing ecological resources like food or shelter, keeping rivals at bay in order to ensure the pair-bond, or safeguarding offspring against attacks by conspecific outsiders.

Daubentoniids – aye-ayes

The aye-aye is the only genus of the daubentoniid family. It is one of the most unusual living primates and has numerous traits that no other primates have. Only one species exists today, and it exhibits an extreme dietary specialization coordinated by unique traits that have evolved in several systems. It eats wood-boring insect larvae. Ayes-ayes have a feature called echolocation, a sonar-like hearing system, and a clawed middle finger with the functionality of an arm, able to flex, extend, move laterally and rotate about its long axis. These are parts of a foraging and food-gathering system used in detecting, snatching, and extracting larvae from an access hole the animal bites into a tree. The process is an extreme example of what primatologists call **extractive foraging** (Figure 3.10), a method seen in several other primate species to be discussed in later chapters.

The nocturnal aye-aye, *Daubentonia madagascariensis*, is the only living primate genus and species assigned to a family of its own. Its highly unusual feeding niche and

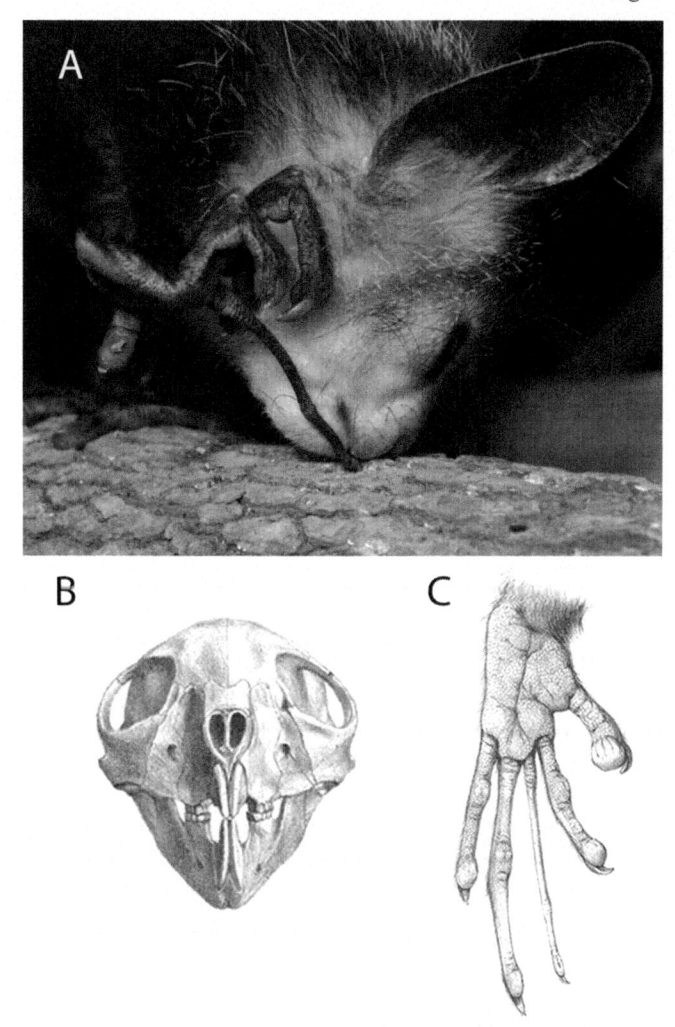

Figure 3.10 (a) An aye-aye tap-scanning with its wiry middle finger, ears cocked to listen and nose pressed to the bark to smell. (b) An aye-aye skull illustrating the specialized gnawing incisors. (c) Close-up of the hand showing finger sizes and shapes, and clawed finger tips.

Image credits: (a), David Haring, Duke Lemur Center. (b), Owen, R. (1866). On the aye-aye (*Chiromys*, Cuvier: *Chiromys madagascariensis*, Desm.; *Sciurus madagascariensis*, Gmelin, Sonnerat; *Lemur psilodactylus*, Schreber, Shaw). *Transactions of the Zoological Society of London*, 5, 33–101. (c), Nash Collection, University of Wisconsin Libraries.

derived anatomy separate the aye-aye ecologically from all lemurs everywhere on the island, including at Andasibe. While its foraging and feeding habits are very particular, they are not restrictive, as the medium-sized aye-aye is highly adaptable and lives in many different habitats in the eastern and western parts of the island, though it does not occur in spiny forests. In Table 3.1, we classify the aye-aye's diet as insectivorous because of the animal's unique physical characteristics directly related to hunting, capturing, and eating larval insects, even though its specialized dentition is also used to eat fruits and other material. They eat fruit meat and seeds that are protected by hard outer coverings, adult insects, fungus, and cancer-like growths (canker) that appear on diseased trees as a result of fungal or bacterial damage. Aye-ayes are basically arboreal quadrupeds that frequently climb tree trunks to access the understory. They use the forest floor, where rotting and dead trees can be found, more than any of Madagascar's primates except the semi-terrestrial ring-tailed lemur.

The aye-aye's specialized feeding adaptation is comprised of:

- A relatively large brain – the largest of the living lemurs – with expanded olfactory and auditory regions
- Huge external ears
- A massively built, short snout
- A powerful neck
- Extremely specialized teeth and jaws
- Hands with long fingers – especially digits III and IV, the middle and ring fingers – tipped with large claws

Aye-ayes target wood-boring beetle larvae that live deep inside trees, which is a complicated process requiring special search methods and anatomical tools to extract the grubs. They eat the seeds of fruits that are protected by a woody covering that is harder than any fruits or seeds consumed by South American primates, to be discussed in Chapter 4, that also specialize in hard-shelled, mechanically defended food, including Brazil nuts. Aye-ayes can even open coconuts. For perspective on the aye-aye's dental and manual apparatuses which make this possible: where people typically use hammers, chisels, screw drivers, or hacksaws to open a coconut to get at the fruit meat inside the aye-aye uses its front teeth. Using its weirdly thin, twistable digit tipped with a claw, the aye-aye can snag a grub by feeling around inside a tree hole, without being able to see its target.

The craniodental anatomy that supports extractive foraging includes:

- A reduced dental formula, 1.0.1.3
- Enormous, continually growing incisors
- A specialized lower jaw joint
- Flat cheek teeth designed for crushing and grinding

The extremely reduced dental formula – with a total of 20 teeth compared to 36 teeth in a mouse lemur – makes room for the long roots of the aye-aye's huge incisors. It is with these huge, well-anchored incisors that the aye-aye is able to bite into extremely hard surfaces like a tree trunk or coconut. Additionally, the incisors grow continuously throughout the life of the individual. This compensates for wear at the tip of the incisors that results from biting into such hard, woody material. The incisor crowns have a thick, hard enamel covering on the outer side, and the teeth wear in a way that produces a beveled, chisel-like sharp edge. The specialized jaw joint permits a very large gape and protects the lower jaw from the twisting it encounters while biting and tearing at wood.

The grubs that aye-ayes eat are in an immature larval stage. They are soft and relatively large, some measuring several inches in length, and have a high protein content. While many insectivorous primates have sharp, pointed cheek teeth to chew the tough exoskeletons of the insects they eat, the aye-ayes have flat cheek teeth that are suitable for crushing and grinding soft insect larvae. The aye-aye's back teeth are biomechanically adapted to processing soft stuff while its front teeth are adapted to gnawing through very tough stuff.

Guided to a feeding spot by an acute sense of smell and/or visually noticing a good prospect, such as a decaying, infested tree trunk, to pinpoint larvae the aye-aye uses a specialized form of hearing to discover the tunnels where grubs live inside. Homing in on the source of a sound by hearing is a typical behavioral process among mammals, but the aye-aye's capabilities are greatly enhanced. The aye-aye's large ears improve hearing by working like dish antennae to catch sound waves and direct them toward the ear drums.

Daubentonia is one of several mammals (others include bats and whales) that has evolved a sound-localizing system called biological echolocation. This primate uses a drumming technique called **tap-scanning**, or percussive foraging, to generate outgoing sound and an incoming echo. The wiry middle finger plays a central role in the process. The aye-aye rapidly taps the claw of its middle finger on the surface of the tree while it cocks its large ears and listens for the reflected sound that signals where the tree is hollow and where the grubs can be found. Next, the aye-aye bites a hole into the tree using the incisors as a pincer to tear away chunks of wood and expose a tunnel where larvae reside, aided by powerful neck muscles. The aye-aye then inserts its middle finger into the hole, finds a grub and pulls it out and eats it (Figure 3.10).

Individual aye-ayes live in semi-solitary groups, which enables them to forage alone, although pairs have been seen foraging together. Small group size is consistent with the species' strong focus on the time-consuming, complex food-access behaviors associated with a diet that incorporates spot foods that are widely distributed but highly localized in space, that offer limited personal access and are exceptionally difficult to obtain.

Paleocommunities

The sites on Madagascar that produce subfossils are often caves (some submerged in fresh water drainage systems), surface or shallow deposits in areas that were once marshy wetlands, and riverbanks, places that offer protection from significant physical damage. They frequently yield large numbers of specimens representing many taxa of animals and plants, extinct and extant.

One cave, Ankilitelo, in the southwest, dated 500–600 years old, yielded 5,000 subfossil bones from 32 species of mammals. The primates in that paleocommunity consisted of at least seven extant and five extinct lemur species. Many of Madagascar's paleocommunities are continuous with today's primate communities. The lemur communities that primatologists see today are contemporary segments of a dynamic evolutionary history. They are a slice in time, like the prehistoric (referring to recent times before written records) slice of time revealed at Ankilitelo.

Madagascar's subfossils provide fundamental information about the diversity of the island's primates during timescales spanning centuries and millennia. They document novel evolutionary adaptations, and the consequences of anthropogenic threats that affected lemurs then and continue to this day. Two elements are key to understanding the demise of lemur species represented in the subfossil record. Essentially all of them were gigantic species, and they all became extinct after the arrival of people on the island several thousand years ago.

- **Diversity**. There are eight extinct genera in the subfossil record – roughly half as many as today's total primate fauna – and 16 species. One subfossil species is included in an extant genus. It is a large extinct aye-aye. The subfossil genera include animals that are sufficiently different from the living forms to be classified in three unique families.
- **Evolutionary adaptations**. The size range, locomotor diversity, and utilized habitats were remarkable within paleocommunities. Body sizes generally ranged from 3.5 oz. (100 g) to 66 lbs. (30 kg), but subfossil remains at one site document a notable outlier. The largest genus, *Archaeoindris*, which lived about 4,000 years ago, was gorilla-sized. As many as 20 lemur species existed sympatrically in paleocommunities. Many subfossils were either nocturnal or cathemeral. Locomotor patterns of the extinct genera included quadrupedalism; deliberate slow-climbing; suspensory behavior; and vertical clinging and climbing behaviors not matched by any living primate but resembling Australia's koala, a marsupial. Ecologically, most genera were fully arboreal but two were semi-terrestrial, more dedicated to ground-dwelling than modern ring-tailed lemurs, with several skeletal adaptations resembling baboons. Dietarily, the range encompassed folivory, frugivory, seed-eating, and extractive foraging in the manner of an aye-aye, among other patterns.
- **Community ecology**. These extinct lemur giants were part of Madagascar's **megafauna**, animals that are particularly large in body mass. The giant lemurs coexisted with other gigantic animals, including so-called dwarf hippopotamus and the largest flightless birds that ever existed, elephant birds, nearly 10 ft (3 m) tall and weighing up to 1,100 lbs. (500 kg). The fact that extant lemur species lived in the paleocommunities together with giant lemurs and elephant birds means that the evolutionary journey of present day lemurs was influenced by the animals that are now extinct. The pattern of niche differentiation exhibited today is part of the continuum that was evolving in a paleocommunity.

The conservation status of Madagascar's lemurs

The lemurs of Madagascar today are the world's most at-risk primates. More than 95% of lemur species are classified as Critically Endangered, Endangered, or Vulnerable. Vulnerable is defined by the IUCN as "… likely to become endangered unless the circumstances threatening its survival and reproduction improve." The most severe threats come from habitat loss due to deforestation, logging, and clearing forests for agriculture.

Like many other large mammals, species of Madagascar's larger-bodied primate families, except for the aye-aye, are in the most jeopardy, partly because they require more food and have longer interbirth intervals, which means it takes them longer to generate and secure reproducing populations that are large enough to withstand serious declines (Figure 3.11). With only one genus consisting of one species in the family, 100% of daubentoniids are Endangered, but not Critically Endangered. The severity of threats to Madagascar's primates is felt even by the smallest species, the cheirogaleids, where well over 50% are Endangered and Critically Endangered.

The indriid family, the largest Madagascar primates and the most easily seen because two of the three genera are diurnal, is particularly vulnerable. One of them, the white-and-brown, dawn-calling Coquerel's Sifaka, *Propithecus coquereli*, is currently regarded one of the 25 most imperiled primate species (Figure 1.11).

IUCN classified the ring-tailed lemur, *Lemur catta*, as Endangered in 2022. Ring-tails are a case study of rapid population decline. In the 1960s and 1970s, they were

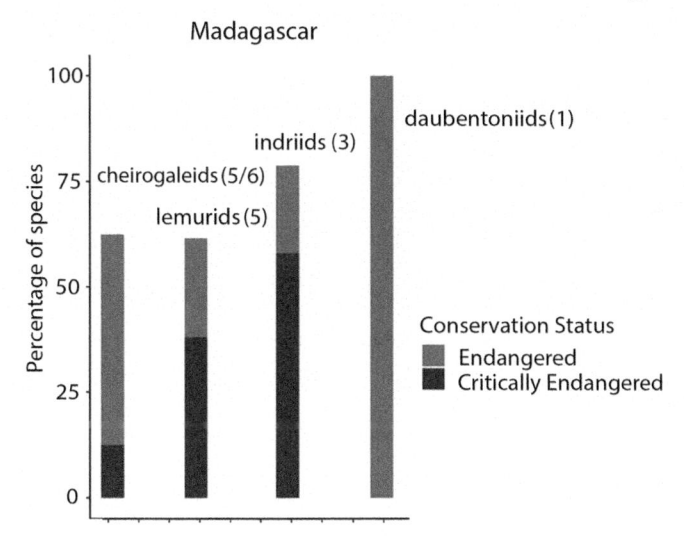

Figure 3.11 The threat levels to the four Madagascar lemur families. Numbers in parentheses refer to the number of genera in each group (genus *Lepilemur* is not included here among the cheirogaleids).

Adapted from Fernández, D., Kerhoas, D., Dempsey, A., Billany, J., McCabe, G., & Argirova, E. (2021). The current status of the world's primates: Mapping threats to understand priorities for primate conservation. *International Journal of Primatology*. https://doi.org/10.1007/s10764-021-00242-2.

flourishing in south and southwestern Madagascar, where these semi-terrestrial animals live in drier, relatively open country. This made *Lemur catta* ideal for primatologists to study intensively, and ring-tails became the foundation for all subsequent ecological and behavioral research on Madagascar's lemurs. However, a detailed survey of 32 sites in their historic range estimated that, as of 2017, the population had declined by 95% since 2000.

What are the reasons for this 95% reduction in population in a roughly 15-year time period? The principal reasons for the decline are:

- Habitat loss
- Hunting
- The illegal pet trade

Madagascar has attracted enormous local, national, and international attention as one of the world's biodiversity hotspots with many endemic and threatened species. There are a range of projects and programs designed to mitigate the existential threats facing lemurs. The result is a large network of nearly 50 national parks and nature reserves spread across the island that are dedicated to protecting fauna and flora, with a particular emphasis on the lemurs. Dozens of non-governmental organizations (NGOs) are committed to conservation projects in Madagascar, and ecotourism is an established method to increase awareness, educate people, and raise funds.

These are among the positive responses to the conservation crisis that inspire optimism among the global leaders of the primate conservation movement, as discussed in Chapter 9.

Madagascar: lemurs
Key concepts

Biodiversity hotspot
Strepsirhines and haplorhines
Olfaction-vision tradeoff
Sexual dimorphism and monomorphism
Microhabitat

Quizlet

1. What are three of the features that make lemurs strepsirhine primates?
2. Why does the fat-tailed dwarf lemur hibernate?
3. What is a toothcomb?
4. What primate feature is measured by the intermembral index?
5. How do lemurs communicate?
6. Why are large-bodied primate families more endangered than smaller ones?
7. What is a subfossil?

Bibliography

Atsalis, S. (2015). *A Natural History of the Brown Mouse Lemur*. Oxfordshire, Routledge.

Ganzhorn, J. U. (1989). Niche separation of seven lemur species in the Eastern rainforest of Madagascar. *Oecologia*, 79(2), 279–286.

Ganzhorn, J. U., Wright, P. C., & Ratsimbazafy, J. (1999). Primate communities: Madagascar. In C. Janson, J. G. Fleagle, & K. Reed (Eds.), *Primate Communities* (pp. 75–89). Cambridge, Cambridge University Press.

Glander, K., Wright, P., Seigler, D., Randrianasolo, V., & Randrianasolo, B. (1989). Consumption of cyanogenic bamboo by a newly discovered species of bamboo lemur. *American Journal of Primatology*, 19, 119–124. https://doi.org/10.1002/ajp.1350190205

Gould, L., Sauther, M., & Cameron, A. (2010). Lemuriformes. In C. J. Campbell, A. Fuentes, K. C. MacKinnon, S. K. Bearder, & R. M. Stumpf (Eds.), *Primates in Perspective* (pp. 58–79). New York, Oxford University Press.

Kappeler, P. (2012). Behavioral ecology of strepsirrhines and tarsiers. In J. C. Mitani, J. Call, P. M. Kappeler, R. A. Palombit, & J.B. Silk (Eds.), *The Evolution of Primate Societies* (pp. 17–42). Chicago, University of Chicago Press.

LaFleur, M., Clarke, T. A., Reuter, K., & Schaeffer, T. (2016). Rapid decrease in populations of wild ring-tailed lemurs (*Lemur catta*) in Madagascar. *Folia Primatologica*, 87(5), 320–330. https://doi.org/10.1159/000455121

Sterling, E. J., & McCreless, E. E. (2006). Adaptations in the aye-aye: A review. In: L. Gould & M. L. Sauther (Eds.), *Lemurs. Developments in Primatology: Progress and Prospect* (pp. 159–184). New York, Springer.

Walker-Bolton, A. D., & Parga, J. A. (2017). "Stink flirting" in ring-tailed lemurs (*Lemur catta*): Male olfactory displays to females as honest, costly signals. *American Journal of Primatology*, 79(12), e22724. https://doi.org/10.1002/ajp.22724

Wright, P. C. (1999). Lemur traits and Madagascar ecology: Coping with an island environment. *American Journal of Physical Anthropology*, Suppl. 29, 31–72. https://doi.org/10.1002/(sici)1096-8644(1999)110:29+<31::aid-ajpa3>3.0.co;2-0

4 South America
New World monkeys

Chapter Contents

South America's geography and climate	69
Platyrrhines are haplorhine primates	71
Platyrrhines are anthropoid primates	72
Platyrrhines and catarrhines	75
The platyrrhine profile	76
Platyrrhines living in communities: diversity, ecology, and adaptation	81
Atelids – frugivore-folivores	82
Pitheciids – fruit-huskers and seedeaters	87
Cebids – predatory omnivorous frugivores	90
Paleocommunities	99
The conservation status of platyrrhines	99
Key concepts	101
Quizlet	101
Bibliography	101

Superfamily Ateloidea – New World monkeys
   Family Atelidae (atelids) – prehensile-tailed platyrrhines
      Subfamily Atelinae (atelines)
      Subfamily Alouattinae (alouattines)
   Family Pitheciidae (pitheciids) – platyrrhine fruit-huskers and seedeaters
      Subfamily Pitheciinae (pitheciines)
      Subfamily Homunculinae (homunculines)
   Family Cebidae (cebids) – predaceous frugivorous platyrrhines
      Subfamily Cebinae (cebines)
      Subfamily Callitrichinae (callitrichines)

South America's geography and climate

This chapter focuses on the 16 genera belonging to three families of primates that live in South America, the New World monkeys, also called the platyrrhines. This group of primates is endemic to the region. There are several platyrrhine species that inhabit the rainforests of Mexico and Central America, as well as continental South America. The name New World monkeys references the centuries-old geographic distinction between Eurasia

DOI: 10.4324/9781003257257-4

and Africa (the Old World) and the Americas (the New World). They are also called neotropical monkeys and ateloids, a simplification of formal taxonomic nomenclature.

South America is about 30 times larger than the island of Madagascar, the region discussed in Chapter 3, yet it has about the same number of primate genera (Figure 1.2). Much of its landmass is situated in the tropical zone, the perpetually warm region that spans the equator and receives more direct sunlight than any other area on earth, as discussed in Chapter 1. At the equator, the sun at noon is aligned overhead throughout the year, so the temperature is stable, there is no true winter, and daylength is consistently divided into nearly equal 12-hour periods of night and day. The equator passes directly through the Amazonian rainforest across a 2,000-mile (3,219 km) trajectory that roughly parallels the east-west course of the Amazon River, and the tropical zone extends about 1,600 miles (2,574 km) north and south of the equator (Figure 4.1). That makes for a vast area of lowlands east of the Andes Mountains, about 60% of the continent, consisting of tropical and subtropical environments. This is a principal reason why Amazonia encompasses the largest rainforest biome in the world, with more land-living plants and animals than any other place on Earth.

The continent is entirely surrounded by oceans, the South Atlantic Ocean, including a portion that is formally called the Caribbean Sea, and the South Pacific Ocean, except for the narrow land link of Panama in the north which is only about 50 miles (80.5 km) wide. The oceans have a strong impact on the continent's climate and habitats because of their influence on temperature, wind, rainfall, and seasonality. Warm Atlantic waters along the eastern coastline contribute to the warmth and humidity of the trade winds

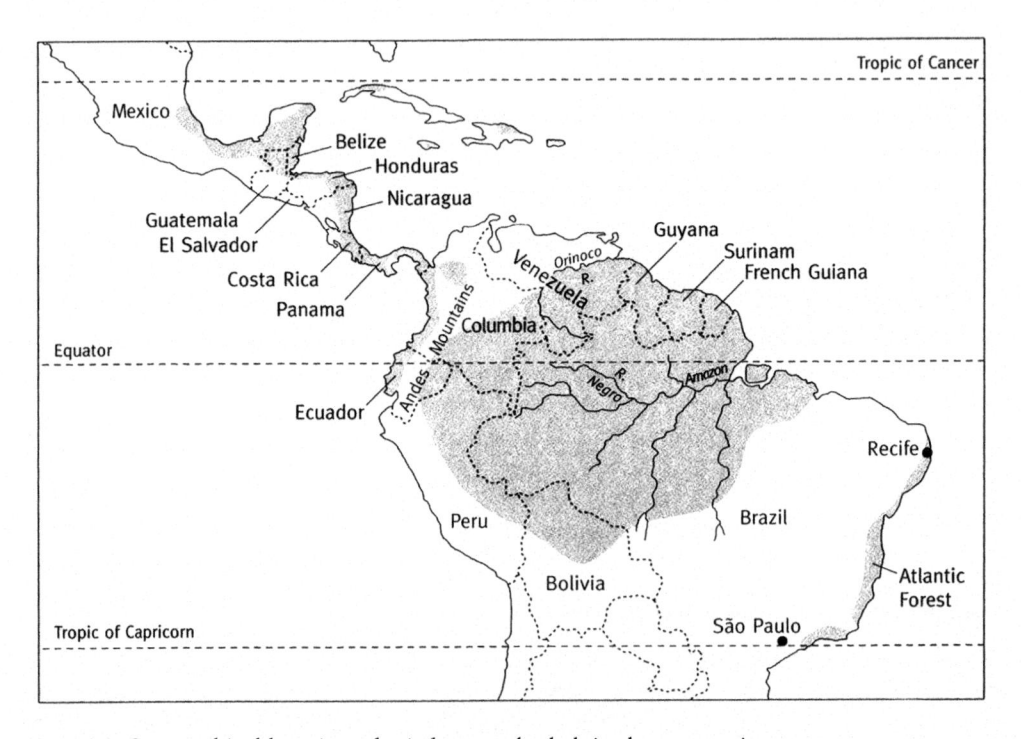

Figure 4.1 Geographical location of rainforests, shaded, in the neotropics.

From Corlett, R., & Primack, R. (2011). *Tropical Rain Forests: An Ecological and Biogeographical Comparison* (2nd edition). New York, Wiley-Blackwell.

blowing off the ocean toward the Amazon basin. The trade winds also send warm, moist rains to sustain a separate rainforest along the Atlantic Coast. The geographical circumstances toward the bottom half of the continent differ. Because of its location, and narrow, tapering shape the south is more responsive to ocean currents that have been cooled by the Antarctic Ocean. This makes it drier, colder, and more seasonal than the northern part of the continent. The environment in the south cannot sustain a rainforest biome.

South America supports two great tropical rainforests where primates are found (Figure 4.1). One is Amazonia, the enormous continuous rainforest situated east of the Andes. It is roughly 2.5 million square miles (6.5 million km^2), equal in size to the 48 contiguous United States. The other is the much smaller Atlantic Forest, about 35,000 square miles (91,000 km^2), located in a long, irregular strip of land bordering the Atlantic Ocean. The two rainforests are widely separated geographically by a massive, drier, more seasonal terrain comprised of tropical savanna, deciduous woodland, shrubland, and thorny forest. Few platyrrhine species inhabit these less resource-dense ecosystems, and the animals are mostly found in narrow bands of gallery forest that grow alongside small rivers and streams.

As many as 12 to 13 primate species and genera can be found in many individual communities spread throughout the Amazon basin. Five genera, including two that are endemics, are found only in the smaller Atlantic Forest in the east. Primates are less abundant taxonomically in the Atlantic Forest because it is far smaller and botanically less rich than Amazonia, but it is formally recognized by international conservation groups as a biodiversity hotspot.

The dominant ecological feature of the Amazon is the Amazon River (Rio Amazonas). It is the centerpiece of the largest river basin in the world. Thousands of other long rivers fan out from it in a large, spidery network. The rivers help to irrigate the land, distribute nutrients that collect in the water, and contribute to the humid weather. During the rainy season flooding converts huge areas of the basin into a flooded, swamp ecology. The widest point of the massive Amazon River, seven miles (11.3 km) across during the dry season, turns into a 25-mile (40 km) span during the wet season. Some platyrrhine primates have a preference for flooded forests, as discussed below.

The vastness of Amazonia and its river system creates its own weather system. It releases an enormous volume of moisture that evaporates into the atmosphere above the forest. Seventy-five percent of the rising moisture is returned to the landscape as rainfall. It is an enormous amount of fresh water, equaling almost 20% of all the world's river water.

Platyrrhines are haplorhine primates

Platyrrhines are haplorhine primates as distinct from the lemurs of Madagascar, discussed in the previous chapter, that are strepsirhines. The haplorhine primates also include tarsiers, Old World monkeys and apes. Haplorhines are the "visual branch" of the olfaction-vision tradeoff.

The major features of platyrrhines and other living haplorhines that distinguish them from strepsirhines include:

- Dry external nose with continuous upper lip and no rhinarium
- Reduced internal (bony) nose
- Reduced vomeronasal organ

South America: New World monkeys

- Reduced olfactory bulb
- Eyes with a fovea, a small indentation of the retina, and no tapetum lucidum
- Robust, high-crowned upper central incisors and vertically oriented lower incisors
- Differences in the anatomy of the placenta

Haplorhines, unlike strepsirhines, are diurnal, with two exceptions, South American owl monkeys and Asian tarsiers. Haplorhine primates exhibit reduced versions of the morphology and behaviors that characterize strepsirhine olfaction. They rely less on detecting environmental odors, and few species depend on scent marking for communication.

Haplorhines have a diurnal eye and keen eyesight, in contrast to the nocturnal strepsirhine eye, which cannot produce a sharp image. Their refined vision is made possible by a dense concentration of photoreceptor cells in the fovea of the retina that are particularly sensitive to bright light, in addition to highly developed visual centers of the brain.

- Owl monkeys and tarsiers, the only two nocturnal haplorhines, have eyes with a fovea. Their eyes lack a tapetum, the light amplifier behind the retina that occurs in strepsirhines. As an alternative nocturnal adaptation, they have large eyeballs to maximize the amount of light that enters the eye. The nocturnal adaptations in these two haplorhines evolved independently as shifts from the adaptive zones occupied by their ancestors, which were diurnal.

The external nose of platyrrhines, as in all haplorhines, is separated from the mouth by a mobile upper lip. The upper lip is supported by a band of muscle that encircles the mouth and enhances its functional potential as an ingestive tool, especially in anthropoids, platyrrhines, and catarrhines, where the lips often work together with the incisors. Mobile lips enable monkeys and apes to efficiently manipulate foods orally, to tug a berry off a stem, fold a batch of leaves into the mouth, or spit out unwanted material. With supple, well-muscled lips, anthropoids are able to shape the mouth to produce sound variations while vocalizing. As visually oriented diurnal primates, many species also communicate by contorting the mouth to make different facial expressions (Figure 4.2).

The structure of the haplorhine placenta is distinctly different from strepsirhines'. In haplorhines the placenta attaches to the uterine wall at one or two sites (discs) and the maternal blood vessels are in closer contact with fetal blood vessels.

Platyrrhines are anthropoid primates

The platyrrhines and catarrhines are anthropoids, today's most widespread and ecologically diversified radiation of primates. Anthropoids are recognizable by their close-set eyes, relatively short, non-protruding and deep faces, wide mouths, and their ability to make facial expressions.

They are found in the neotropics – platyrrhines – as well as in Africa and Asia – catarrhines, the Old World monkeys and apes. Platyrrhines typically inhabit tropical rainforests; catarrhines inhabit diverse biomes that include tropical rainforests, open grasslands, semi-deserts, and temperate forests where snowy winters occur. Most anthropoids live in trees; a few catarrhines live on the ground. Some anthropoids are small; a few are super-large. Anthropoid brains are larger than strepsirhine brains relative to body size. This corresponds with the larger amount of visual input that is stored and processed by the anthropoid brain's well developed visual centers.

Figure 4.2 South American capuchin monkeys can shape their mobile lips and mouth to produce a variety of facial expressions.

From Rosenberger, A.L. (2020). *New World Monkeys: The Evolutionary Odyssey*. Princeton, Princeton University Press.

A relatively large brain size is also correlated in many cases with advanced cognitive skills. South American capuchin monkeys, for example, use rocks as tools to smash open hard-shelled fruits and to dig for drinking water in areas that are seasonally dry and lack juicy fruits, which is typically the main source of water for a primate.

The most clear-cut, diagnostic hard-anatomy features shared by all anthropoids include:

- A complete bony eye socket, which is shaped roughly like a cone with the apex at the rear, called orbital closure. Anthropoids are the only primates and only mammals that have this morphology. Other primates have a more primitive pattern, an open orbit that is framed on the lateral side by a slim bony strut called the **postorbital bar** (Figure 4.3).
- Upright, robust, tall (high-crowned) incisor teeth. Anthropoids are the most committed primate frugivores. They have well-developed upper and lower incisor teeth suited for biting fruits of varying sizes, shapes, and hardness.
- A rigid lower jaw. The **mandible**, lower jaw, is distinctive in that the right and left sides are fused in the midline into a single U-shaped bone that is usually wide at the front to accommodate a row of four robust incisors. A single-bone mandible enhances the functional efficiency of masticatory muscles that attach to it on either side of the head, for both biting and chewing. In contrast, in most strepsirhines, all tarsiers, and many other mammals the mandible is formed by two bones, one on the right side and the other on the left, that are connected anteriorly by a joint.

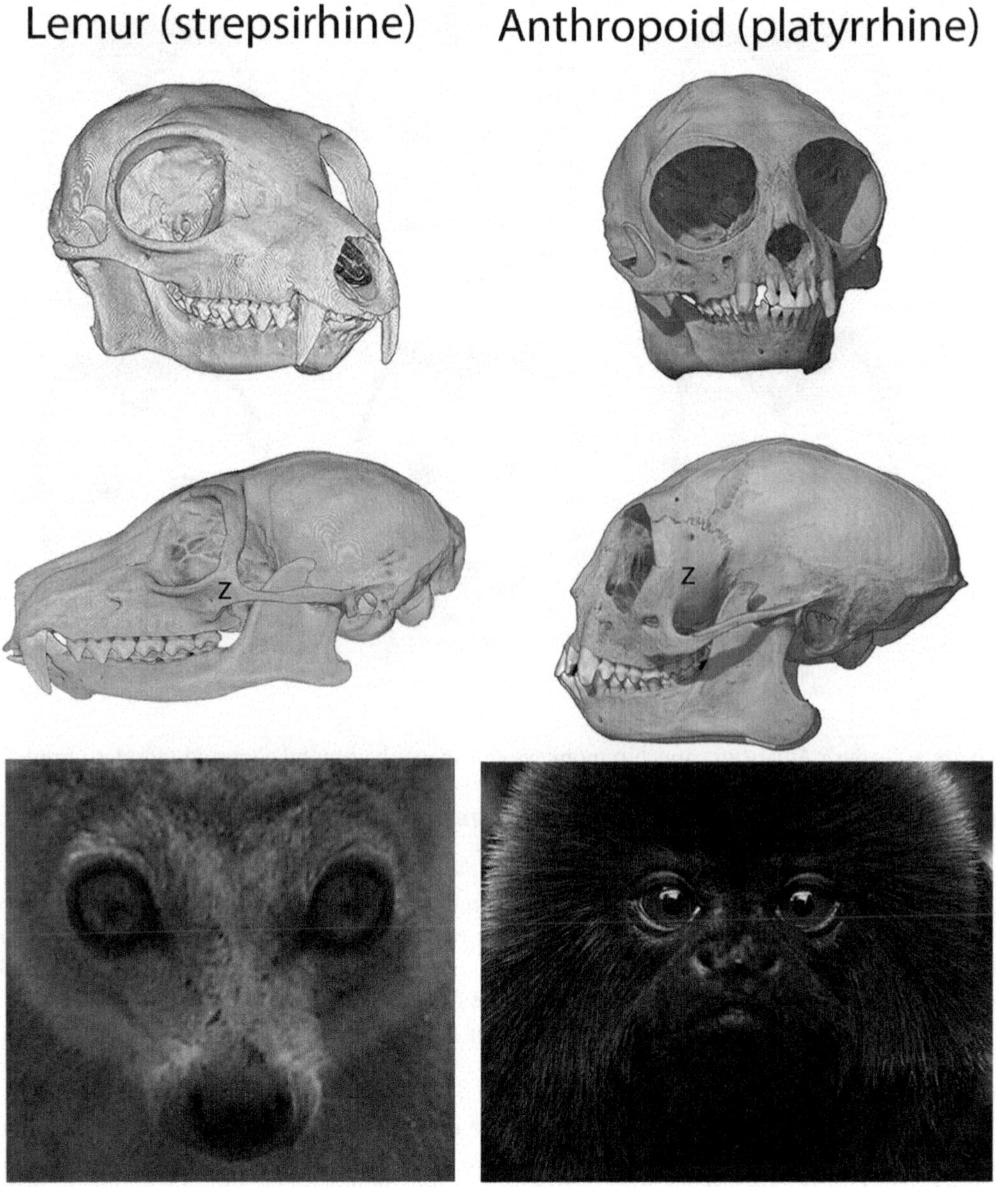

74 *South America: New World monkeys*

Lemur (strepsirhine)   Anthropoid (platyrrhine)

Figure 4.3 Skulls of a crowned lemur (left) and a platyrrhine, the Goeldi's monkey (right), contrasting an open strepsirhine orbit with postorbital bar and an anthropoid's complete orbital closure. Faces of the same species show the large, bulging lemur eyes under the contour of skin and fur behind the pupils, and the small, recessed anthropoid eyes. The slender zygomatic bone (z) forms a simple bar of bone in strepsirhines and is wider to close off the eye socket in anthropoids.

Credits: Crania derived from Digimorph (http://digimorph.org). Photo portraits: *Eulemur coronatus* female, by Frank Wouters (CC-BY-SA 2.0); Goeldi's Marmoset, by Cam MacMahon (Wikimedia, Pubic Domain).

- Larger relative brain size. Living anthropoids have relatively large brains. The ways that functional areas of the brain are proportioned and wired are also different. For example, in keeping with the olfaction-vision tradeoff, the olfactory lobe at the front of the brain is considerably smaller in anthropoids than in strepsirhines, while the occipital lobe at the back, where visual information is processed, is larger.

Platyrrhines and catarrhines

Outwardly, as noted in Chapter 1, platyrrhines are distinguished from catarrhines by nose shape (Figure 4.4). Platyrrhine nostrils are set wide apart, separated by a broad fleshy strip of tissue, and the nostrils open to the side. Catarrhine noses have closely spaced, downward-facing or forward-facing nostrils with a narrow fleshy dividing strip.

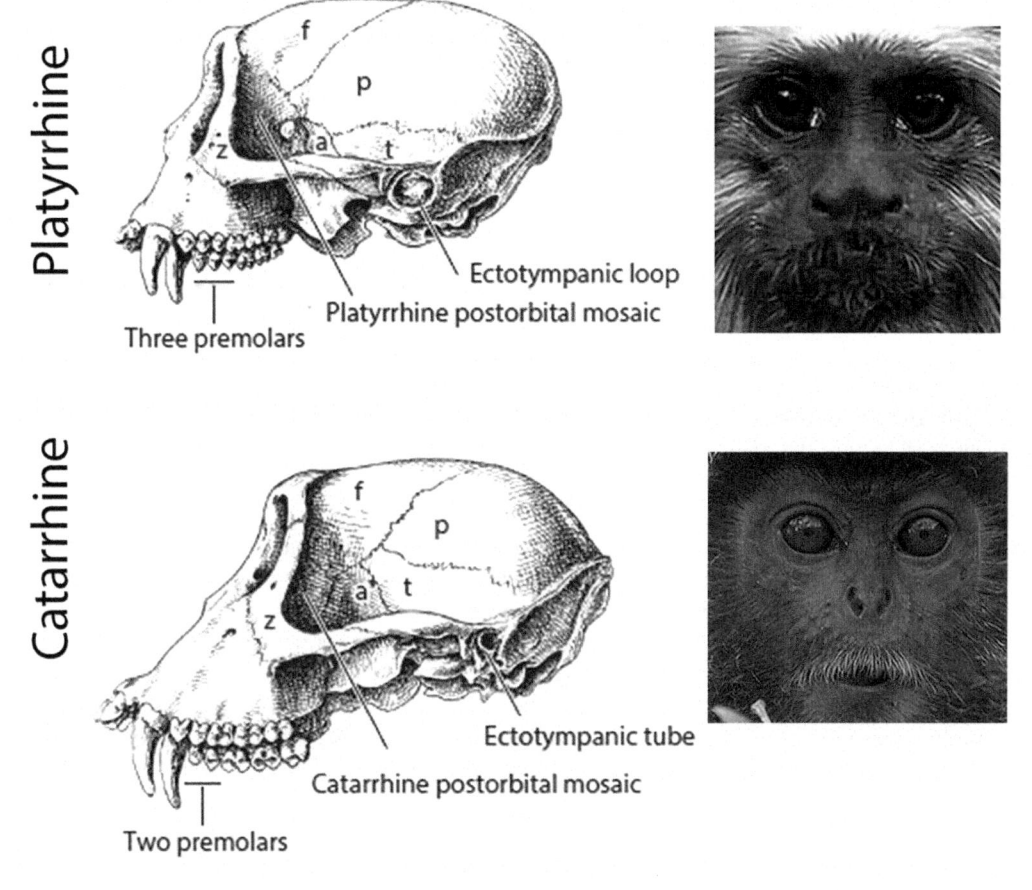

Figure 4.4 Crania (top, capuchin monkey; bottom, rhesus macaque) and faces of a platyrrhine and catarrhine illustrating nose shape and several other distinguishing characteristics of the cranium and dentition. The labeled bones are: a, alisphenoid; f, frontal; p, parietal; z, zygomatic.

Left panel adapted from Rosenberger, A.L. (2020). *New World Monkeys: The Evolutionary Odyssey*. Princeton, Princeton University Press. Photo credits (top and bottom): *Leontopithecus chrysomelas* (6337871306).jpg by Ruben Undheim (CC-BY-SA 2.0); Silvered Leaf Monkey (*Trachypithecus cristatus*) (8127485334).jpg, by Bernard Dupont (CC-BY-SA 2.0).

76 South America: New World monkeys

Internally, there are other differences in cranial and dental morphology that distinguish living platyrrhines and catarrhines (Figure 4.4), including:

- The **postorbital mosaic** is the arrangement of bones that meet to close the braincase near the cranio-facial junction behind the orbit. In platyrrhines the mosaic is formed by the zygomatic and parietal bones joining to separate the frontal and alisphenoid bones. In catarrhines, the zygomatic and parietal bones are separated by the frontal, temporal, and alisphenoid bones.
- The **ectotympanic bone** is the structure that frames the eardrum. In platyrrhines the ectotympanic is essentially a flat, horseshoe-shaped loop oriented vertically. It is attached to the side of the cranium at the opening to the ear. In catarrhines, the ectotympanic bone is elongated, shaped like a tube. The eardrum is located at the inner end of the tube, as opposed to being close to outside opening of the ear.
- The **dental formula** differs, fundamentally because platyrrhines have one more premolar. Their dental formula is 2-1-3-3 – it is the third number in the formula that refers to how many premolars there are. Catarrhines, with only two premolars, have a dental formula of 2-1-2-3. It is important to note that marmosets and tamarins differ from other platyrrhines in having only two molars – the fourth number in the formula – and their dental formula is 2-1-3-2.

The platyrrhine profile

Living New World monkeys are highly varied in their dietary, locomotor, and behavioral adaptations. There are distinctive patterns that define platyrrhines' phylogeny, anatomy, behavior, and lifestyle. Many of the descriptive parameters are interconnected and work together to form adaptative complexes. For platyrrhines it is important to consider how the subfamilies within families are related (Figure 4.5). For illustrative purposes, six of the sixteen genera are shown in Figure 4.5.

The following characteristics constitute the platyrrhine profile:

- **Phylogeny**. The nearest relatives of the platyrrhines are catarrhines, the Old World monkeys and apes. Together, the platyrrhines and catarrhines comprise the living anthropoids, a monophyletic group, descending from a single common ancestor.
- **Body mass**. Platyrrhines exhibit a wide range of body sizes, from the small 3.5-ounce (~100 g) pygmy marmoset, *Cebuella*, to the large 20-plus pound (~10 kg) muriqui, *Brachyteles*, which is roughly 100 times heavier. Most platyrrhines are in the small and medium size classes. The pygmy marmoset is the world's smallest living anthropoid. The smallest living primate is the lemur, Madame Berthe's lemur, which weighs about 1 oz. (28 g), discussed in the previous chapter.
- **Activity rhythm**. All platyrrhine genera are diurnal except for the owl monkey, *Aotus*, which is nocturnal throughout its geographical range in the Amazon. An exception is one population that is cathemeral. It lives far south of the equator where habitats are semi-arid, the nights are colder, and food supplies not as abundant.
- **Ecological domain**. All platyrrhines are arboreal and only rarely come to the ground. None have morphological adaptations relating to locomotion on the ground.
- **Diet**. Platyrrhines are essentially omnivorous frugivores and eat a seasonally varied diet that encompasses a range of vegetation that is commonly eaten by primates. In addition to the one genus that is notably folivorous, the howler monkey, *Alouatta*, there are several patterns that are highly distinctive.

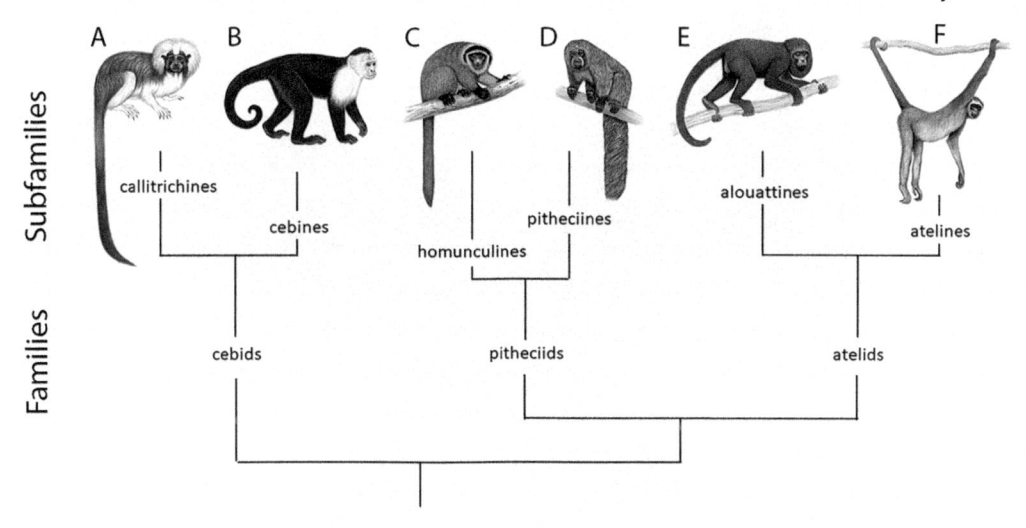

Figure 4.5 Cladistic relationships of the three major living platyrrhine families and six subfamilies represented by genera: (a) tamarin, *Saguinus*; (b) capuchin, *Cebus*; (c) titi monkey, *Callicebus*; (d) saki, *Pithecia*; (e), howler monkey, *Alouatta*; (f) muriqui, *Brachyteles*. Not drawn to scale.

Adapted from Rosenberger, A.L. (2020). *New World Monkeys: The Evolutionary Odyssey*. Princeton, Princeton University Press. Artwork by Stephen D. Nash, courtesy of the IUCN SSC Primate Specialist Group.

- The cebids constitute an entire platyrrhine family that is omnivorous and particularly predaceous. Their diets include significant proportions of fruit, gums, insects, and small vertebrates. While prey is often an incidental food for many other anthropoids, cebids pursue animal matter persistently.
- Among the pitheciids, the saki and uacari monkeys feed intensively on the seeds of fruits that are encased in very hard shells. They have unique craniodental adaptations to open them up. The pitheciid family is the only major taxonomic group or radiation of primates that is comprised of genera that feed on fruits with tough rinds and several that are craniodentally specialized seedeaters.
- The **muriqui** is a platyrrhine genus that depends on eating leaves in addition to fruits. But, with the exception of its dentition, the muriqui does not exhibit the other anatomical or behavioral correlates of folivory that are found among most other leaf-eating primates.

- **Locomotion.** Platyrrhines are extremely varied in their locomotor styles. Two important clades have evolved unique specializations. The smallest platyrrhines, the callitrichine subfamily, have clawed digits that are vertical-clinging adaptations. The largest platyrrhines, the atelid family, have a long, grasping, fully **prehensile tail** – the only primates that do – with a long, specialized gripping pad covered in friction skin. It extends along the endmost third of the tail (Figure 4.6). The prehensile tail is used by these monkeys in travelling, climbing, and hanging from branches when they assume suspensory postures to feed. Another platyrrhine genus, the capuchin, has a relatively short, strong, and furry tail that is semi-prehensile, as discussed below.

Muriqui and spider monkeys employ the prehensile tail in a novel way, using it in a form of modified brachiation. Brachiation is a highly specialized, two-armed, hand-over-hand,

Figure 4.6 Spider monkeys have a long, fully prehensile tail with an extensive, belt-like gripping pad on the underside covered in friction skin.

Photo credit: *Ateles fusciceps robustus* moving2.jpg, by Patrick Müller (CC-BY-SA-3.0).

swinging style of below-branch locomotion. It is extensively used by the gibbons and siamang, the lesser apes of Asia that we will discuss in Chapter 6. The muriqui and spider monkeys propel themselves hand-over-hand through the trees with the tail acting as a third limb, holding on with the tail as well as the forelimbs.

The pitheciine subfamily presents other noteworthy examples of platyrrhine locomotor and postural diversity. Their postural specializations facilitate food-handling. The bearded saki, *Chiropotes*, and the uacari, *Cacajao*, are seed-eating genera that habitually hang by their feet while handling the hard-shelled fruits that they crack open with their teeth. The third genus of the subfamily, the saki monkey, *Pithecia*, is the only habitual vertical-clinging-and-leaping platyrrhine monkey.

- **Craniodental morphology.** Craniodental morphology is more varied in New World monkeys than in any other major primate group.

  - The pitheciines have a complete masticatory system – incisors, canines, premolars, molars, lower jaws, faces, and musculature – that is designed to breach hard-shelled fruits, to expose, extract, and eat what is inside, including the seeds.
  - Howler monkeys have one of the primates' most unusual heads, accommodating an enormous vocalization mechanism that is situated in the throat below the base of the skull, as discussed in this chapter and further in Chapter 7.
  - Cebines, like the capuchin skull shown in Figure 4.4, have short-faced, rounded crania with large braincases that house exceptionally large brains relative to their body size.

- The nocturnal owl monkeys have large orbits to hold their large eyes.
- Two genera of atelids have each independently evolved leaf-eating dentition.
- Two callitrichine genera have specialized gum-gouging anterior teeth.
- Four out of five callitrichine genera exhibit a reduced dental formula with only two molars, as noted above.

- **Pelage.** New World monkeys have a wide range of pelage (fur) types, in color, texture, length, and distribution on the body. Colors range from dull to bright. Some pelage patterns are species-specific identifiers ensuring that conspecifics recognize one another; some pelage serves as camouflage; other patterns contribute to visual displays used in communication. In species that exhibit sexual dimorphism, pelage may also be a feature distinguishing males and females (Figure 4.7). Platyrrhine faces may be accentuated by leathery skin, ranging from tan to beige to black to red, or they may be striped by white-on-black markings. The forehead may be accentuated by a plush horizontal band of fur above the eyes. Tails may have thin, short fur, or long, thick, bushy fur. Some species are monochromatic – black or red or off-white.
- **Sexual monomorphism and dimorphism.** Platyrrhines exhibit the full range of species-specific possibilities with respect to monomorphism and male-biased dimorphism in body mass, canine size, and coat color patterns. In white-faced saki monkeys sexual dimorphism is only expressed in the pelage. Males and females are the same in body size, canine size, and other skeletal traits (Figure 4.7).
- **Communication.** To communicate, platyrrhines use a wide range of visual, auditory and olfactory signals, and facial expressions, body postures, and vocalizations (Box 4.1). Communication via scent is important in several genera, particularly among marmosets and tamarins, and smell is vital to the nocturnal owl monkeys. These platyrrhines have various glands used in scent-marking behavior to communicate by depositing odors. They also have a functional vomeronasal organ to facilitate pheromonal communication.

Box 4.1 Vocal communication

Vocalizations are the most common and important means by which platyrrhines and other anthropoid primates communicate. They make a great variety of sounds – chirps, shrieks, screams, grunts, barks, coos, howls, and more, in styles that are species-specific. Some species use specialized anatomies to generate calls. Howler monkeys have an unusual hyoid bone and larynx, and a strong diaphragm, that enables them to roar loudly. Gibbons and siamang, the lesser apes in Asia, have large throat sacs which they fill with air and expel to produce their loud calls.

Communication plays a vital role in regulating behavior within social groups. Experiments in the lab and in the field demonstrate that acoustic signals convey information that is understood by the receivers.

Of the many types of call signals used within groups, there are:

- Contact calls, used to maintain mutual awareness among individuals that are spread out while foraging.
- Moving calls, that alert troop members and coordinate a shift in foraging movements.

- Feeding calls, to share the location of a good feeding site among troop members.
- Alarm calls, to signal danger, such as the presence of a specific type of predator (eagle, snake, or leopard) that can elicit a recognizably different call in vervet monkeys.
- Loud calls, long distance calls which may be calls directed at other social groups.

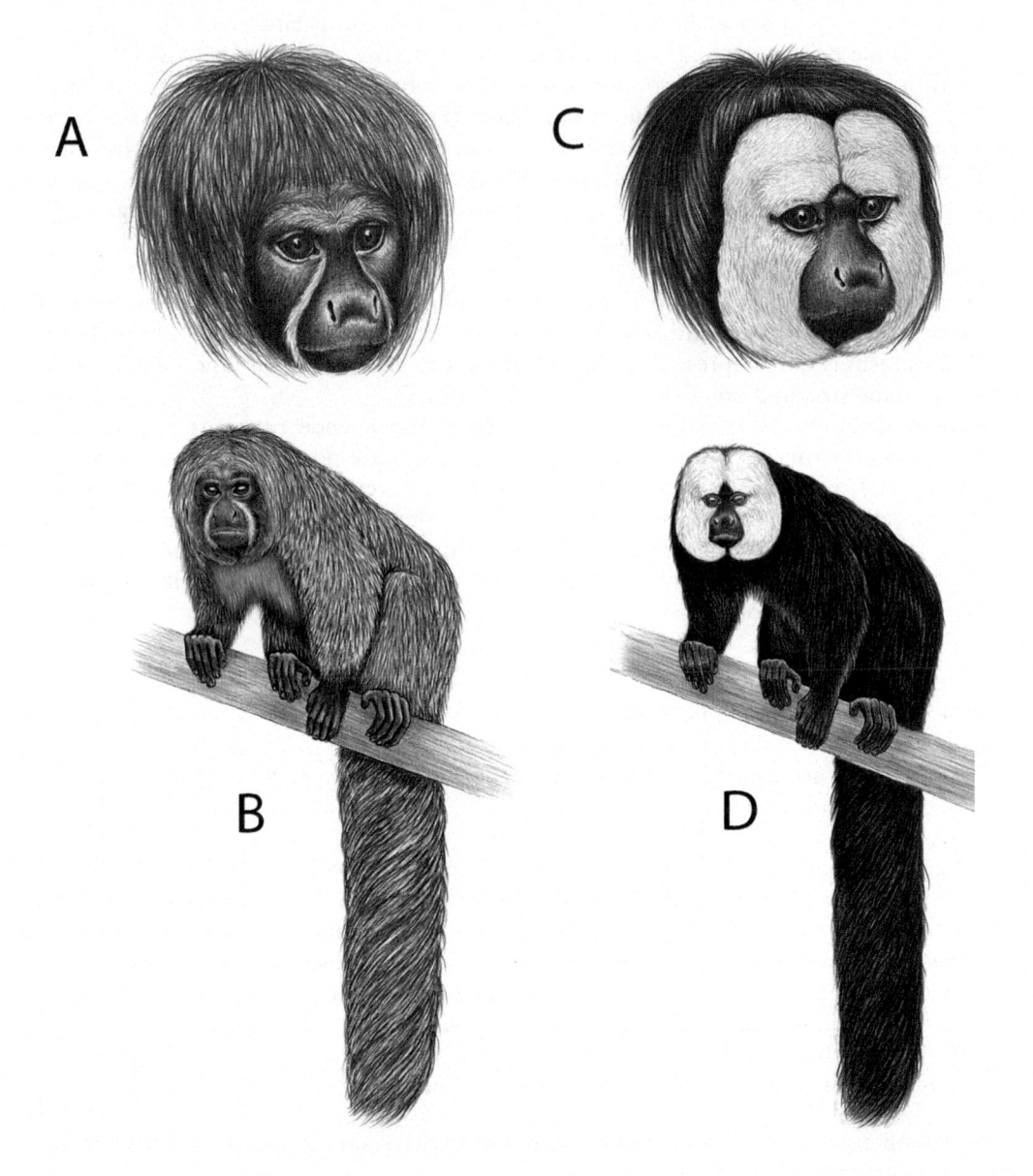

Figure 4.7 Female (a, b) and male (c, d) white-faced saki monkeys are sexually dichromatic in pelage.

Artwork by Stephen D. Nash, Courtesy of the IUCN SSC Primate Specialist Group.

- **Social systems and mating.** New World monkeys exhibit nearly all of the basic social systems and mating patterns found among primates. An exception is the semi-solitary category. Platyrrhine social systems include pair-bonds, unimale-multifemale troops, multimale-multifemale troops, social groups that are cohesive but split up temporarily, male and female dominance hierarchies, and co-dominant societies. There are other patterns as well, such as the sexually segregated groups that occur outside of the breeding season in some species of squirrel monkeys. In the complex social and mating systems of some marmosets a high-ranking female may breed with several males while subordinate females are blocked from breeding, largely by pheromonal means. In this system, troop members also help to rear offspring.
- **Reproductive pattern.** The majority of platyrrhines have litters of singletons but marmosets and tamarins have twins.
- **Parenting styles.** Parenting styles are diverse. Mothers are typically the main caregivers, but two other arrangements are noteworthy in the extent of fathers' and siblings' roles in giving care to infants and young. In pair-bonded monogamous species like owl and titi monkeys, fathers play important supportive roles in rearing young. In marmosets and tamarins, mothers produce and nurse twins twice a year, meaning a reproducing female bears the energetic burden of simultaneously lactating one litter while gestating another. Marmoset and tamarin adult males and siblings help by communal rearing, carrying the twins soon after birth and helping to feed them as they are weaned, sharing food, and providing protection.
- **Encephalization.** The proportional size of the brain is called encephalization. The platyrrhine radiation exhibits three patterns:
  - A normative situation, where the brain size of species conforms to the standard platyrrhine proportions that correlate with body size, as in some marmosets and tamarins.
  - High encephalization, where brain size exceeds the volume that is predicted by body size, as in squirrel monkeys and capuchins. The clever, tool-using capuchin, a 3–9 lb. (1.4–4 kg) monkey in the medium size class, has brain-to-body-mass proportions that make it highly encephalized, exceeding the ratios of large-brained modern apes.
  - De-encephalization, where brain size is less than the size predicted by body mass, as in the mostly leaf-eating howler monkey, *Alouatta*.

Platyrrhines living in communities: diversity, ecology, and adaptation

Platyrrhine biodiversity is distributed in various concentrations of species in certain geographic areas and ecological zones. The size of terrain that a species occupies varies as well. Some species are widely distributed, and others are found concentrated in fewer areas. A notable geographical feature of the platyrrhine radiation is a disjunct distribution. This is the broad geographical separation between monkey communities that live in the two great South American rainforests, Amazonia, and the Atlantic Forest.

The rich biotic communities in many parts of Amazonia are packed with primates. Some areas are inhabited by a dozen or more New World monkey genera, including pairs of sympatric species belonging to the same genus. To examine how ecological niches are partitioned by the monkeys, we begin by focusing on platyrrhines that live in Manu National Park in Peru, an area that has been designated a globally important

United Nations Educational, Scientific and Cultural Organization (UNESCO) Biosphere Reserve. It is a relatively intact, evergreen Amazonian ecosystem. Research has been ongoing there since the mid-1970s.

Eleven of the 16 living platyrrhine genera recognized in this book exist at Manu, while five do not. Two genera are endemic to the disjunct Atlantic Forest, the lion marmoset, *Leontopithecus*, also known as the lion tamarin, and the muriqui, *Brachyteles*. Both are among the world's most endangered primates. Two other genera occur in the Amazon in habitats far from Manu, the uacari, *Cacajao*, and the bearded saki, *Chiropotes*. They live in areas that correspond with the geographical distribution of the particular trees on which they rely. The fifth genus not at Manu is the marmoset *Callithrix*. It is widespread in diverse habitats east of the Manu region, including rainforest, tropical savanna, grassland, and shrubland.

The 11 genera present at Manu represent all of the family-level platyrrhine clades. All are arboreal and all are diurnal, except for the nocturnal the owl monkey. Monkeys belonging to the same subfamilies live similar lives. The main ways in which niches are partitioned among them are based on several ecological correlates, including body mass, diet, locomotion, and microhabitat preferences (Table 4.1). Another factor in how the niches of closely related species are partitioned is shifts in feeding strategy, how species cope with seasonal depletions in resources due to seasonality and the natural botanical rhythms of the forest. Each of the three major family-level clades occupies a distinctive adaptive zone in the forest based on its unique sets of adaptations relating to these parameters. Their phylogeny and adaptations are connected, as Table 4.1 shows. Each family, representing a clade, shares a pattern of similar adaptive traits and habitat preferences.

Atelids – frugivore-folivores

The four genera of the atelid family are by far the largest-bodied platyrrhines. Even in the smallest, most sexually dimorphic atelid genus, the howler monkey, females weigh about twice as much as the largest species in the other families, the pitheciids and cebids.

Three of the four atelid genera occur at Manu. They are widespread in the Amazon basin and are often found in sympatry. The fourth, *Brachyteles*, only occurs in the Atlantic Forest, as we have seen. There, the muriqui is sympatric with howler monkeys. But South American spider monkeys and woolly monkeys only live in the Amazonian rainforest.

Atelids are the only primates that have evolved a fully prehensile tail. They have a diverse set of musculoskeletal and neurological characteristics that make powerful, secure usage of the tail possible, including a long patch of hairless, finger-like skin on the underside that resists slippage (Figure 4.6). In spider monkeys, *Ateles*, for example, the tail ends in a tip with tactile abilities and motor control that functionally resemble the dexterity of a human finger.

All four atelid genera share the prehensile-tail complex, a derived homologous characteristic, evidence that this feature was present in their last common ancestor. There are a variety of ways in which atelids use their tails, from serving as an occasional overhead hook during the slow quadrupedal locomotion of howler monkeys, *Alouatta*; to sliding on an overhead branch during rapid quadrupedalism in woolly monkeys, *Lagothrix*; to using the tail as a fifth limb that is incorporated into the hand-over-hand, suspensory, brachiation-like locomotion of spider monkeys, *Ateles*, and muriqui, *Brachyteles*.

Table 4.1 Ecological characteristics of platyrrhines at Manu

Family	Common name	Genus and species	Size	Activity cycle	Locomotion	Diet	Habitat	Forest stratum
Atelidae	Red-faced black spider monkey	*Ateles paniscus*	Large	Diurnal	Quadrupedal	F,L,N	Forest	Canopy
Atelidae	Humboldt's woolly monkey	*Lagothrix lagothricha*	Large	Diurnal	Quadrupedal	F,L	Forest	Canopy
Atelidae	Colombian red howler monkey	*Alouatta seniculus*	Large	Diurnal	Quadrupedal	F,L,U	Forest	Canopy, subcanopy
Pitheciidae	Humboldt's night monkey	*Aotus trivirgatus*	Medium	Nocturnal	Quadrupedal	F,I,L,N	Forest	Canopy, subcanopy, understory
Pitheciidae	Red-bellied titi	*Callicebus moloch*	Medium	Diurnal	Quadrupedal	F,U,L,I	Waterside, forest	Subcanopy, canopy
Pitheciidae	Monk saki	*Pithecia monachus*	Medium	Diurnal	Quadrupedal	Seeds, F	Forest	Canopy, subcanopy
Cebidae	Tufted capuchin	*Cebus apella*	Medium	Diurnal	Quadrupedal	F,I,S,N,P,V	All types	Subcanopy, canopy
Cebidae	Humboldt's white fronted capuchin	*Cebus albifrons*	Medium	Diurnal	Quadrupedal	F,I,S,N,V	All types	Subcanopy, canopy, ground
Cebidae	Guianan squirrel monkey	*Saimiri sciureus*	Small	Diurnal	Quadrupedal	F,I,N	All types	Subcanopy, vines, canopy
Cebidae	Goeldi's monkey	*Callimico goeldii*	Small	Diurnal	Clawed quadrupedalism	Fungus,F,I	Bamboo thickets	Vines, subcanopy
Cebidae	Emperor tamarin	*Saguinus imperator*	Small	Diurnal	Clawed quadrupedalism	F,I,N,G	Forest, openings	Vines, subdanopy
Cebidae	Spix's saddle-back tamarin	*Saguinus fuscicollis*	Small	Diurnal	Clawed quadrupedalism	F,I,N,G	Forest, openings	Vines, subcanopy, canopy
Cebidae	Pygmy marmoset	*Cebuella pygmaea*	Small	Diurnal	Clawed quadrupedalism	G, I,N,F	Waterside, swamp	Subcanopy, vines

Compiled from Janson, C., & Emmons, L. (1990). Ecological structure of the non-flying mammal community at Cocha Cashu Biological Station, Manu National Park, Peru. In A. H. Gentry (Ed.), *Four Neotropical Rainforests* (pp. 314–338). New Haven, Yale University Press and other sources.

Dietary items are listed in their order of importance according to sources, abbreviated as: F, Fruit pulp; Fu, Fungus; I, Insects and invertebrates; S, Seeds; N, Nectar; P, Pith; V, Terrestrial vertebrates; G, Gums; U, Unripe fruit.

- **Diet and locomotion**

    The large atelids are canopy-dwellers. As a group, they are considered frugivore-folivores. This descriptive categorization is important because it clearly separates their ecological position relative to other platyrrhines on the basis of diet at Manu and elsewhere, specifically what forest products are crucial sources of protein.

    Atelids principally rely on leaves, preferably young ones, for protein. Cebids mainly rely on insects and small vertebrates for their protein. Pitheciids either rely on seeds (pitheciines) or varying proportions of insects and leaves depending on place and season (homunculines). Only woolly monkeys among atelids are known to forage and feed on insects to a significant degree, but that still accounts for less than 10% of their diet. The relatively large body mass of atelids correlates with leaf-eating, but the ecological separations and diverse dietary adaptations of atelids are not captured by the simple frugivore-folivore categorization.

    The howler monkey, the only living genus of its subfamily, is the most divergent atelid in many important ways that are all connected to diet. It is the most folivorous form. It also likes to eat figs, like all primates, and may prefer various fruits over leaves when available. Howlers prefer young leaves that are more easily digested than mature leaves. Yet, howlers have capacious guts that are adapted to processing older leaves. Its large, crested molars are a classic biomechanical example of the shearing teeth commonly found among folivores, but they are also morphologically suited to chewing fruits by crushing and grinding. Long-term field studies conducted at many sites in various habitats have found that more than half of the time howlers spend feeding – and sometimes as much as 80% and more – is spent eating leaves.

    Alouatta is the most geographically widespread platyrrhine genus. One reason is the capacity to exist by efficiently feeding on leaves, the most ubiquitous source of food produced by a tropical and subtropical forest, whether evergreen or deciduous, wet or dry. Howlers typically have color vision, which sharpens their ability to locate younger leaves, which are lighter shades of green and may begin growing with a reddish hue. *Alouatta* is the only platyrrhine genus having **trichromatic vision**, which is the ability to see combinations of red, green, and blue. It is the only platyrrhine genus in which that ability is present in both males and females.

    Howlers are habitat generalists of the frugivore-folivore adaptive zone by virtue of their ability to subsist on a narrow range of plentiful foods. They live throughout the Amazon, in woodlands and savanna gallery forests that extend outside the rainforest basin, in the Atlantic Forest, and in Central American forests as far north as southern Mexico. In contrast, the more omnivorous capuchins, which belong to the cebid family, are habitat generalists that occupy the frugivorous-faunivorous adaptive zone. They are also pioneering species, with a comparable geographical distribution, and they live under an even wider range of environmental and botanical conditions.

    The stoutly built howler's essential ecological strategy is that of an **energy minimizer**, a pattern that markedly distinguishes howlers from the other atelid subfamily, the atelines, comprised of woolly monkeys, spider monkeys, and muriqui. Leaves are a relatively poor source of nutrition that must be eaten in bulk when serving as a staple. It is a diet that is conducive to a lifestyle that discourages locomotor exertion and other energy-intensive activities. So, howlers travel little to forage – leafy food is all around them – and they spend much of their day resting, as much as 80%, which gives the gut time to detoxify and digest the mass of leaves that fills their stomachs.

Their sedentary pattern is reflected in a simple style of quadrupedal locomotion which contrasts with the more acrobatic atelines. Howlers have very small brains relative to body size, resembling other anthropoid folivores and herbivorous mammals generally. They are de-encephalized, as an adaptation to conserve energy; the brain is the organ that consumes the most energy. The most encephalized platyrrhines are frugivores and predaceous omnivores.

Spider monkeys, *Ateles*, lithe and long-limbed, are at the opposite end of the atelid spectrum in terms of energy expenditure and consumption. *Ateles* is an **energy maximizer**, taking in relatively large amounts of nutritious foods and also expending much effort in the foraging process. They eat quick-energy, sugary ripe fruits. Spider monkeys range widely to get to trees with these fruits, moving rapidly above and below branches in a climbing-and-clambering style of quadrupedalism. With a burst of energy, they swing through the trees with their prehensile tail providing both support and propulsion. At some sites where spiders have been well studied, 80%–90% of feeding time is spent eating fruit. Roughly 10%–20% of feeding may be spent eating leaves but they forage widely for them because spider monkeys are highly selective in choosing young leaves.

Ateles cheek teeth are blunt and relatively small, and its incisors are large. This is an effective pattern for biting and mashing large, soft fruits but not for cutting or shearing mature leaves that are tough. The woolly monkeys, *Lagothrix*, are more robustly built quadrupeds that do not locomote in the highly acrobatic fashion of spider monkeys. They also rely heavily on ripe fruits. But while living sympatrically with spider monkeys, woolly monkeys are ecologically separated from them by selecting fruits that are different from those eaten by *Ateles*.

- **Social systems and mating**

Group sizes and social systems vary widely among atelid genera, even between congeners, species of the same genus. Social groups can include as few as six individuals and as many as 60.

Alouatta tend to live in some of the smallest troops, which are often unimale-multifemale units, but some howler species can have multimale-multifemale groups of 30–40 individuals. Group members act cohesively, and are organized in social-dominance hierarchies. The animals forage, eat, and rest together. When reaching reproductive age, both males and females disperse from the natal group to find mates. Dominant males have exclusive access to reproducing females, so in multimale groups aggression among males can be fierce. The aggressive, domineering behaviors exhibited by howlers are associated with the degree of sexual dimorphism in various features that characterizes the genus. For example, males are physically much larger, weighing 1.2–1.6 times the weight of females in various species.

The most severe episodes of male aggression in howlers occur when a number of non-breeding, roving males attempt to drive out an alpha male in order to take over the females that he controls. The hostile animals may also be infanticidal; they are a cause of high rates of infant mortality in howlers. After takeovers, infanticidal males are soon able to sire offspring with reproductively active females because those whose infants were killed begin ovulating when they are no longer nursing. As with capuchin monkeys, such takeovers happen episodically during the lifetime of a social group. Alpha males do not hold their dominant positions permanently.

Howlers are energy minimizers, as noted, conserving energy where they can, including in defense of females from outside males. Loudly roaring is the method they use to warn outside males to keep away. This is the howler's most unusual morphological and behavioral adaptation. These loud roars can be heard across miles of forest, at any time of day or night. A roaring howler male often ignites a chain of call-and-response howls of neighborhood troops answering one another. As mentioned, the calls are produced through a complex set of adaptations, including a large hyoid bone.

Loud roaring is a form of expression that gives an impression of body size. It alerts non-resident *Alouatta* males and neighboring troops to the likelihood that a large, vigorous adult male in good condition is present, willing, and able to guard and defend his group. When joined by females, who also howl but have a smaller vocal mechanism, it announces group size, cohesion, and stability. As an adaptation, a long-range vocal broadcast is energetically efficient and less risky than face-to-face confrontations to keep outside males at bay, or to discourage subordinate resident males from attempting to displace their superior.

Spider monkeys, in contrast, live in large multimale-multifemale groups including dozens of individuals. These groups are distinctive in that smaller, fluid subgroups of varying sizes and compositions get together on a daily basis to forage. This is a fission-fusion social system. The *Ateles* foraging parties often consist of females and their offspring. Ecologically, small foraging blocs minimize competition among group members when important resources, like ripe fruits, are widely distributed in the forest but concentrated in small patches. The absence of male-biased sexual dimorphism, or the fact that adult females of some spider monkey species are larger than males, eliminates the prospect that foraging parties composed of females and their offspring will be challenged by male parties.

Woolly monkeys, are dimorphic sexually with males weighing 1.2–1.6 times more than females. They also live in groups that can be composed of dozens of individuals. They prefer ripe fruit but tend to approach feeding differently. They generally forage as cohesive units with individuals spread far apart in the forest, keeping in touch by vocalizing but will, at times split up into discrete, temporary foraging parties.

Brachyteles, the muriqui

The Atlantic Forest's muriqui is another exceptional atelid. It is the largest genus of the family and the largest platyrrhine. Its adaptations do not conform to several of the features typically associated with either frugivory or folivory. *Brachyteles* is characterized by a unique combination of fundamental adaptive features – a spider monkey-like body with a howler monkey-like dentition. Muriqui are somewhat more robust than spider monkeys and have larger muzzles, otherwise they are similar in their physiques, postcranial skeletons, and locomotor behaviors. Muriqui and spider monkeys are also similar in some of their habitual patterns of play – they hang by their tails face-to-face, arms wrapped around one another, vocalizing with twittering sounds. The anatomical and behavioral resemblances are complete down to small details. Both of them lack a thumb or have only a vestigial nubbin. Only the teeth are different and the correlating structure of the face.

Spider monkeys have large biting incisors and small, nondescript cheek teeth with blunt cusps, a frugivorous pattern. Muriqui have small incisors and large crested molars, a folivorous pattern functionally in line with *Alouatta* but different in the details. While long-term studies have documented an annual diet that comprises more than 50% leaves, and possibly as much as 67% in one instance, digestion in muriqui more closely resembles the fast throughput that characterizes spider monkeys and other frugivores. The digestive strategy of muriqui does not follow the *Alouatta* pattern of slow, efficient leaf digestion, which requires long food-retention times and gradual detoxification. *Brachyteles* digests its food quickly and does not process leaves as efficiently. But its large body mass and correspondingly large gut may compensate by allowing the animals to often consume large amounts of leaves.

Brachyteles behaves like an energy-maximizer, like *Ateles*. Both genera are highly acrobatic and move in the same ways when negotiating the canopy. Despite the muriqui's folivorous dental adaptations, it does not have a small brain like *Alouatta*. The neural requirements of acrobatic, tail-assisted locomotion, as well as the complexity of their social systems, result in their retaining a high degree of encephalization commensurate with their body size.

Brachyteles resembles *Ateles* in exhibiting fission-fusion social systems and a female social dispersal pattern, and in the expression of secondary sexual characteristics. Both genera are either sexually monomorphic, or, in some spider monkey species, females are slightly larger. In muriqui, there are no male-female differences in physique or canine size that could be factors during male-female interactions, or in male-male **agonistic** exchanges, behaviors that involve aggressive and submissive interactions. Their canines are small as well as being monomorphic. This is highly unusual in an atelid. *Ateles* canines are large even though they are monomorphic, and the other genera, *Alouatta* and *Lagothrix*, have large *and* dimorphic canines.

The lack of a prominent canine to be used in a threat display is related to the muriqui's uniquely peaceful social system. Males and females are co-dominant, and males form strong bonds with one another.

Pitheciids – fruit-huskers and seedeaters

Three of the five genera of the pitheciid family exist at Manu: the saki, *Pithecia*; the titi monkey, *Callicebus*; and the owl monkey, *Aotus*. Other pitheciids living outside the region are the bearded saki, *Chiropotes*, and the uacari, *Cacajao*. All five pitheciids are medium-sized monkeys, generally quadrupedal. Members of the pitheciine subfamily, the sakis and uakaris, are identifiable by a hyper-platyrrhine nose, nostrils spaced very far apart and separated by a very wide fleshy strip (Figure 4.7). They also have distinctly bushy tails, even *Cacajao*, which has a very short, stubby tail. Some have plush, luxuriant coats and big beards and tufts of fur around the head. The bald uacari, *Cacajao calvus*, is as its name suggests, bald. It has a red, leathery face. The homunculine subfamily (named after a fossil genus, *Homunculus*, that belongs to this taxonomic group) includes the owl monkey, *Aotus*, and the titi monkey, *Callicebus*.

Pitheciines are the only specialized seed-eating adaptive radiation among all the living primates. The saki monkeys, bearded sakis, and uacaris have a divergent craniodental morphology, which is exaggerated in *Chiropotes* and *Cacajao*. Their anatomy is adapted to their reliance on seeds that grow inside hard-shelled fruit, more specifically, to the challenge of breaking open the shells. The two homunculines, *Aotus* and *Callicebus*, are each distinctive cranially, particularly the nocturnal owl monkey, with its very large orbits. Both genera have well-developed incisors that reflect their habit of using these teeth to peel the tough rinds of particular fruits or to scrape off the fleshy material that coats a large seed.

- **Diet, locomotion, and activity rhythm**

    The pitheciids are all diurnal, except for the nocturnal owl monkeys. A population of one species, the Azara's night monkey, *Aotus azare*, is cathemeral, as mentioned. They are found in the southernmost region of the genus's distribution, in a colder, semi-humid, non-evergreen ecosystem where nutritious foods are less plentiful. The activity cycle of Azara's night monkey is sensitive to the amount of moonlight that is available. When there is no moonlight and the amount of nighttime light is reduced, the monkeys spend daytime hours foraging for food.

    Pitheciids are quadrupeds but, as is often the case with arboreal primates, there are variations in locomotor and postural behaviors. For example, the saki spends more time in the understory and commonly uses vertical-clinging-and-leaping, while the larger forms, the bearded saki and uacari, predominantly use the canopy and often hang upside-down, suspended by their feet. This ability is associated with food-handling, especially the challenge of splitting open seed-containing fruits with hard protective coverings.

- **Social systems and mating**

    Aotus and *Callicebus* have similar social and mating systems. Their combination of social system with related morphologies and behaviors is different from all other platyrrhines, and all other primates. They live in stable, family-like groups, pair-bonded monogamous units consisting of no more than an adult pair and their young. There is no sexual dimorphism in body size or in canine size. The fact that their canines are very small, hardly projecting above the level of adjacent teeth, is a unique feature. It minimizes their use in threat displays and is associated with low levels of aggression within a social group.

    In both owl and titi monkeys, the bond between the adult male and female is enduring, potentially lasting for years. Their relationship is highly cooperative. They share food with each other, which mutually reinforces their social bond, and is particularly beneficial when a female is lactating. The rearing method of *Aotus* and *Callicebus* is classified as biparental care, but females do relatively little caregiving after giving birth, other than nursing the young. Fathers are responsible for transporting, feeding, grooming, playing, and protecting infants and juveniles. Fathers maintain a close relationship with offspring for years, until it is time for the young to disperse, to leave the natal group to find a mate. Both males and females disperse, which is a common pattern among platyrrhine species with small social groups.

    Aotus and *Callicebus* both defend their small territories from conspecifics by displays, vocalizing, and chasing away outsiders. Owl monkeys may become quite aggressive and fight. Territorial defense may be a means to control habitat and food resources, to serve as a form of **mate-guarding** to keep outsiders from co-opting a mate, or to discourage group takeovers by outsiders. The arrival of a "floater," either an adult male or female that is not paired but is seeking social and reproductive opportunity, can result in the replacement of the resident same-sex adult.

    Titi monkey groups maintain spatial separation from other groups by announcing their location frequently, almost daily. The adults let out a vocalization called a **dawn call** soon after leaving their sleeping tree in the early morning. The vocalization is a loud, staccato cry which carries far into the forest. The dawn call is often made by the male and female together, in a duet. The call is much more complex acoustically, and delivered at a much higher pitch, than the deep, loud roars of howler monkeys discussed above.

Tail-twining is another social activity that is typical of both titi monkeys and owl monkeys. It begins with the adults but often also involves the young. It is a fixed action pattern, an innate behavior that is species- and context-specific. The animals sit side-by-side in close contact, and the adults twist their tails together (Figure 4.8). This is a daily behavior in *Callicebus* that lasts for long periods in the late afternoon, once the group has settled down on a large branch where they will sleep at night. No primates other than *Callicebus* and *Aotus* twine their tails.

Among the pitheciines, the saki monkeys also live in small social groups, have small home ranges, and defend their territories like the homunculines. The bearded saki and the uacari are different, though less is known about their social systems. They live in multimale-multifemale groups with counts as large as 44 in the bearded saki and more than 150 individual members in the uacari. *Chiropotes* and *Cacajao* are fast-moving animals with large home ranges that often include seasonally flooded forests.

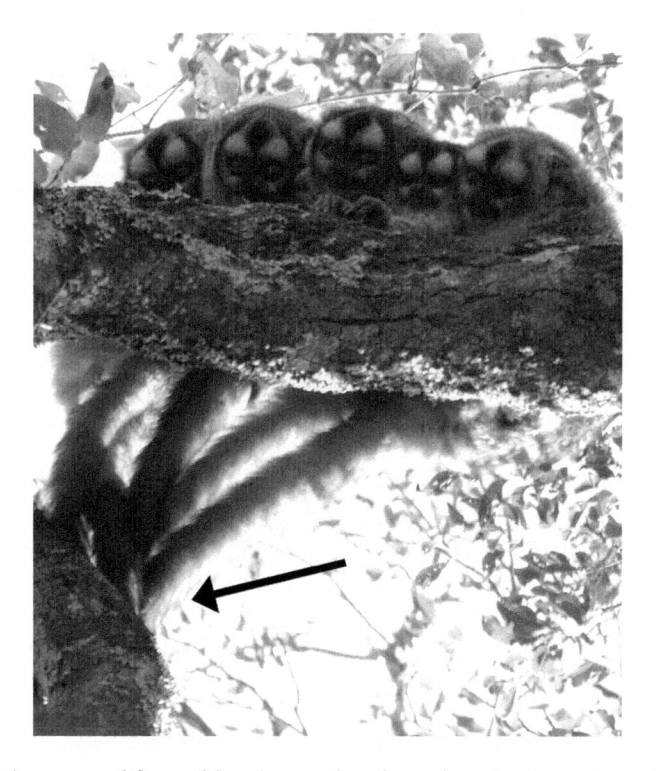

Figure 4.8 A family group of five wild cathemeral owl monkeys in Argentina twining tails (at the arrow).

Courtesy of Margaret Corley and the Owl Monkey Project.

90 South America: New World monkeys

Cebids – predatory omnivorous frugivores

Among platyrrhines, cebids are the most taxonomically diverse family-level clade, with seven genera, more than any other family. Of the seven cebid genera, six live in Manu. They have similar resource needs because they are closely related phylogenetically, but they have also diverged to occupy distinct ecological niches within the family's adaptive zone (Table 4.1). At Manu two of the genera are each represented by sets of sympatric species. There are two capuchin species belonging to the genus *Cebus*, and two tamarin species belonging to *Saguinus*.

Cebids have adaptively differentiated within a frugivorous-faunivorous adaptive zone in a manner that makes them unique in a number of ways. They are the only anthropoid clade that:

- Exhibits a large range of body sizes, with the largest species, the capuchins, weighing roughly 35 times the mass of the smallest, the pygmy marmoset
- Occupies an adaptive zone in which predation on insects and small vertebrates is a fundamental ecological characteristic
- Exploits ecological opportunities in the understory microhabitat, below the continuous forest canopy
- Has species falling in the low end of the primate body mass range, including the pygmy marmoset
- Has a combination of ecological adaptations in each of the two main subfamily clades that include highly significant size increases, **gigantism**, as well as decreases, **dwarfism**
- Presents a great variety of social and mating systems, including patterns revolving around sexual monomorphism and dimorphism; female dominance and male dominance; unimale-multifemale and multimale-multifemale units; suppression of reproduction by females; communal rearing; and a sexually segregated social system
- Includes smaller monkeys whose relative brain sizes exceed those of very large apes

The seven genera of living cebids are classified in two subfamily clades, the cebines and callitrichines. At first glance, they may seem very different considering their body size disparities and how they locomote as a consequence. The small callitrichines scurry about on twigs and are facile in navigating the understory, climbing narrow vines and stationing themselves on wide tree trunks while feeding on gums. The much larger capuchin monkeys locomote through the canopy quadrupedally and hang from their semi-prehensile tails while feeding. They may stretch away from a tree trunk by adopting a tripod stance, braced by two feet and the grasping tail. The locomotor differences between callitrichines and cebines are alternative methods to exploit frugivorous, predacious dietary niches at different body-mass scales.

Despite these differences, cebids are a coherent ecological radiation of predaceous, omnivorous frugivores that is predicated on a small body size, with the exception of one genus, *Cebus*, that dramatically deviates in size yet continues to occupy the same adaptive zone. Cebids are all alike in that they search for prey visually, probing for them manually. A tiny pygmy marmoset pursues a grasshopper on a leaf; a lion marmoset, five times larger than the pygmy marmoset, hunts down a frog that is immersed in a pool of water in the base of a pitcher plant; a squirrel monkey, seven times larger than the pygmy marmoset, flips through a batch of dead leaves where crickets hide; or a capuchin, 35 times larger than the pygmy marmoset, tears through a bird's nest looking for eggs.

Fruits and insects found in the canopy are very important to all cebids, as is the nectar produced by flowers. Tree gums, more prevalent below the canopy, on tree trunks, are a major source of food for callitrichines but not for cebines. Cebines lack the postural adaptations needed to access gum-producing sites efficiently, and don't have the ability to cling to vertical supports, face-forward, for sustained periods of time. Also, for cebines with larger body masses, and living in larger social groups, spot-feeding on gums is not strategically or nutritionally advantageous. They spend their time more productively foraging for their preferred foods in the richer canopy.

Cebids live in a variety of social systems with varied mating strategies and life-history patterns. These include unimale-multifemale and multimale-multifemale systems. Based on data from Manu and elsewhere in Peru, the seven cebid genera exhibit a variety of troop sizes. The midpoints of the average number of individuals observed in wild groups ranges from five in the emperor tamarin to 35 in squirrel monkeys. Feeding opportunity, as always, is an influential factor. Groups are size-limited when operating in the sub-canopy, where the majority of supports are vertically oriented and present physical and spatial constraints. Canopy-feeding is less limiting and provides more opportunities, so larger groups can be sustained.

Cebines – capuchins and squirrel monkeys

The capuchins and squirrel monkeys belong to the cebine subfamily. Both genera are predaceous, omnivorous frugivores, but *Cebus*, the capuchin, and *Saimiri*, the squirrel monkey, differ considerably in body mass; *Saimiri* is roughly one-third the size of *Cebus*. While they both belong to the medium size class, the difference is a key aspect of their biology that produces a distinct ecological separation between them.

Anatomically, cebines have relatively short faces, very close-set orbits, and oval braincases. The size and shape of their braincases, with high foreheads and large, rounded occipital regions in the back, accommodate large brains. Cebines have large, sexually dimorphic canine teeth. Their cheek teeth are distinctly different from one another. Squirrel monkey molars have pointy, sharp cusps while capuchin molars are relatively flat with blunt cusps. Their limb proportions are unremarkable as both genera are basically quadrupedal animals.

- **Diet and locomotion**

    Capuchins are one of the most widespread platyrrhines in terms of geography, ecosystem ecology, and habitat utilization. In addition to living in Amazonia, they inhabit the fragmented Atlantic Forest, and narrow gallery forests in drier areas. Capuchins eat from more than 200 species of plants, including a wide variety of fruits and vegetation, and consume well over a dozen types of invertebrates. Diversified feeding patterns make capuchins habitat generalists with considerable niche breadth. This means they go almost anywhere in the forest and take advantage of a large array of food options that benefit their survival and reproduction. They are considered a pioneer species, among the first to colonize a new area when it becomes physically accessible and botanically feasible. In contrast, the niche of the squirrel monkeys is much narrower. They live in lowland, humid forests, are not found in the drier savannas, and do not occur in the Atlantic Forest.

    The dietary profile and habitat preferences of squirrel monkeys overlap with capuchins, but *Saimiri* are more reliant on figs, in particular. About 50% of *Saimiri*'s annual diet consists of figs, small, soft fruits that can be densely concentrated in places. This enables more individuals to feed simultaneously. Large troops consist of dozens of individuals. Fig trees are very widely dispersed in the forest, so *Saimiri* have very large **home ranges**, areas in which a social group concentrates its foraging, feeding, and resting. Larger and stronger, capuchins are more versatile feeders, so they are able to exist in smaller home ranges.

Capuchins do not generally eat leaves, even though their body mass exceeds other platyrrhines' that rely on leaves to an important extent. These monkeys are driven by phylogeny to eat insects and small vertebrates as a source of protein to complement fruit. They spend about 50% of their time looking for and eating prey, and 40% of their time traveling and feeding on fruit and other sorts of non-leafy vegetable matter. The vertebrates that capuchins consume include rodents, opossums, young birds, lizards, frogs, and more, in addition to a wide variety of larvae and insects that they come across or uncover by rummaging through dead foliage or by stripping bark from trees.

The larger capuchins are stockier in build than the smaller, thinner squirrel monkeys. Capuchins muscle their way through the treetops and encounter few obstacles that hinder their search for something to eat. They break tough fruits apart by beating them against a branch until they crack, to eat the seeds inside. They use rocks as tools to smash open hard fruits and nuts. They bite through the hard stalk of a palm frond to get at the edible pith. They snap off the ends of dead tree limbs to expose insect infestations. The foraging methods of *Cebus* combine strength and brains in frugivorous or predaceous feeding. The capuchin's semi-prehensile tail is used in climbing and is critical during feeding. A *Cebus* can hang or extend its upper body away from a tree so it can grab and manipulate food with its hands.

Saimiri, on the other hand, glide quadrupedally through the forest canopy picking at things to eat, flushing out barely concealed insects or chasing down the more mobile ones. Squirrel monkeys use more gap-leaping in their quadrupedal modes. Their long tails are used for balance while leaping, and can be bent around a branch for steadiness while the animals are holding food in their hands.

The capuchins and squirrel monkeys exploit a variety of microhabitats year-round in the canopy, understory, and for *Cebus* even the forest floor when they eat palm fruits that have fallen to the ground. Yet, capuchins are very large-brained and squirrel monkeys have a high basal metabolic rate. Both genera have energy needs that can be satisfied best by foraging and feeding in the canopy, which produces the most plentiful foods.

- **Social systems and mating**

In cebines, social groups and breeding are coordinated in a number of ways. As sexually dimorphic species, cebines generally live in multimale-multifemale groups, but of very different kinds. *Saimiri* species exhibit stark variations in their social systems. In some species, troop members do not behave aggressively and there is no evidence of dominance hierarchies. Troop members in other species are aggressive, females dominate males, and even juveniles may harass adult males.

Costa Rican squirrel monkeys, *Saimiri oerstedii*, exhibit one of the most unusual social systems found among primates. They live in sexually segregated subgroups without coherent dominance hierarchies. Females and immature offspring tend to remain spatially separated from male groups year-round and have little interaction with them until the two- to three-month mating season begins. Then, as in most species of

Saimiri, males undergo a dramatic, hormonally driven change called the fatted-male syndrome, which is maintained throughout the mating period. Their weight increases by more than 20% due to water retention, their shoulders become enlarged, and their fur becomes fluffy. Their testosterone levels peak, and fighting breaks out as males compete for the attention of individual females when they enter estrous for a brief two days. This is when females become sexually active and reproductively fertile. Fighting separates the dominant and subordinate males. Subordinate males generally do not breed.

Capuchins are organized differently, in heterosexual groups. There are dominance hierarchies among the males and females. Adult males are dominant over females but females are known to form protective coalitions when necessary. The circumstances when *Cebus* males become highly aggressive occur when roving males living in the forest unaffiliated with a stable social and reproductive group attempt a group takeover, an effort to oust an alpha male and commandeer the females. When this happens, females may band together to vigorously protect their offspring and sometimes help the resident alpha male fend off attackers. When takeovers succeed, overthrown males can be seriously injured, banished to live wounded in the forest where their chances to survive are diminished, leaving only the females to protect the offspring. Infanticide often ensues, perpetrated by the marauding males. If the intruders succeed in killing young monkeys, the mothers' reproductive cycles restart and they again become fertile. Long-term studies in the wild have documented that established social groups may experience takeovers every few years. So, during her lifetime, an adult female will live in a group that is controlled by several different male hierarchies.

Callitrichines – marmosets, tamarins, and Goeldi's monkey

There are many species of callitrichines, and they are especially noteworthy for the many ways they look different, including features that function as species-level identifiers so the animals can recognize conspecifics (Figure 4.9). Their faces are remarkably diverse. The head of one tamarin species is topped with a stiff, wide central brush of white hair that can be flattened or raised as a communications gesture. Several closely related species of the tamarins have prominent and varied mustaches. In some the mustache is a short-haired rectangular patch of white below the nose that extends onto the cheeks; in others it is a thin line that spreads to encircle the mouth; one has long whiskers shaped like downturned handlebars. Tufts or tassels of hair about the ears accentuate the head in one group of species belonging to the marmoset *Callithrix*. The genus *Leontopithecus* has a lion-like mane. In callitrichines, fur on the back may be a solid color or mottled. Legs and arms may be offset from the trunk by contrasting colors.

- **Diet and locomotion**

    As a correlate of small body size, one of the most distinctive anatomical features of callitrichines is their clawed digits (Figure 4.9a). Callitrichine claws are a locomotor, postural, and feeding adaptation particular to life in the understory. It enables these small monkeys to move up and down tree trunks by grappling vertical supports that are too wide to be grasped with a foot or hand. Holding such a posture is advantageous when scanning for and ambushing prey beneath the canopy, and when feeding on gums that exude from the tree trunks. This is particularly important for

Figure 4.9 Four callitrichines: (a) black-tailed marmoset; (b) emperor tamarin; (c) cotton-top tamarin; (d) common marmoset.

Photo credits: (a) Black-tailed Marmoset (*Mico melanurus*).jpg, by Bernard Dupont (CC-BY-SA-2.0); (b) Tamarin portrait 2 edit2.jpg, by Brocken Inaglory (CC-BY-SA-3.0); (c) Cotton Top Tamarin by Tricha.jpg., by Ltshears (CC-BY-SA 3.0); (d) Common marmoset (*Callithrix jacchus*) cockroach.jpg, by Bart van Dorp (CC-BY-SA 2.0).

the smallest genera, *Cebuella* and *Callithrix*, that spend significant amounts of time hanging on while gouging holes in the bark with their anterior teeth to generate a flow of tree gum.

Most of the smaller callitrichines need less food and space than cebines to satisfy their dietary needs. Their foraging focuses on the discontinuous lower strata of the forest and areas more rarely used by other monkeys. Callitrichines frequent microhabitats at the edges of the forest where dense canopy cover has been disturbed, such as treefalls. These areas can be overgrown with shrubs and tangles of vegetations containing concentrations of potential prey, like insects and small vertebrates. Dense tangles are also a safer place for a small monkey to forage than the perimeters of large tree crowns that are more exposed to predaceous birds flying overhead.

While gum-eating is important to all callitrichines, it is especially vital to the pygmy marmoset, *Cebuella*, and the marmoset, *Callithrix*. These animals have functionally specialized anterior teeth and jaws that enable them to produce a flow of gum by scraping and gnawing away tree bark. Being smaller in size than the others, they are able to subsist on a narrower range of nutritionally adequate foods. Other callitrichines lack these dental specializations for gouging, and they must rely on gum sites they happen to find after trees have been damaged by insect activity or other means.

Callitrichines locomote in the canopy by scurrying and leaping quadrupedally. But in the subcanopy where vertical supports are predominant as vines, saplings, and tree trunks, they use a leaping motion by jumping horizontally or in ways reminiscent of vertical-clingers-and-leapers. This is especially true of Goeldi's monkey, *Callimico*, a genus that habitually uses trunk-to-trunk leaping between medium- or large-diameter supports, where they perch while scanning for prey living in the leaf litter on the forest floor below. In areas where *Callimico* are sympatric with tamarins, both species often forage together but the Goeldi's employ trunk-to-trunk leaping twice as often as the tamarins.

The pygmy marmoset, *Cebuella*, and Goeldi's monkey, *Callimico*, are each ecologically distinguished from all other primates by unique diets. *Cebuella* is the most gumivorous primate in Amazonia and they occupy microhabitats that are rarely used by other genera, especially river banks – they are habitat specialists. *Cebuella* eats little fruit, but does eat insects. They are tiny monkeys and sometimes rely on a single gum-producing tree in a very small home range. When that tree is tapped out or destroyed by damage, a *Cebuella* troop simply moves to another site with a suitable gum tree.

Goeldi's monkey, *Callimico*, are also habitat and dietary specialists. They have been well-studied outside of Manu in parts of Bolivia and Brazil. They live in rainforest habitats, but are most abundant where there are stands of bamboo. Much of their diet is composed of fungi: bamboo fungus and jelly fungus that tend to grow in wet habitats on decayed or fallen trees and tree limbs. These monkeys spend a lot of their time traveling to find fungus. Both types of fungi are relatively rare in the rainforest, and inconsistently available. Fungus can take days or months to regrow after being harvested by the animals.

- **Social systems and mating**

    Callitrichine social groups tend to be small, although groups of up to 20 individuals have been observed. Their social organization is based on multimale-multifemale configurations. The identity and relatedness of group members can be very complex. This has been investigated in tamarins by employing paternity tests using DNA taken from hair follicle cells from individuals living in geographically adjacent social groups that were captured and then released. The study showed that resident males may not be the fathers of the young in a cohesive social group, and older and younger adult females may not be mother-daughter pairs, as in the social groups of many other primate species.

There is a wide variety of mating patterns among callitrichines. Alpha females may mate with one or several males; one male may mate with multiple females; and mating may occur between multiple males and multiple females. Despite this variety, callitrichine social groups share a general pattern of care-giving behavior in which various group members contribute to rearing offspring. This unique communal rearing system revolves around other exceptional conditions shared by nearly all callitrichines. They give birth to twins, sometimes at the high rate of twice each year. The dominant breeding female suppresses reproduction in other resident females by way of pheromonal and possibly agonistic behavioral means, which is why they are available care-givers. Only one female breeds. There are a few exceptions to the pattern. Goeldi's monkey, *Callimico*, is different. It gives birth to singletons, and two females in a group may be active breeders at the same time. Lion marmosets, *Leontopithecus*, a genus that has litters of twins, may also have multiple breeding females in a social group.

Dry season niche partitioning

The seasonality of food production in rainforests is associated with markedly different feeding strategies and niche partitioning among Manu's cebids. Their diets vary more during the dry-season months than at other times of the year. There is a shortage of choice fruits during the four or five dry months at Manu, when the forest may produce roughly 20% of the amount of fruit that is found during the wettest months. Cebids also eat insects. But because insects rely on forest foods, their numbers diminish during the dry season, too. Consequently, the amount of these major nutritional resources for cebids falls greatly. Studies have shown that monkeys lose weight during this period. For example, saddle-back tamarins, *Saguinus fuscicollis* (sometimes called *Leontocebus weddelli*) have been found to lose about 5% of their body weight during the dry season.

The potential for food competition increases during the dry season, and the species-specific behaviors of the animals change as they contend in different ways with the food shortages.

The squirrel and capuchin monkeys shift their feeding preferences by focusing on very different species of fruit that reflect the animals' contrast in size and strength.

The squirrel monkey, *Saimiri*, concentrates on soft, ripe figs, even though locating choice fig trees requires extensive travel. The capuchin, *Cebus* – bigger, stronger, more dexterous – relies on harvesting hard palm nuts for food, which are entirely outside the squirrel monkey's physical abilities. For capuchins, either figs, palm fruits or nuts become fallback foods. **Fallback foods** are foods that primates rely on when commonly eaten ones are unavailable. They are typically more challenging to access and/or process; require more handling time before feeding; exert more wear-and-tear on the dentition; may be more difficult to digest; and may be less valuable nutritionally.

The two sympatric capuchins at Manu, *Cebus apella* (*C. apella*), the tufted capuchin (also known as *Sapajus macrocephalus*) and *Cebus albifrons* (*C. albifrons*), Humboldt's white-fronted capuchin, are most distinctly separated ecologically during the dry season when both turn to foods that they barely eat at other times of the year. The more powerfully built, robust *C. apella* seeks out the large, hard nuts of *Astrocaryum* palms that become abundant and ripe during the dry season. For about a month at that time, these animals do not eat any other types of fruit. *C. apella* can crack the nuts open with a single bite to get at the nutritious plant tissue that coats the seed inside, the endosperm. They also eat pith from the interior of the *Astrocaryum* palm frond's very hard central stem.

Cebus albifrons also eats *Astrocaryum* endosperm as a fallback food. But because it has less powerful jaws than *Cebus apella*, it cannot bite the nuts open, and so it cannot harvest them directly from the trees. Instead, they must wait for the nuts to ripen, fall to the forest floor, and begin to rot. The animals then come to the ground to rummage through the debris, scanning and sniffing to find nuts that have been weakened because they were damaged by beetles, forest-floor species that feed on *Astrocaryum* as well.

For *C. albifrons*, the process of selecting these rotted nuts requires time and effort to look for those that have turned a lighter color, with the right smell, with a beetle drill hole, and one whose endosperm has not been fully eaten by the beetles. The monkeys spend a great deal of time each day on the ground searching, sometimes inspecting more than 20 fallen nuts before choosing one. Even then they may not be able to bite it open. They carry it back into a tree and smash it against a branch to try to open the nut, but that does not always work either. For *C. albifrons*, selecting and handling each nut that is finally eaten can take 5–10 minutes.

The *C. albifrons* approach is much less efficient than that of *C. apella*, the arboreal, fresh-fruit, grab-crunch-and-eat method. This difference underscores how the niches of the two species are different. The tufted capuchin, *C. apella*, is able to satisfy more of its energy needs in a smaller home range than the white-fronted capuchin, *C. albifrons*, because every patch of *Astrocaryum* palms is a food bonanza, in nuts and pith. But for *C. albifrons*, the labor- and time-intensive requirements of palm-nut foraging are so demanding that they must rely on a second fallback food, figs, during the dry season, like squirrel monkeys. This means *C. albifrons* must travel more and maintain larger home ranges than *C. apella* during the dry season.

The saddle-back tamarin, *Saguinus fuscicollis*, and emperor tamarin, *Saguinus imperator*, the two sympatric tamarin species at Manu, have very similar feeding habits. Groups of both species forage together in the same microhabitats, mostly in the subcanopy, forming mixed-species associations. They feed on the same fruit species and many of the same kinds of insects, including grasshoppers and crickets that comprise the bulk of their prey. What distinguishes the two *Saguinus* species ecologically is the manner in which they hunt, their targets, and locomotor styles. Saddle-backs search for brown insects, such as grasshoppers and locusts that are found in bark crevasses or tree knot-holes located on vertical substrates, which they negotiate by clinging-and-leaping. Emperor tamarins prefer more mobile green insects which they grab from the surfaces of leaves that cluster in dense patches, using a quadrupedal, lunging prey-capture technique.

Non-Amazonian cebids

Two genera of callitrichines are found in other ecosystems. One is located in a vast area that makes up roughly the eastern half of the Amazon basin below the Amazon River, and continues across tropical savanna to reach the Atlantic Forest. This is *Callithrix*, the marmoset. Different species of this genus are found in the wet rainforests and dry habitats encompassed by these ecosystems. The other genus is *Leontopithecus*, the lion marmoset. It is confined to the disjunct rainforest of eastern Brazil, the Atlantic Forest, where these two genera can be found living sympatrically.

Table 4.2 Comparison of the 16 genera of New World monkeys that inhabit the Amazon basin and the Atlantic Forest in Brazil, and their roughly approximated body weights and body-size classes. Note that four genera are found in both regions. *Leontopithecus* and *Brachyteles* only occur in the Atlantic Forest

Body weight (approximate)	Size category	Amazon	Atlantic Forest
100 g (3.5 oz.)	Small	*Cebuella*	
250–300 g (0.7–0.5 lb.)	Small	*Saguinus, Callithrix*	*Callithrix*
500 g (1.1 lb.)	Medium	*Callimico*	*Leontopithecus*
1,000 g (2.2 lb.)	Medium	*Callicebus, Aotus, Pithecia, Saimiri*	*Callicebus*
3,000 g (6.6 lb.)	Medium	*Cebus, Chiropotes, Cacajao*	*Cebus*
5,000 g (11 lb)	Large	*Alouatta*	*Alouatta*
10,000 g (22 lb.)	Large	*Ateles, Lagothrix*	*Brachyteles*

The Atlantic Forest is a unique South American biome with a composition of primates that is quite distinct from the Amazonian basin and from the community at Manu (Table 4.2). This is significant for the adaptations and ecological positions of the animals. Members of all three platyrrhine families are found in the Atlantic Forest, including four genera that occur in the Amazon basin, though they are much less diverse taxonomically.

Each of the platyrrhine genera in the Atlantic Forest consists of a single species in the conservative taxonomy used here, while in the Amazon basin most of the genera are comprised of two or more species. The owl monkey, *Aotus*, an ecologically flexible genus, is notably absent from the Atlantic Forest though it is otherwise widespread and occupies a distinctive nocturnal niche that is not available to other New World monkeys. Seed-eating sakis and uakaris are also not found in the Atlantic Forest.

The Atlantic Forest platyrrhines that are congeneric with those in the Amazonian basin are ecologically differentiated from one another locally by phylogeny and adaptation, as in all primates. The two unique Atlantic genera are distinctive in several ways, as we see below. Neither of them has an adaptive counterpart that lives in the Amazon basin. One is an endemic cebid, *Leontopithecus*, that occupies an unusual predatory feeding niche, and the other is an endemic atelid, the muriqui, *Brachyteles*, that is a unique type of frugivore-folivore, as discussed above.

Body size differences strongly influence niche differentiation among the monkeys of the Atlantic Forest (Table 4.2). Each of the genera is well separated from others by size. In the Amazon, in contrast, multiple genera overlap in body weight. Two to four genera fall in the same body mass categories.

Ecologically, the Atlantic Forest's lion marmoset, *Leontopithecus*, stands out as a super-predator compared to other callitrichines. It is morphologically distinguished by its larger size, bigger teeth, and relatively long forearms and hands, an adaptive complex that enables the animals to find, catch, and eat small vertebrates, like frogs, in an unusual way. In addition, this facilitates a broad range of frugivory, which includes eating fruit of various sizes, nectar, and gums.

Lion marmosets are specialized foragers in a distinct microhabitat where their morphology is uniquely advantageous. It searches for insects and vertebrates that live in tank bromeliads, which are a characteristic feature of the Atlantic Forest where they grow abundantly high up in the trees. Bromeliads are plants that grow on other plants (epiphytes, also known as air plants) without taking root in the ground. They have a self-watering mechanism consisting of a large basin-like center in the middle of the plant from which long, stiff leaves grow outward in a circular arrangement, forming a water reservoir. The pool of water attracts a large variety of aquatic insects as well as frogs, salamanders, and snakes. *Leontopithecus* forages in these bromeliads, using its long, slender hands and forearms to reach into them and probe the tanks for prey.

The marmoset *Callithrix* is the smallest primate in the Atlantic Forest ecosystem. *Callithrix* is a specialized gum-eater, like its closest genus-level relative, the pygmy marmoset from Amazonia. This feeding adaptation distinguishes them from the other sympatric monkeys by body size and microhabitat. They prefer to forage and feed in the subcanopy. Specialized bark-gouging anterior teeth also enable *Callithrix* to exploit tree gum and inhabit many drier and botanically poorer areas within the tropical savanna, grassland, and shrubland that is situated between the Amazon and Atlantic Forest.

The social organization and breeding systems of *Callithrix* and *Leontopithecus* resemble the patterns seen in *Saguinus* as discussed above. Groups are small, alpha females suppress the reproduction of other females, one or more males may be mating partners, and groups engage in communal rearing.

Paleocommunities

In South America, there is fossil evidence of an important paleocommunity from a 12–14 million-year-old site in the Andes of northwestern Colombia, known as La Venta. An arid desert today, La Venta documents a rich collection of remains of forest-loving terrestrial and marine vertebrates that were then part of the greater Amazonian rainforest fauna. There are nearly a dozen genera and species of fossil primates, many related to extant platyrrhines, in sizes that range from a small tamarin to a large howler monkey. All these fossils are known from teeth and jaws, a few from postcranial remains as well. The dietary adaptations represented include frugivory, seed-eating, folivory, and insectivory. There is a specimen with a well-preserved mandible, dentition, and a piece of the lower face from a nocturnal species of owl monkey that is barely distinguishable from, and is classified in, the same genus as the extant owl monkey. It is called *Aotus dindensis*.

At least a half-dozen fossils of primate species that lived at La Venta can be directly linked phylogenetically and adaptively with taxa alive today in Amazonian basin communities. Three or four of the La Venta genera are also linked with 20 million-year-old platyrrhine fossils discovered in Argentina, which supports the idea that subtropical forests existed there before grasslands set in about 14 million years ago.

No native primates live anywhere in the Caribbean today. But four genera of subfossil platyrrhine primates have been discovered in the Caribbean among the islands of Cuba, Hispaniola, and Jamaica. These animals were all endemic to the Caribbean and had been there for millions of years. Because they existed as small island populations, they were also subject to intense threats due to natural and anthropogenic hazards. One of the most interesting subfossils is *Xenothrix*, from Jamaica. *Xenothrix* existed until at least 900 years ago.

The conservation status of platyrrhines

New World monkeys are severely threatened (Figure 4.10). The habitats where platyrrhine primates live in South America are disappearing with alarming speed, due in large part to deforestation. For example, by some estimates more than 278,000 miles^2 (~720,000 km^2)

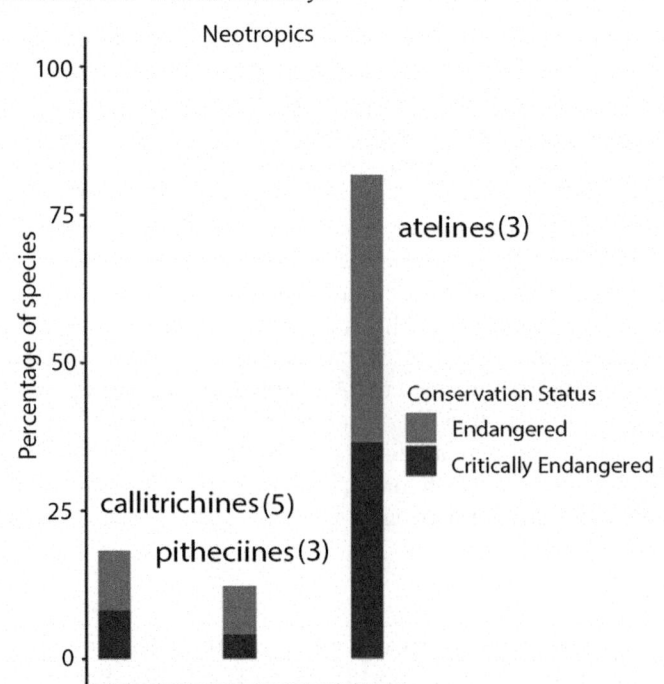

Figure 4.10 The threat levels to three of six subfamilies of New World monkeys. Numbers in parentheses refer to the number of genera in each group.

Adapted from Fernández, D., Kerhoas, D., Dempsey, A., Billany, J., McCabe, G., & Argirova, E. (2021). The current status of the world's primates: Mapping threats to understand priorities for primate conservation. *International Journal of Primatology*. https://doi.org/10.1007/s10764-021-00242-2.

of the Brazilian Amazonian rainforest, were lost since 1970, which is about 20% in the last 50 years. All experts agree that the rate of deforestation is increasing.

In 2019, conservationists warned that the Amazon basin, which is the heart of neotropical primate life, may have reached an ecological tipping point. Dry seasons had become longer, drier, and hotter, and record-breaking droughts occurred in 2005, 2011, and 2015. If deforestation is not mitigated, mainly by reforesting areas that have already been cleared, scientists predict the rainforest will devolve into a savanna biome, where platyrrhines cannot live.

Seven of the 25 most imperiled primate species identified in 2022 by the IUCN SSC Primate Specialist Group are South American monkeys (Figure 1.11). The largest platyrrhines, the atelines, are at the highest risk because they require more space, reproduce slowly, and they either require mature forests where commercial logging is now intense, as with spider monkeys, or are now confined to living in small habitat fragments that limit population size and growth for various reasons, which is the case with muriqui (Figure 4.10).

There are long-term efforts designed to increase the survivability of platyrrhine species. Two successful projects located in the Atlantic Forest are especially noteworthy. Projeto Muriqui de Caratinga focuses on the muriqui, the atelid *Brachyteles arachnoides*. Associação Mico-Leão-Dourado is devoted to the golden lion marmoset (also called golden lion tamarin), the callitrichine *Leontopithecus rosalia*. The muriqui and golden lion marmosets are endemic species of the Atlantic Forest whose populations have declined to very low numbers as their habitats have become fragmented and degraded.

The main objective of both projects has been to secure sufficient space for a population of each species that can be protected in their natural habitats. Each project led to the establishment of a biological reserve, local educational programs that encourage their conservation, and to the continuous presence of researchers studying the animals and their habitats, conducted by Brazilian and foreign scientists. See Chapter 9 for more on threats to primates globally and conservation efforts.

Key concepts

Phylogeny and classification
Niche breadth
Seasonality and fallback foods
Energy minimizers and maximizers
Habitat generalists and specialists

Quizlet

1 What makes anthropoids the most widespread and ecologically diversified radiation of primates?
2 Give three features that make platyrrhines haplorhines.
3 Describe the different shapes of the platyrrhine and catarrhine noses.
4 What is the activity rhythm of platyrrhines? What is the one exception?
5 Describe two of the varied locomotor styles of platyrrhines.
6 Describe the atelid family's prehensile tail and how it is used.
7 Describe the social organization and breeding system of one of the platyrrhine genera.

Bibliography

Butler, R. A. (2020). Calculating Deforestation in the Amazon. *Mongabay.* https://rainforests.mongabay.com/amazon/deforestation_calculations.html

Fernández-Duque, E., Fiore, A., & Huck, M. (2012). The behavior, ecology, and social evolution of New World Monkeys. In J. C. Mitani, J. Call, P. M. Kappeler, R. A. Palombit & J. B. Silk (Eds.), *The Evolution of Primate Societies* (pp. 43–64). Chicago, University of Chicago Press.

Fragazy, D. M., Visalberghi, E., & Fedigan, L. M. (2004). *The Complete Capuchin: The Biology of the Genus Cebus.* Cambridge, Cambridge University Press.

Garber, P.A., Porter, I.M., Spross, J & DiFiore, A. (2016) Tamarins: insights into monogamous and non-monogamous single female social and breeding systems. *American Journal of Primatology,* 78(3), 298–314.

Janson, C., & Emmons, L. (1990). Ecological structure of the nonflying mammal community at Cocha Cashu Biological Station, Manu National Park, Peru. In A. H. Gentry (Ed.), *Four Neoropical Rainforests* (pp. 314–338). New Haven, Yale University Press.

Lovejoy, T. E., & Nobre, C. (2019). Amazon tipping point: Last chance for action. *Science Advances,* 5(12), eaba2949. https://doi.org/10.1126/sciadv.aba2949

Porter, L. M. (2006). *The Behavioral Ecology of Callimicos and Tamarins in Northwestern Bolivia.* London, Pearson.

Rosenberger, A. L. (2020). *New World Monkeys: The Evolutionary Odyssey.* Princeton, Princeton University Press.

Rosenberger, A. L., Tejedor, M. F., Cooke, S. B., & Pekar, S. (2009). Platyrrhine ecophylogenetics in space and time. In P. A. Garber, A. Estrada, J. C. Bicca-Marques, E. W. Heymann, & K. B. Strier (Eds.), *South American Primates: Comparative Perspectives in the Study of Behavior, Ecology, and Conservation* (pp. 69–113). New York, Springer.

Rylands, A. B. (Ed.). (1993). *Marmosets and Tamarins: Systematics, Behaviour, and Ecology.* Oxford, Oxford University Press.

Terborgh, J. (1983). *Five New World Primates.* Princeton, Princeton University Press.

5 Africa
Lorises, galagos, Old World monkeys, and great apes

Chapter Contents

Mainland Africa's geography and climate	104
The African radiation: lorises, galagos, Old Word monkeys, and great apes	105
Lorisoids – lorises, pottos, angwantibos, and galagos	106
The lorisoid profile	107
Catarrhines – Old World monkeys and apes	109
Teeth	111
Color vision, olfaction, and communication	111
Cercopithecoids – Old World monkeys	112
Bilophodont molar teeth	112
Cranial and postcranial characteristics	112
Cheek pouches and guts: cercopithecine and colobine specializations	112
Ischial callosities	115
The cercopithecoid profile	115
Hominoids – apes	117
The hominoid profile	119
African primates living in communities: diversity, ecology, and adaptation	122
Lorisoids – nocturnal frugivorous-gumivorous predators	124
Cercopithecids – omnivores, frugivores, and folivores	126
Other cercopithecids – the long-faced papionins	130
Hominids – semi-terrestrial frugivore-folivores	133
Other African apes – gorillas, the largest living primates	138
Paleocommunities	140
The conservation status of African primates	141
Key concepts	142
Quizlet	142
Bibliography	143

Superfamily Lorisoidea (lorisoids) – lorises and galagos
   Family Lorisidae (lorisids) – lorises
   Family Galagidae (galagids) – galagos
Superfamily Cercopithecoidea (cercopithecoids) – Old World monkeys
   Family Cercopithecidae (cercopithecids) – Old World monkeys
      Subfamily Cercopithecinae (cercopithecines) – cheek-pouched monkeys
      Subfamily Colobinae (colobines) – leaf monkeys

DOI: 10.4324/9781003257257-5

Superfamily Hominoidea (hominoids) – apes
  Family Hominidae (hominids) – great apes
    Subfamily Homininae (hominines) – African great apes

Mainland Africa's geography and climate

The primates found in Africa include four of the seven major groups of primates: lorises, galagos, Old World monkeys, and apes. All but the galagos are also found in Asia. Galagos are endemic to mainland Africa. Africa and Asia are the only landmasses where multiple major radiations occur. In Madagascar we find only lemurs. In the neotropics we find only New World monkeys. And only in Asia do we find tarsiers, along with lorises, Old World monkeys, and apes. The apes that occur in Africa are great apes, chimpanzees, bonobos, and gorillas. The other great ape, the orangutan, and the lesser apes, the gibbons and siamang, are endemic to Asia, while both Old World monkey subfamilies are Afroasian.

This means the taxonomic composition and biogeographical context of the African primate fauna are very different from the regions discussed in Chapters 3 and 4. Madagascar's lemurs and the neotropical New World monkeys each represent individual, coherent phylogenetic and taxonomic groups found exclusively on these landmasses and nowhere else. In contrast, Africa's lorises, Old World monkeys, and great apes belong to the same family-level clades as the lorises, Old World monkeys, and great apes of Asia. They have the same basic anatomical and behavioral patterns, but there are taxonomic and adaptative differences, at the genus and species levels, that distinguish the African and Asian forms.

We continue to refer to this region, mainland Africa, as Africa even though Madagascar is also a part of Africa geographically. The mainland is enormous, only about 4% smaller than the largest continent, which is Asia. The equator transects the middle of the African continent, meaning much of it is located within the tropical zone, making the temperature relatively hot (Figure 5.1). A sizeable area within north Africa's tropical and subtropical zones consists of the Sahara Desert, where it rarely rains and the temperatures average over 100°F (37.8°C) for months each year. At the southern end of the continent there is a second major desert, the Kalahari. Together, these deserts comprise about one-third of the mainland, making Africa the hottest continent on Earth.

Less than 10% of mainland Africa consists of continuous, dense evergreen tropical rainforest, located in its midsection. Yet, because the surface area of Africa is so large the rainforest is enormous, over 2.2 million square miles (~5.7 million square kilometers). The Congo Basin of Central Africa is the heart of the African rainforest. It is fed by the Congo River, the second largest river in the world after the Amazon River in South America. A topographically flat region that is rimmed by sloping terrains, the Congo Basin contains the largest swamp forest in the world.

Primate communities figure prominently in Africa's rainforests, and each community may include more than a dozen sympatric primate species. Moving away from the rainforests, the numbers of primate species decline in different biotas to the east, north, and south, were there are botanically different habitats, less rainfall, and markedly varying seasonal temperatures. In woodlands, where trees are less dense, there are more deciduous species, and more shrubs, herbs, and grasses that grow in the understory. There, four primate species may coexist sympatrically, including three arboreal forms and one terrestrial. Where there are trees in the resource-poor savanna grasslands, there may be

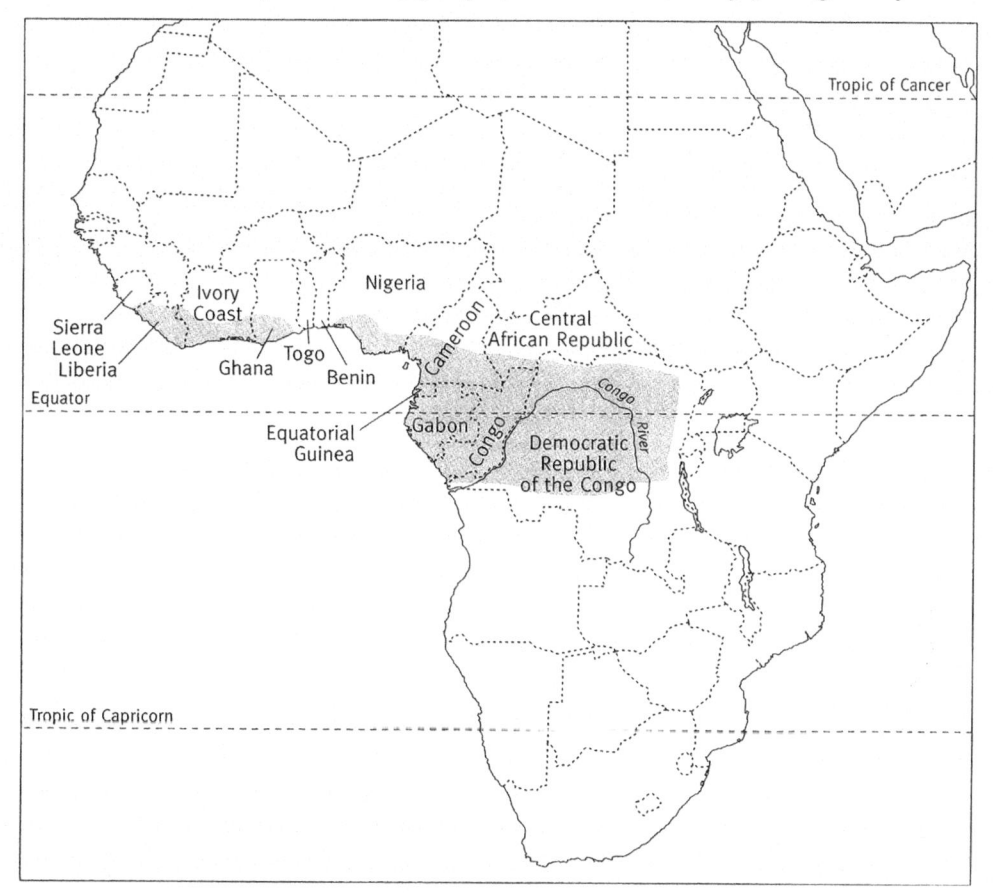

Figure 5.1 Geographical location of rainforests, shaded, on mainland Africa.

From Corlett, R. & Premack, R. (2011). *Tropical Rain Forests: An Ecological and Biogeographical Comparison* (2nd edition). New York, Wiley-Blackwell.

only two or three primate species, semi-terrestrial and terrestrial. In some areas, only one monkey species occurs.

African primates have flourished in habitats that might seem inhospitable to a forest-loving mammalian order like the primates. There are six terrestrial baboon species geographically distributed in enormous tracts across open savannas and woodlands. They range from the southernmost tip of the continent to the edge of the Sahara in the north and into the rocky deserts and hills of the Arabian Peninsula (which is technically Asia).

The African radiation: lorises, galagos, Old Word monkeys, and great apes

As noted, the primate radiation of mainland Africa differs from the primate radiations of Madagascar and South America in a number of ways. Primates in Africa comprise:

- A phylogenetically disparate fauna, composed of both strepsirhines and anthropoids
- Nocturnal and diurnal radiations

- An unparalleled spectrum of habitats occupied, ranging from rainforests to deserts
- Taxonomic groups that evolved semi-terrestrial and terrestrial adaptations
- Many extra-large species (20–60 lbs., 9–27 kg) and several super-large species (70–400 lbs., 32–181 kg)
- Few truly faunivorous, insectivorous, or gumivorous specialists
- Communities with the highest numbers of coexisting species, up to 17, including some with four congeneric sympatric species
- Species with uniquely complex social systems involving enormous groups composed of hundreds of individuals
- Species with the greatest cognitive abilities of any primates
- The most carnivorous anthropoids that are also likely the most intelligent ones – chimpanzees and bonobos – whose lifestyles reflect an entirely unique ecological role

Chimpanzees and bonobos are ecologically unique because they are team-hunting predators and meat-eaters, but they are not comparable to flesh-eating mammals known as carnivorans, such as large cats (felids – like lions, tigers, and leopards) that are obligatory hunters needing to eat meat for their survival on an almost daily basis. Chimpanzees and bonobos are frugivorous omnivores that hunt and eat mammals, especially monkeys that live in the same forests, as a part of their broad-spectrum diet. In the Tai Forest of the Côte d'Ivoire (Ivory Coast) that we discuss below, chimpanzees hunt seven of the sympatric monkey species there as well as a coexisting strepsirhine. They hunt almost every week, and at least once a day during a three-month hunting season.

As a result of their hunting habits, chimpanzees stand apart as community members. Their hunting of co-located primates imposes a larger, more unusual selective load on the other species than the more benign types of indirect competitive interactions that drive ecological differentiation and enable sympatric species to obtain resources while staying out of each other's way.

The distinctive features of the African primates cut across very different clades with different lifestyles. That contrasts with the more homogenous patterns that characterize the radiations of Madagascar and the neotropics, each of which constitutes a single clade with a foundational set of adaptations: lemurs are a clade of strepsirhines, platyrrhines are a clade of anthropoids. The African primates include both strepsirhines and anthropoids in a single region, and species belonging to both of these clades live in the same communities.

Lorisoids – lorises, pottos, angwantibos, and galagos

The lorisoids (Figure 5.2) consist of two distinct groups that are assigned taxonomically to two families, the lorises (also called lorids and lorisids) and galagos (also called galagonids and galagids). While fruit, gum, and animal matter are primary food resources for lorisoids, there are consistent differences between these families that reflect alternative predatory strategies. These are most evident morphologically and behaviorally in the locomotor system. Lorises are stealthy, pursuit predators, and move quadrupedally. Galagos are ambush predators that move by vertical-clinging-and-leaping. Each group exhibits radical specializations relating to these different styles of predation.

The lorises are the more diverse family morphologically. The African lorises include two species of the potto genus, *Perodicticus*, and one species of the angwantibo, *Arctocebus*.

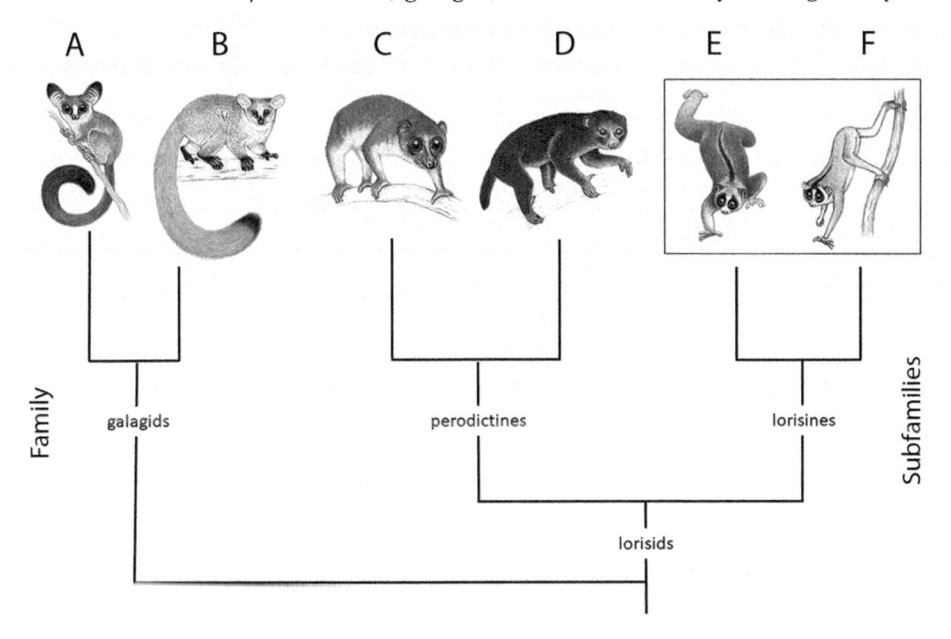

Figure 5.2 Cladistic relationships of the three major groups of living African and Asian (in the rectangle) lorisoids represented by (a) the dwarf galago, *Galagoides*; (b) greater galago, *Otolemur*; (c) angwantibo, *Arctocebus*; (d), potto, *Perodicticus*; (e) slow loris, *Nycticebus*; (f) slender loris, *Loris*. Not drawn to scale.

Artwork by Stephen D. Nash, courtesy of Stephen D. Nash and the IUCN SSC Primate Specialist Group.

Both genera are classified in the perodictine subfamily. Asian lorises, the slender genus, *Nycticebus*, and the slow loris, *Loris*, are classified in the lorisine subfamily.

The genus- and species-level taxonomy of galagos is a matter of debate. Three galagid genera are recognized by all researchers: *Galago*, the lesser galago, *Euoticus*, the needle-clawed galago, and *Otolemur*, the greater galago. Some scientists recognize two others, *Paragalago*, the eastern dwarf galago, and *Sciurocheirus*, the squirrel galago. As many as 18 galago species have been identified.

The lorisoid profile

In summarizing the biological profile of lorisoids, as in other chapters, we emphasize that the features described separately for introductory purposes are usually parts of functionally interconnected systems.

- Phylogeny. Lorisoids are most closely related to Madagascar's cheirogaleids based on the fact that they share derived cranial features and adaptive hypothermia, a trait discussed in Chapter 3. Not all primatologists agree with this hypothesis because the molecular data presents an alternative view, that lorisoids are most closely related to Madagascar's lemurs as a whole, not to any one individual family, as will be discussed when we review the fossil record in Chapter 8.
- Body mass and shape. Lorisoids are small strepsirhines, weighing less than one pound (~0.45 kg), but a few species of each family fall into the medium-size class. The largest

lorisoid is the potto, *Perodicticus potto*, weighing roughly three pounds (1.4 kg). Lorises have large, furry round heads; forward-facing large eyes that produce a flat-faced appearance; small ears; either a stubby tail or none at all; large, wide hands and feet; and, nearly equal forelimb-to-hindlimb proportions. Galagos weigh roughly 1.8 oz. to 2.6 lbs. (50 g–1.2 kg); have longer snouts; very large, mobile ears; a long tail that is often bushy; narrower hands; longer feet, and distinctly long hindlimbs.

- **Activity cycle**. Both lorises and galagos are nocturnal.
- **Ecological domain**. Lorises and galagos are arboreal and rarely come to the ground.
- **Diet**. The species primarily feed on varying proportions of fruits, gums, and animal matter. Several galagos and lorises are strongly gumivorous. They spend more than 50% of their activity budget on accessing and eating gum. Some species have very strong toothcombs. In feeding, most use the toothcomb as a scraper to remove existing gum deposits, but some use the toothcomb as a gouge to stimulate tree gums to flow.
- **Locomotion**. The differences in lorisid and galagid locomotor styles are adapted to the types of prey they prefer, and the methods the animals use to search for and capture prey. Lorises feed on slow-moving insects, including caterpillars and beetles. They rely on vision and olfaction to find prey and approach their targets stealthily using quadrumanous, i.e., four-footed, walking and climbing to avoid rustling leaves. Galagos tend to eat noisy, fast-moving insects. These primates rely on hearing and vision to detect prey, and catch them by pouncing.
- **Craniodental morphology**. As strepsirhines, all lorises and galagos have a toothcomb. The toothcombs of galagos that rely more on eating gums have longer, more robust crowns. The distinctive upper incisors of lorises and galagos have a peg-like shape. They are set wide apart in the midline. Pottos and galagos have relatively long snouts and small braincases.
- **Pelage**. Lorisoid fur is drably colored in greys and browns, providing nocturnal camouflage.
- **Sexual monomorphism**. Like other strepsirhines, lorisoids are sexually monomorphic.
- **Communication**. As nocturnal animals, lorises and galagos rely on olfactory and vocal communication. Species-specific vocalizations are important among galagos in social contexts. Certain calls are used as an antipredator alarm, which attracts conspecifics in an effort to mob and discourage a predator. The slow-moving lorises rely less on vocalizations and more on olfaction for communication. Both groups distribute odors using scent-marking glands and by depositing urine.
- **Social systems and mating**. Lorisoids are classified as semi-solitary, but they are not asocial. In some species individuals spend time together or in close proximity. The most interactive species spend nearly 50% of their time in social settings. In some galagos, adult females tend to affiliate with one another; in others, a female may associate with several males. "Group sleeps" are common in lorisoids, when animals nest together during daylight hours in tangles of vegetation or in tree hollows.
- **Reproductive pattern**. Most galagos and the African lorises give birth to singletons, once or twice a year. The lorises are unusual in producing infants with relatively low birth weights that require a lengthy lactation period.
- **Parenting styles**. Mothers carry newborns and infants in their mouth for short distances, holding the young by the neck, although in some species an infant is able to cling to the fur on its mother's back. Non-independent offspring are frequently parked

in nests while a mother forages. She may shuttle her infants to several different parking spots each night while she travels to feed.

The predatory adaptations, concealing coloration, and slow movements of lorises in particular are associated with a profoundly distinctive morphology and a lifetime of stealth. They are quiet to avoid being heard by tree-climbing predators, and to avoid being heard by their prey. Their hands and feet – with thumb and large toe positioned opposite the three lateral fingers and toes (second digits are vestigial) – are highly modified to provide strong, pincer-like grips. The touch pads on the palms, soles, and digits are enhanced to amplify traction. The trunk is flexible and can become rigid with muscular control. All these features enable lorises to move on top of, below, and between branches without rattling them, to snatch their prey by surprise and elude predators.

In a threatening situation, with a predator approaching head-on, a potto may take up a face-to-face defensive posture, freezing in place and deeply bending its head and neck downward between the arms. This bent-down maneuver exposes the potto's defensive shield, a raised area of thickened skin on the back of the neck that is reinforced by specializations of the spine and shoulder blades. The potto's shield offers some protection against a neck-biting attack from a carnivoran, for example. Observers have also seen pottos use their shield proactively: firmly anchored by the strong hands and feet, they thrust the shield against a would-be attacker with sufficient force to cause it to lose balance and fall. In dire circumstances, the potto may simply escape by taking a nose-dive to the forest floor.

Catarrhines – Old World monkeys and apes

Catarrhines are a tremendously successful adaptive radiation of living primates with a geographical distribution that is unmatched by any other primates. This is due largely to the ecological dominance of Old World monkeys, the cercopithecoids. The other catarrhine group consists of apes, the hominoids. Both groups live in Africa and Asia. The cercopithecoids occupy the broadest ranges of climate and habitat, partly because they have evolved terrestrial lineages. Populations of a few ground-dwelling species have become accustomed to living in urban areas as well, like Asian langurs and macaques that will be discussed in Chapter 6.

Some of the catarrhines' crucial adaptations are highly sophisticated when compared with the majority of other primates, which provide various ecological and behavioral advantages. For example, catarrhines are the only clade in which true red, green, and blue color vision, trichromacy, is universal, affording the animals a high-resolution, fine-grained visual comprehension of the environment. Catarrhines have the largest relative brain sizes as a group (as opposed to a single genus, the platyrrhine capuchin monkey that has quite a large brain). Many genera appear to be quite advanced cognitively. The potential impact of large brains is well illustrated by chimpanzees that make and use tools in a variety of ways.

The living cercopithecoids include one family, the cercopithecids, that consists of two subfamilies, each with representatives in Africa and Asia (Figure 5.3). Cercopithecines are known as the cheek-pouched monkeys. They are generally omnivorous with a preference for fruit. Being anatomically specialized folivores, the colobines are called leaf monkeys, but seeds are also important in their diets.

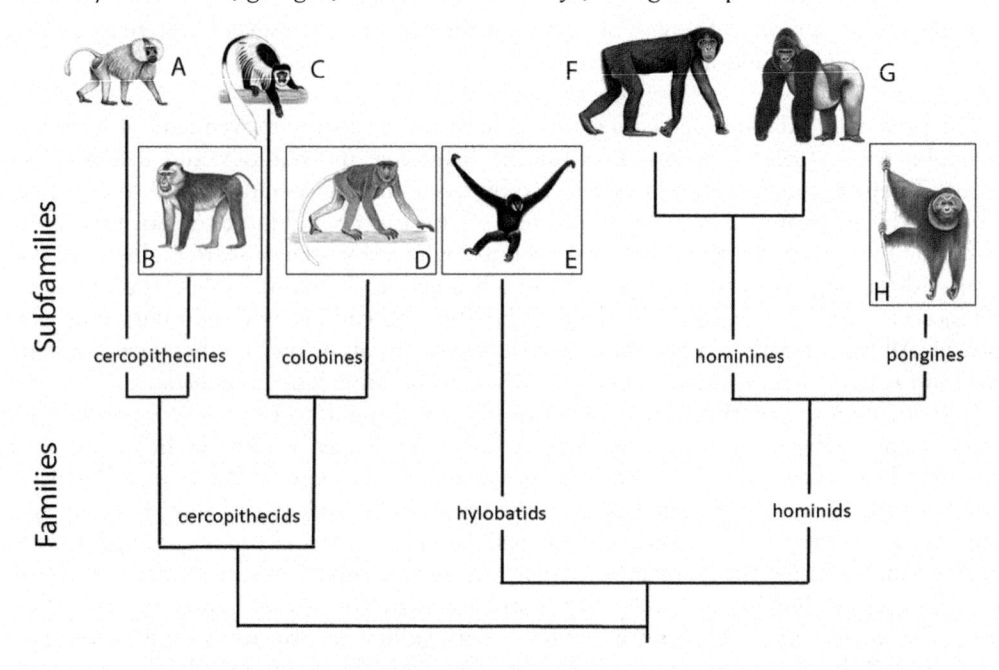

110 *Africa: lorises, galagos, Old World monkeys, and great apes*

Figure 5.3 Cladistic relationships of the three living catarrhine families and subfamily clades, with Asian taxa enclosed in rectangles. They are: (a) a baboon, *Papio*; (b) macaque, *Macaca*; (c) guereza, *Colobus*; (d) proboscis monkey, *Nasalis*; (e) siamang, *Symphalangus*; (f) *Pan*, chimpanzee (bonobo); (g) gorilla, *Gorilla*; (h) orangutan, *Pongo*. Not drawn to scale.

Artwork by Stephen D. Nash, courtesy of Stephen D. Nash and the IUCN SSC Primate Specialist Group.

The African cercopithecines are a sub-Saharan group, with the exception of macaques, genus *Macaca*, the most widely distributed cercopithecid. One species, the Barbary macaque, lives in montane forests in northwest Africa. Other species of macaques are found throughout tropical and semi-tropical Southern and Southeast Asia, including islands in Indonesia and the Philippines. The distribution of the genus also extends northward into temperate zones of northern China and Japan.

The seven genera of living hominoids are divided into two families, hylobatids, consisting of the lesser apes, and hominids, including the great apes. As suggested by these common names, there are significant body size differences between the families, and there are geographical differences as well. Hylobatids live exclusively in Asia while hominids live in both Africa and Asia.

There is a marked difference in the taxonomic abundance of hominids and cercopithecids. Today's Old World monkeys are the most prolific catarrhines, with (conservatively) more than 20 genera versus the seven genera of living apes. This was different in the distant past. As will be discussed in Chapter 8, ape species and genera were plentiful and taxonomically dominant millions of years ago, before major changes in African ecology began to favor the diversification of monkeys and negatively influenced the biodiversity of apes.

The facial and cranial anatomy of living catarrhines was introduced in Chapter 4 in contrasting the platyrrhine and catarrhine patterns (Figure 4.4). Those features, and others, are summarized in Table 5.1, and briefly reviewed here.

Table 5.1 A summary of the distinguishing traits of living catarrhines and platyrrhines

Catarrhines	Platyrrhines
Narrowly spaced nostrils	Widely spaced nostrils
Frontal-alisphenoid postorbital mosaic	Zygomatic-parietal postorbital mosaic
Ectotympanic tube-shaped	Ectotympanic horseshoe-shaped
Two premolars	Three premolars
Canines usually large	Canines variable in size
Sexual canine dimorphism common	Taxonomically variable
Single-cusped, modified anterior premolar (p3)	No
Molar-like posterior premolar (p4)	Rarely
Vomeronasal organ highly reduced	Vomeronasal organ moderately reduced
Color vision	Rarely
Ischial tubersosities and callosities	No
Opposable thumb	No
Medium to super-large body mass	Small to large body mass

Teeth

The canine teeth of catarrhines, particularly the uppers, are generally large and sexually dimorphic. As in other primates, they play a major role in gestural communication. Canines can be used to flash at conspecifics, threateningly or in appeasement, during agonistic encounters, social interactions that reflect conflict and may lead to aggression. With well-developed facial muscles, the catarrhines use a variety of facial expressions during which the upper and lower canines are exposed. Of course, large canines can also be lethal offensive or defensive weapons.

To accommodate the tall, dagger-like canine crown during occlusion, the lower anterior premolar, termed p3, has a distinctive, single-cusped morphology. It is called a sectorial tooth, a reference to its narrow "cutting" shape, and is most pronounced in males as part of the sexual dimorphism complex. The crown is narrow and large to accommodate the large upper canines when the teeth come together. Grinding the upper canine tooth against p3, a behavior called canine honing, keeps the back edge of the upper canine sharp. The premolar behind it, p4, is different. It resembles the molars and functions like one.

Color vision, olfaction, and communication

With color vision, the diurnal catarrhines have a superior visual system that can extract a large amount of information from the environment. Trichromacy is advantageous in foraging by enabling a primate to discern reddish fruits from a background of green foliage, and young light-colored leaves from those that are older and a darker green. Associated with the development of color vision is the variety of pelage colors found among cercopithecoid species, including, reds, blues, and purples. Colors can be species-specific. In some species, patches of skin color and patterns of fur color are sexually dimorphic and play important roles in signaling reproductive status. Or, they serve as attention-getting features that influence mate choice, in mandrills, for example.

Catarrhines are less reliant on olfactory communication than platyrrhines and strepsirhines. They have a highly reduced vomeronasal organ and do not scent mark. The skin glands of apes, especially those located in the armpits and pubic region, produce odors that appear to be pheromonal. For social and reproductive purposes, catarrhines combine olfactory, acoustic, and visual information in complex ways, supported by an advanced cognitive processing system.

Cercopithecoids – Old World monkeys

Old World monkeys are widely distributed, live in very different habitats, eat a variety of foods, and range from being arboreal to terrestrial, yet all of the genera are remarkably uniform anatomically. They are classified into one family, the cercopithecids, which is divided into two subfamilies, cercopithecines and colobines.

The biodiversity count of African cercopithecids strongly favors one subfamily. The colobines consist of only two or three genera that live in rainforests, while there are ten cercopithecine genera ranging throughout the Sub-Saharan region, in forest and non-forested habitats. The cercopithecine subfamily is also subdivided into two distinct phylogenetic and taxonomic groups, called cercopithecins and papionins.

Bilophodont molar teeth

The one feature that distinguishes all Old World monkeys from apes and all other primates is their molar morphology. The molar teeth of all cercopithecoids conform to a phylogenetically derived crown pattern called **bilophodonty**, meaning two-ridged or two-crested (Figure 5.4).

The bilophdont pattern of upper and lower molars consists of four cusps, two outer (buccal) and two inner (lingual), that are arranged like the corners of a rectangle. Parallel transverse ridges connect the paired front and back cusps. The upper and lower molars are almost mirror images of one another. In all other primates upper and lower crowns have different numbers of cusps and different shapes overall, though they are designed to reciprocate one another in action. Several examples of this are shown in Chapter 8 (Figure 8.1).

Cranial and postcranial characteristics

Cercopithecid crania are quite similar, though there are taxonomic differences. The African baboons have evolved distinctively long faces. In the postcranial skeleton, the uniformity continues. All cercopithecids are either arboreal or terrestrial quadrupeds with limited variation in the proportions of the long bones, hands, feet, and tail, though the limbs of the more terrestrial forms are relatively longer than the arboreal species. Their elbows are designed for stable quadrupedal locomotion. Overall, this marked skeletal regularity contrasts with the diverse anatomies and locomotor behaviors seen among strepsirhines and platyrrhines.

Cheek pouches and guts: cercopithecine and colobine specializations

The main morphological and ecological distinctions between the two subfamilies of Old World monkeys relate to diet and microhabitat. Cercopithecines are frugivorous omnivores and many feed on the ground, where they collect fruit, seeds, roots, bulbs, or grass blades. All cercopithecines, but no colobines, have a set of cheek pouches, expandable

Lower molars

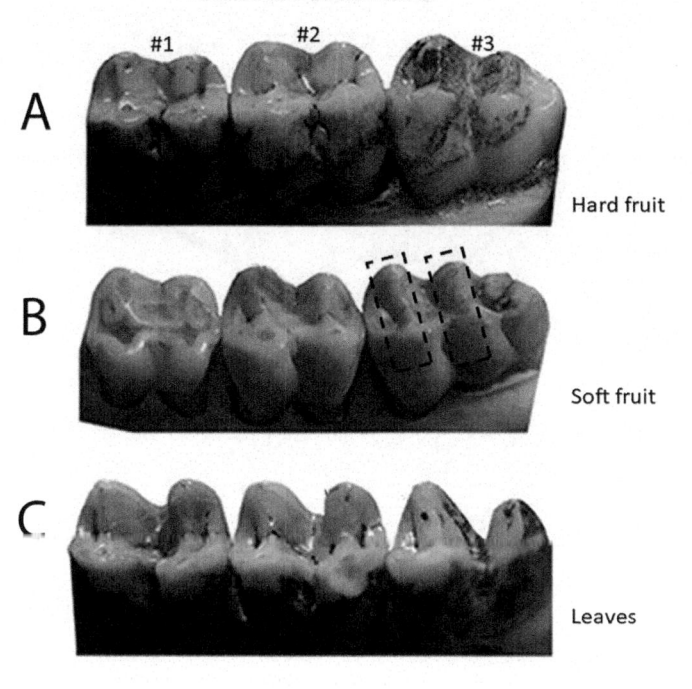

Figure 5.4 Inner (lingual) views of three right lower molars (left to right; first, second, and third) of Old World monkeys illustrating the bilophodont crown pattern. The characteristic transverse ridges (lophs) are framed by the rectangles in B. The species show a gradient of crown shapes associated with various diets: (a) a frugivorous mangabey with blunt cusps and thick enamel, for chewing hard fruits, (b) an omnivorous-frugivorous macaque with large crushing basins between the ridges, to eat relatively soft fruits; (c) a folivorous langur with tall, pointy cusps and long outer shearing crests to shred leaves.

Adapted from Ungar, P. S. (2015). Primate teeth and plant fracture properties. *Nature Education Knowledge*, 6(7), 3. https://www.nature.com/scitable/knowledge/library/primate-teeth-and-plant-fracture-properties-135718653/

pockets that develop within the inside wall of the cheeks and extend into the neck (Figure 5.5).

Cheek pouches, which have evolved in a number of other mammals, are used to temporarily store or transport food before eating it. While pouched, some pre-digestion of food also occurs as a reaction to chemical enzymes that are present. A storage pouch conserves time and energy by enabling a monkey to collect as much food as possible in a single sitting without moving much.

In cercopithecines the pouch promotes terrestrial feeding efficiency, especially under the influence of **contest competition**. The cheek pouches are beneficial when several individuals forage in close proximity and a preferred feeding spot is potentially contestable, when the food source can only physically accommodate one or a few individuals at a time, or when the amount is insufficient to feed more than one monkey. Filling up the pouch rapidly and moving out of the way can avoid altercations, especially for low-ranking animals living in social groups with dominance hierarchies, where access to food is often regulated by agonistic threats. The cheek pouch is also an advantage

Figure 5.5 The cheek pouches (arrow) of this bonnet macaque are filled with fruit.

Photo credit: Bonnet macaque DSC 0893.jpg, by T. R. Shankar Raman (CC-BY-SA-4.0).

among ground-feeders while out in the open. It allows the animals to secure more than a normal mouthful of food and then retreat to the safety of trees before being detected or stricken by predators. Similar benefits would also be derived while feeding on fruits in the treetops.

Living colobines are arboreal leaf- and seed-eaters, though they have long been known informally as leaf monkeys because of their highly specialized stomachs and shearing cheek teeth. Colobine stomachs are divided into several chambers that contain microbiota to aid digestion by fermenting food, a process seen also among folivorous lemurs on Madagascar and New World howler monkeys in the neotropics. However, colobines are the only primates that ferment leaves in the stomach, a part of the gastrointestinal tract called the foregut, as opposed to the hindgut, which is downstream of the stomach anatomically and includes the intestines and caecum, which is where other leaf-eating primates ferment material to be digested. The colobine dentition is also designed to facilitate leaf-harvesting and mastication, with relatively small incisors and relatively pointed, crested cheek teeth. Even the few colobine populations that are habitually terrestrial have the same stomach morphology.

It is important to underscore the significance of seeds in the colobine diet. Many colobine species spend large proportions of their feeding time eating seeds: 25%–35% is not uncommon and some species devote 50%–60% of their time to it. Their specialized stomachs are well adapted to detoxify the compounds in seeds and to ferment seed coats, which, like leaves, are also difficult to digest.

Ischial callosities

All Old World monkeys have ischial callosities on their rumps. Among apes, only gibbons and siamang exhibit them. Ischial callosities are fibrous, fatty cushions on the buttocks that are covered in thick, no-slip skin. They are tightly attached to underlying pelvic bone, the ischium. The callosities are an adaptation to sitting securely on branches while feeding, resting, and sleeping in place for many hours, especially on branches that are relatively narrow and potentially present balancing issues to a relatively large primate. Cercopithecid sitting postures involve sitting on the rump with the torso held vertically and the feet placed in front of the body, with knees bent. In that position, most of the body weight falls directly on the buttocks. The callosities prevent slippage and reduce the pressure on the ischium.

In great apes, extra-large and super-large body mass makes long periods of sitting on small branches an impractical approach to feeding or resting in the trees. They lack ischial callosities. At night, the great apes build tree nests by bending leaves and branches into platforms, and they sleep on their sides or backs. The largest apes, gorillas, may sleep on the ground, also in nests.

The cercopithecoid profile

- **Phylogeny.** The closest relatives of cercopithecoid Old World monkeys are the apes.
- **Body mass and shape.** Most cercopithecoids fall into the large and extra-large body mass categories, although smaller cercopithecines are medium-sized. By far the smallest species is the arboreal talapoin monkey, weighting about 2.2 pounds (1 kg). Cercopithecoid tails are almost always relatively long, except for a few macaque species that have short tails, and their limbs are about equal in length.
- **Activity rhythm.** All cercopithecoids are diurnal.
- **Ecological domain.** Arboreal, semi-terrestrial, and terrestrial.
- **Diet.** Cercopithecines are omnivores that prefer fruit while colobines are primarily leaf- and seed-eating folivores. Terrestrial cercopithecids eat a variety of foods found on the ground, including grasses, herbs, and plant parts that need to be dug out, like roots and bulbs. A few species consume important amounts of animal matter, including invertebrates and small mammals, and a few rely on gums and other exudates, mostly in non-forested environments.
- **Locomotion.** Quadrupedalism is the standard form of locomotion. Arboreal colobines frequently use quadrupedal leaping and some species are adept at using some degree of suspensory locomotion during feeding.
- **Craniodental morphology.** Cercopithecines tend to have longer snouts than colobines, most notably in the baboons. Baboons' sexually dimorphic features and gestural displays, like baring huge canines with open-mouth threats, strongly influence facial structure. Cranial and dental differences between the subfamilies are otherwise subtle. As frugivorous omnivores, cercopithecines have relatively large incisors and molars with relatively blunt cusps. The folivorous colobines have relatively smaller incisors and molars with taller, sharper cusps.
- **Pelage.** With three-color, trichromatic vision, the coat colors and color patterns of the diurnal cercopithecoids are richer and more varied than those of the nocturnal strepsirhines and the dichromatic, diurnal platyrrhines. For example, mandrills have vibrant pelage colors and colored facial skin that is associated with sexual dimorphism. Dominant adult

males have faces with skin that is bright blue, red, and yellow/white, a color combination also seen on their rumps. Their fur is blue- and gold-tinged at the chin and around the head. As further discussed in Chapter 7, these patterns play an important role in mating.
- **Sexual monomorphism/dimorphism.** Cercopithecids vary in the degree and anatomical expression of sexual dimorphism. Among cercopithecines the scope of male-female body mass differences ranges from being modest to extreme. For example, males may be 20% heavier than females in macaques, but more than 3.5 times the mass of females in mandrills. The differential in male canine size also ranges from marked to extreme. In outward appearance, the males of several African cercopithecines are also strongly differentiated from females by manes or capes of fur around the shoulders. The largest, most ornate males often gain priority access to females.
- **Communication.** Visual gestures based on well differentiated facial muscles and auditory signaling are highly developed, while olfactory communication plays a limited role (Figure 5.6).
- **Social systems and mating.** Group sizes and social systems vary in cercopithecids, ranging from small family units consisting of a mated pair and their offspring, which is rare, to complex multi-tier systems involving hundreds of individuals with subsets of unimale-multifemale mating groups, and unattached bachelor male groups as well. Social structure is often shaped by dominance relationships, both same-sex and heterosexual. Male dominance is mediated by highly aggressive behaviors in some African cercopithecines but not among colobines. Female coalitions, driven by kinship

Figure 5.6 A female gelada showing submissive behavior by executing the lip flip gesture.
Photo credit: ZooZürich – *Theropithecus gelada* 15.JPG, by Albinfo (CC-BY-SA_3.0).

ties, help determine female status in large troops among most cercopithecine species, but female colobines do not form coalitions. There, rank is determined by individual behaviors, which in some cases allows subadult females to achieve an alpha status above the ranks of males.
- **Reproductive pattern.** All cercopithecids give birth to singletons that are spaced apart by a year or more; twins occur rarely.
- **Parenting styles.** Mothers are the main caregivers but males may provide indirect care by protecting their likely offspring from harassment by other troop members.

Hominoids – apes

The chimpanzees, bonobos, gorillas, and other hominoids are apes. They are not monkeys, as they are sometimes erroneously called. Apes are easily identified by one outstanding feature – they do not have a tail. Almost all monkeys have a tail.

There are two family-level hominoid clades – hominids and hylobatids (Figure 5.3). Falling mostly into the large body mass category, about 10–20 pounds (4.5–9 kg), the lesser apes are Old World monkey-sized. The great apes are the largest living primates, weighing roughly 70–400 pounds (32–181 kg).

There are three great ape genera, each with two species: *Pan*, chimpanzees and bonobos; *Gorilla*, western and mountain gorillas; and *Pongo*, Bornean and Sumatran orangutans (a second Sumatran species is recognized by some scientists). *Pan* and *Gorilla* are exclusively African while *Pongo* is Asian. The hylobatids, the four lesser ape genera, are also Asian. They are among the primates discussed in Chapter 6.

All apes share relatively long forelimbs and a skeletal morphology that is functionally related to climbing and suspensory behaviors (Figure 5.7). During quadrupedal locomotion, this results in a somewhat more erect, or **orthograde**, bearing. A more horizontal, or **pronograde**, orientation of the body is exhibited by monkeys. Forelimbs and hindlimbs are especially important when climbing tree trunks vertically, a critical locomotor behavior in the semi-terrestrial chimpanzees, bonobos, and gorillas. The most arboreal apes, hylobatids, and orangutans, exhibit the more extreme variations in forelimb and hindlimb mobility, suspensory capabilities, and grasping adaptations. African ape anatomy blends arboreal and terrestrial adaptations.

The skeletal morphology shared by apes is a functional-adaptive complex with a range of distinctive traits, including the following highlighted in Figure 5.7. The numbers in brackets refer to body parts indicated in the figure.

- Long arms to facilitate climbing and suspensory postures [1].
- Broad shoulders, large shoulder blades, and wide, barrel-shaped chests. These traits are functionally related to powerful arms and mobile shoulders, especially lateral and over-the-head arm-raising maneuvers [2].
- The elbow is built to allow full extension of the forearm and the rotation of the radius that moves the hand into the palm-up and palm-down positions. These features enable an ape hanging by its arm to change direction and spin the body around a fixed handhold, as most effectively demonstrated by brachiating gibbons and siamang [3].
- Wrists are also flexible and facilitate extensive side-to-side movement of the hand [4].
- A short, stiff lower back (lumbar region) with wide hips to which the back and strong leg muscles attach. This provides skeletal stability during climbing and suspensory locomotion, and effective use of the hindlimbs in climbing [5].

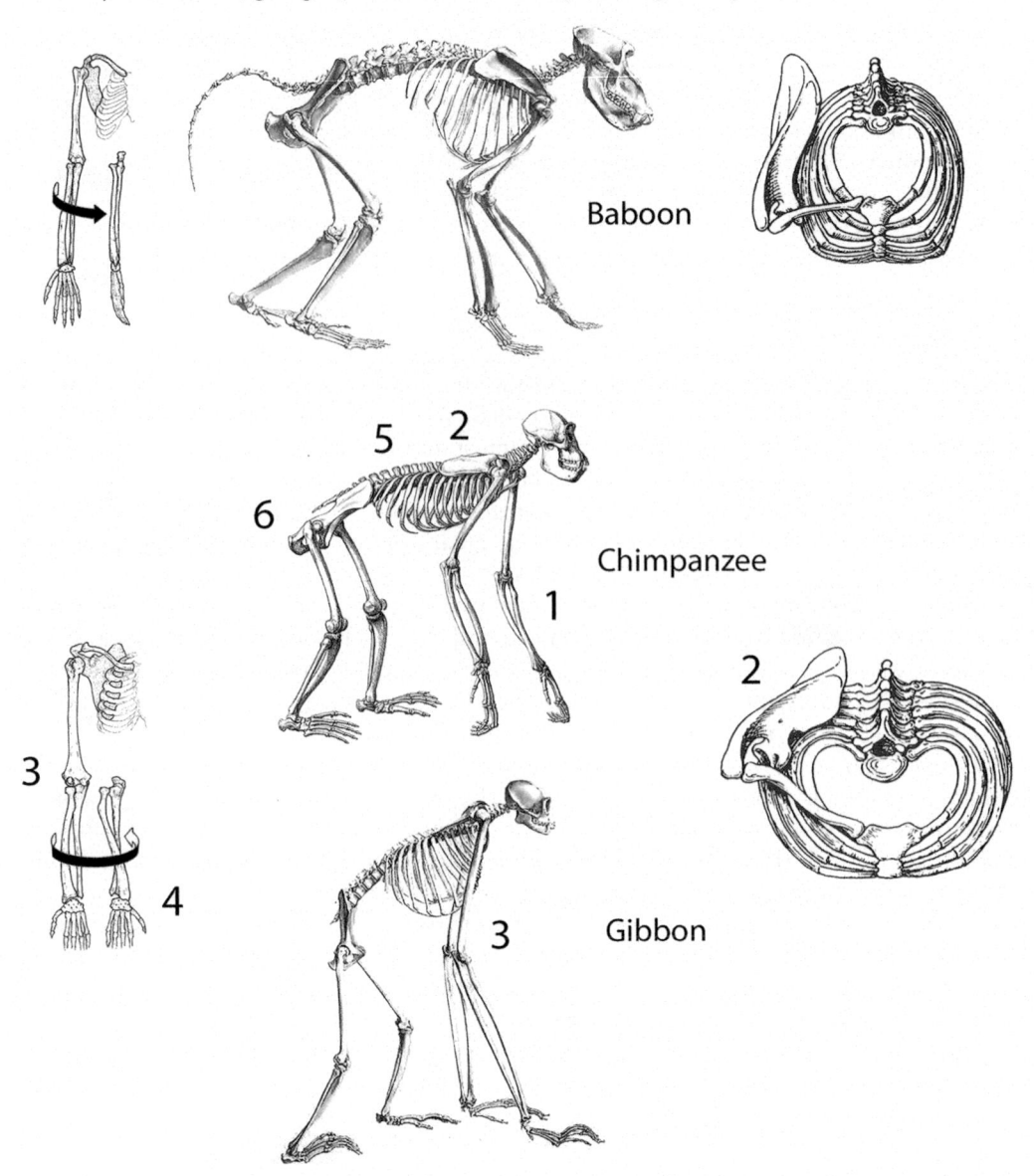

Figure 5.7 Skeletons of a terrestrial baboon, semi-terrestrial chimpanzee, and arboreal gibbon brought to the same standing heights in quadrupedal postures. The numbers highlight ape characteristics, keyed to the text. Note the flexed-fingered, knuckle-walking hand posture of the chimpanzee and the digital or palmar push-off postures of the baboon and gibbon.

Adapted from: de Blainville, H.M. D. (1839–1864). Ostéographie, ou, Description iconographique comparée du squelette et du système dentaire des Mammifères récents et fossiles: pour servir de base à la zoologie et à la géologie. Paris; Schultz, A. H. (1969). The skeleton of the chimpanzee. In G. H. Bourne (Ed.). The Chimpanzee (Vol. 1, pp. 50–103). Basal, Karger; Hunt, K. D. (2016). Why are there apes? Evidence for the co-evolution of ape and monkey ecomorphology. Journal of Anatomy, 228(4), 630–685. https://doi.org/10.1111/joa.12454.

- Lack of an external tail and development of the coccyx in the pelvic anatomy that forms a base of support for internal organs to accommodate the ape's generally orthograde posture while climbing and clambering through the trees [6].

The hominoid profile

- **Phylogeny.** The hominoids, great and lesser apes, are most closely related to cercopithecoids, Old World monkeys.
- **Body mass and shape.** The great apes are powerful, forelimb-dominant animals, though they do have strong hindlimbs as well. They have a large body mass, large heads, broad shoulders and chests, a short back, long muscular forelimbs, and no tail, together producing a compact, stout trunk. Hylobatids, the lesser apes, have exceptionally long arms, a much smaller body-size scale than apes, and a relatively small head, making for an even more compact appearance.
- **Activity cycle.** All apes are diurnal.
- **Ecological domain.** Hylobatids are exclusively arboreal. Orangutans have evolved an extensive skeletal system of arboreal specializations that enable them to negotiate the treetops, yet they do come to the ground, especially large males. *Pan* and *Gorilla* are semi-terrestrial. Their ground-dwelling specializations blend with arboreal anatomies that enable them to be efficient on the ground and also in the trees, in spite of their heavy mass.
- **Diet.** Ape diets are varied. Gibbons are categorized as frugivorous and siamang are folivorous, yet heavily reliant on fruits also. Orangutans prefer ripe fruit, unripe fruit, leaves, bark, and invertebrates. Gorillas eat fruit, leaves, stems, herbaceous plants, and pith, with the proportions varying according to habitat. Some gorillas wade into ponds to eat aquatic plants in seasonally inundated areas. Chimpanzees and bonobos favor ripe fruits but they have an exceptionally diverse diet also consisting of leaves, seeds, flowers, stems, roots, wood, and more. Like gorillas, they also eat the pith of aquatic plants.

  Chimpanzees and bonobos are avid predators. The chimpanzee's favorite prey are red colobus monkeys, which the animals regularly hunt as a group. Chimpanzees also fashion branches into pointed spears, used to probe the hollows of trees in search of nesting galagos, which they stab, kill, and eat. Bonobos hunt red colobus monkeys too, usually in groups, while lone bonobos have been seen hunting and eating small antelopes called duikers.
- **Locomotion.** The ape body plan is basically a design to enhance the climbing and/or suspensory capabilities of a quadrupedal musculoskeletal system, but several divergent specializations have evolved. Hylobatids are an extreme case. They employ a highly acrobatic form of brachiation. If hylobatids are on the ground or on top of large branches taking a few steps, they move bipedally, often holding their arms above their heads because they are so long that they might otherwise contact the ground or the branch that they are walking on if the forelimbs are hanging down at their sides.

  The arboreal orangutans are different in several ways from all the other apes. They have a style of four-footed, quadrumanous locomotion which is a method to negotiate the canopy while holding on to as many supports as possible at all times, to maintain balance and prevent a fall. This is made possible by very large, strong hands and feet, and tremendous mobility at the hip and shoulder joints. On the ground, orangutans move quadrupedally using a variety of hand postures ranging between a palm-down placement to fist-walking, where the load is carried by curled fingers and palm or the outer side of the hand.

African apes habitually knuckle-walk. This involves anatomical reinforcement of the fingers, finger joints, wrists, and elbows facilitating a stiff-armed hyperextension (Figure 5.7). During quadrupedal walking and running, they fold their fingers so their hands contact the substrate along the outer surfaces of enlarged middle phalanges of the second through fifth digits. It is a stable hand position that protects the fingers from hyperextending during locomotion.

- **Craniodental morphology.** In connection with their large body size, great apes have large crania that accommodate relatively large brains and large dentitions, which influences facial size. Gorillas present the most extreme condition, with large jaws and teeth. In adult males there is a massive brow ridge above the orbits. They also have a raised longitudinal ridge or midline bony crest, the **sagittal crest,** which forms on top of the braincase. The sagittal crest anchors powerful masticatory muscles. In the rear of the cranium, strong neck muscles are attached to a similarly strong transverse crest (Figure 5.8). These muscles balance a huge, jutting head that is connected to the vertebral column at the rear of the cranium.

  The molar teeth of all apes have a distinctive shape. It is called the Y-5 crown pattern. The term refers to the five principal cusps of the first and second lower molars that are spaced out and separated by grooves or creases in the enamel that intersect to form a Y.

- **Pelage.** African apes have black coats. Adult alpha male gorillas develop silvery-colored fur on their backs, which is why they are called **silverbacks**. Species of hylobatids may have sexually dimorphic coats, with males usually darker than females, colored in shades of brown to black, while female coats range from beige to blond.

- **Sexual monomorphism/dimorphism.** Most great ape species are highly sexually dimorphic in body mass and canine size. In the great apes, canines are large and

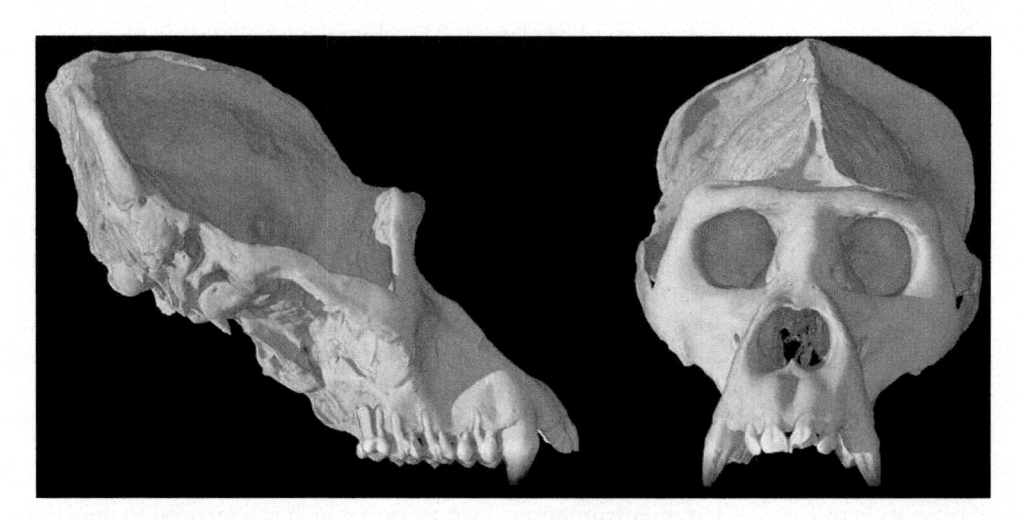

Figure 5.8 Cranium of an adult male gorilla showing the massive face, brow ridge, and well-developed crests on the top and back of the head.

Derived from http://digimorph.org/

robust, more so in males than females. Some, but not all adult orangutan males, are uniquely dimorphic in developing large, fleshy, flange-like pads of soft tissue that extend from the cheeks and frame the face. In hylobatids, there is little body size dimorphism and canines are monomorphic. Both sexes have same-sized, tall, dagger-like canines resembling those of cercopithecoids, even in species that are dimorphic in pelage.

- **Communication.** Great apes have a large repertoire of grunts, growls, barks, whoops, shrieks, and other sounds used in vocal communication. They produce visual displays and combine them with vocalizations. Orangutans **piloerect** their fur, raise and stiffen it, while loud calling and throwing branches. Chimpanzees rain-dance before or during a storm by noisily racing back and forth agitatedly, climbing into and out of trees at speed, and snapping branches. They drum with their hands on fallen tree trunks. They throw stones at specifically selected trees that produce long, deep echoes on impact. **Chest-beating** is a combined audiovisual sign used by adult male gorillas to show off their physical status and advertise their competitive ability, by slapping the hand against a large patch of skin on the upper chest between the shoulders while assuming a bipedal stance. Gibbons have impressive long-distance vocal displays. Both males and females have enlarged throat sacs that they fill with air and then expel to produce complex, staccato loud calls, daily, often at dawn. Mated pairs do this together as a duet, to advertise their pair-bond and defend territory and mates against conspecifics.
- **Social systems and mating.** Apes exhibit a wide range of social systems. Hylobatids form pair-bonded groups and small extended family groups but mating occurs outside the family unit as well, in extra-pair copulations. Gorillas are typically organized in unimale-multifemale groups, in which an alpha or silverback male has exclusive reproductive access to several adult females, but the groups often also contain subordinate males. Chimpanzees live in multimale-multifemale, fission-fusion groups in which females may mate with multiple partners, though a male, or coalition of males, may attempt to restrict sexual access by other resident males. Orangutans are semi-solitary, with males generally living separately from females and her offspring. Some males are able to establish stable ranges that overlap with multiple females for breeding purposes, and other males roam opportunistically in efforts to find breeding partners.
- **Reproductive pattern.** All apes give birth to singletons. The spacing between birth intervals does not vary according to body size. For example, chimpanzee and bonobo intervals are every 4.5–6.0 years, and gorillas are 4.0–4.6 years, while orangutans give birth every 7.2–9.3 years. The long interbirth intervals reflect the slow maturation and long lifespans of these large primates. Great apes take 8–13 years to reach physical maturity and may require a few more years until they first reproduce. Their lifespan can be 30–50 years and more.
- **Parenting styles.** There is a wide range of parenting styles among apes. Orangutan males are not involved in parenting. Hylobatid males, like other pair-bonded primates, assist in rearing by carrying offspring for a period of time. Chimpanzee fathers associate with, play with, and groom their own offspring, especially while they are infants, when infanticide by intruders is a potential threat to young. Adult male gorillas allow the young to stay close, carrying and playing with them, and deterring harassment by others. Alpha gorilla males play an important role in protecting infants, when they are most vulnerable to attack.

African primates living in communities: diversity, ecology, and adaptation

Many African primates live in arboreal-terrestrial communities that include ground- and treetop-dwellers and span an enormous range of body sizes (Table 5.2). This is unlike the canopy communities of lemurs and platyrrhines. Because the African forms include anthropoids and non-anthropoids, depending on habitat, primate communities may be composed of diurnal and nocturnal species.

The biodiversity count of African cercopithecids strongly favors one subfamily. The colobines consist of only two or three genera that live in rainforests, while there are ten cercopithecine genera ranging throughout the Sub-Saharan region, in forest and non-forested habitats. At the Tai Forest of the Côte d'Ivoire (Ivory Coast) that is our case study, there are five cercopithecine genera and three colobine genera. The cercopithecine subfamily is also subdivided into two distinct phylogenetic and taxonomic groups, called cercopithecins and papionins.

Across Africa, the cercopithecins are represented by a large number of species, more than 30 according to some accounts. Cercopithecins are consistently smaller in size and more slender in build than the papionins, and their snouts are shorter. The smallest living anthropoid, the talapoin monkey, *Miopithecus*, which weighs less than three pounds (1.4 kg), is part of the cercopithecin group. The cercopithecins are generally more arboreal than the papionins and more reliant on insects than other cercopithecids. The papionins are distinctive in having unusually long snouts, particularly in the baboons, *Papio*, and mandrills, *Mandrillus*.

One of the outstanding features of the long-tailed, cercopithecin monkeys is the great diversity in the species-specific appearance of their faces, which are distinguished by marking highlights, varied types of beards on the cheeks and chin, differences in eyebrows, eye patches, ear tufts, and more. Their body and facial colorings, the most elaborate of any mammal, are expressed in green, beige, brown, blue, rust, red, grey, yellow, white and black, arranged on the body in small patches, large fields, or stripes. By comparison with papionins, their striking, bright-colored appearance corresponds with a greater reliance on visual gestures and vocal signals to communicate, particularly when the animals are operating in dense canopy foliage or in the shadowed lower strata of the forest, where visual highlights may be beneficial. To regulate behavior and maintain order outside the forest where visibility is better, papionins use gestures like glaring at another individual, which is an act of dominance, and stronger vocalizations. Their coats are more uniform and less brightly colored in browns and grays and blacks, blending with their surroundings and providing a measure of protection from predators.

The visible markings of cercopithecins enable individuals to soundlessly monitor one another, coordinate movements, and maintain group cohesion. At the same time, the different markings displayed by cercopithecin species may relate to another unusual behavior among primates, the tendency of many cercopithecin species to tolerate proximity with other species.

While foraging, cercopithecins frequently form polyspecific associations, the gathering and coordination of two or more species, which may offer mutual benefits to individuals of each species in terms of locating food resources and avoiding predators. Coordinating species recognize each other's predator-specific alarm calls. In that context, it is advantageous for individuals to watch for and recognize conspecifics when they are dispersed while foraging in the difficult visual environment of the treetops, or the dimly lit understory.

Table 5.2 Ecological characteristics of Taï Forest primates

Subfamily	Common name	Genus and species	Body mass (Male)	Body mass (Female)	Activity cycle	Locomotion	Microhabitat	Diet
Perodictinae	West Africa potto	*Perodicticus potto*	M	M	Nocturnal	Slow-climbing quadrupedalism	Arboreal	Animals, F
Galaginae	Demidoff's dwarf galago	*Galagoides demidovii*	S	S	Nocturnal	Quadrupedalism	Arboreal	Animals, F, Gu
Cercopithecinae	Sooty mangabey	*Cercocebus atys*	ExLarge	L	Diurnal	Quadrupedal	Semi-terrestrial	F, In, Gu
Cercopithecinae	Campbell's monkey	*Cercopithecus campbelli*	L	M	Diurnal	Quadrupedal	Semi-terrestrial	F, L, Gu
Cercopithecinae	Diana monkey	*Cercopithecus diana*	L	L	Diurnal	Quadrupedal	Arboreal	F, In, L
Cercopithecinae	Putty-nosed monkey	*Cercopithecus nictitans*	L	M	Diurnal	Quadrupedal	Arboreal	F, In, Gu
Cercopithecinae	Spot-nosed monkey	*Cercopithecus petaurista*	L	M	Diurnal	Quadrupedal	Arboreal	L, F, G
Colobinae	Upper Guinea red colobus	*Procolobus (Piliocolobus) badius*	L	L	Diurnal	Quadrupedal	Arboreal	L, F, Fl
Colobinae	Olive colobus	*Procolobus (Procolobus) verus*	L	M	Diurnal	Quadrupedal	Arboreal	L, F
Colobinae	King colobus	*Colobus polykomos*	ExLarge	L	Diurnal	Quadrupedal	Arboreal	F, L
Homininae	Chimpanzee	*Pan troglodytes*	SupLarge	SupLarge	Diurnal	Knuckle-walking and climbing	Semi-terrestrial	F, L, Animals

Body mass abbreviations: S, small; M, medium; L, large; ExLarge, extra-large; SupLarge, super-large. Diet abbreviations: F, fruit; Fl, flowers; Gu, tree gum; In, insects; L, leaves. Not all food items eaten are listed, and the categories are presented in order of importance. The animal category includes insects, other invertebrates, and small vertebrates, like reptiles, and small mammals.

The Tai Forest of the Côte d'Ivoire (Ivory Coast) is in the Tai National Park, a World Heritage Site that includes one of the last tracts of undisturbed tropical rainforest in western Africa with a rich flora and fauna. Following the taxonomy employed by fieldworkers up until the early 2000s, the Tai Forest's 11 primate species had been allocated to seven primate genera, as in Table 5.2; some primatologists recognize an eighth genus. The total number of sympatric primates at Tai compares favorably with large primate communities of Madagascar and South America, where about a dozen species may coexist.

Lorisoids – nocturnal frugivorous-gumivorous predators

There are two lorisoids in the Tai Forest belonging to two contrasting families, the galagos (galagids) and lorises (lorisids). One is Demidoff's dwarf galago, *Galagoides demidovii*. It is one of the forest galagos. Other galago species are found in relatively open shrubland where there are few trees. The Tai Forest lorisid is the West African potto, *Perodicticus potto*. The same genus also occurs in central African rainforests. A second similar potto genus is also found elsewhere in western Africa, the angwantibo, *Arctocebus*.

Tai's lorisoids, representing two families of small-bodied, nocturnal arboreal strepsirhines, each have an ecological role that is distinctly different from all the other sympatric primates, which are much larger, diurnal, and may be semi- or fully terrestrial. *Perodicticus potto* and *Galagoides demidovii* feed on animals and plants in the canopy that are categorically similar, but the animals are ecologically differentiated from one another by very different adaptations, beginning with their different sizes. The galago is small-sized, weighing about two ounces (57 g), while the potto weighs about two pounds (0.9 kg), making it a middle-sized strepsirhine.

- **Diet and locomotion.** Conspicuous outward differences between the two families reflect the locomotor skeleton. Galagos have long legs and feet, biomechanical adaptations for specialized bipedal leaping. The intermembral index of *Galagoides demidovii* is 68 (see Box 3.1). The forelimb is roughly two-thirds the length of the hindlimb. Forelimbs are functionally irrelevant in springing up and out during vertical-clinging-and-leaping.

  The loris skeleton is very different. It is biomechanically designed to facilitate small-step quadrupedal locomotion, with the body stretching, turning, and twisting its way forward to advance using all four limbs in concert, a quadrumanous pattern of locomotion described as "creeping and crawling." The potto's intermembral index is 88: the hindlimb is only about 10% longer than the forelimb. Progression in lorises is basically four-handed, using all-fours to grip the substrate whenever possible. The locomotor style uses exceptionally mobile joints at shoulders, hips, wrists, and ankles, which facilitate acute angular and rotational movements of the limbs, and specialized hands and feet to provide secure, four-point grasps (Figure 5.9).

Pottos and galagos have different prey targets, foraging styles, and methods of capture. Pottos feed on slow-moving insects and hunt them silently, moving stealthily in a slow, quadrumanous manner. The invertebrates that pottos eat are found in the canopy where they spend equal amounts of time on vertical, horizontal, and oblique substrates.

Demidoff's dwarf galago forages in dense foliage in the understory below the closed canopy, where they mostly use vertical and oblique supports. They are more prone to leaping. They may flush insects out into the open by rapidly walking, running, and

Figure 5.9 A potto stretching for a handhold while moving quadrupedally in the small-branch setting of the canopy.
Courtesy of A. P. Levantis Ornithological Research Institute.

leaping. Once a target is located, the galago snatches the flying insect lunging at it. With feet firmly attached to a branch, the galago pitches itself forward to grab the prey and then draws itself back into a sitting position to eat it. It is a swift, sudden variation of a pattern exhibited by lorises, that execute it in relative slow motion.

Pottos, like many other strepsirhines, also eat fruit, gum, leaves, and buds. Ants, often discovered in abundance while they move along a branch in a large column, are lapped up as the potto walks slowly, nose-to-branch. Land snails, frogs, and small birds are also eaten. But the potto's most idiosyncratic foods are noxious insects that are well protected from predation, including caterpillars that may be physically irritating. A potto will grab a caterpillar in its hand or it may immobilize a caterpillar by squishing it against a branch. Then, before killing it with a bite, the potto rubs the caterpillar to remove the stinging hairs or spines that cover it. After eating a caterpillar, the potto may spend more than a minute cleaning its face and hands on a branch to remove any irritants.

- **Social systems and mating.** The social systems and mating patterns of pottos and galagos are not well known. Both the potto and Demidoff's galago are semi-solitary. Adult pottos do not sleep together, but adult galagos share sleeping sites. It is likely that individuals of both taxa monitor conspecifics using vocalizations and scent marking, like other semi-solitary, nocturnal strepsirhines.

Cercopithecids – omnivores, frugivores, and folivores

The Tai primate fauna is ecologically distinctive in supporting sympatric species of the two subfamilies that belong to the cercopithecid family, five species of the cheek-pouched cercopithecines, and three species of leaf-eating colobines (Figure 5.10). It is unusual that these eight closely related species are able to maintain ecological separation. They are all roughly similar in size, eat fruits and leaves, and have similar locomotor styles, factors that typically separate the niches of primates elsewhere. Nevertheless, we find that they do, as discussed below.

- **Size, diet, and locomotion.** The main ecological factors associated with diet and locomotion that contribute to niche differentiation of seven Tai monkey species are summarized in Figure 5.11. Each species exhibits a distinctive combination in the amount of time spent on the ground, eating fruit, and eating leaves. Body mass, however, differs little among the congeneric species. The three species of *Cercopithecus*, Campbell's monkey, the spot-nosed monkey, and the Diana monkey, are all the same size. The males of these sexually dimorphic species weigh roughly 10–11 lbs. (4.5–5 kg). The sooty mangabey, which belongs to a different genus, *Cercocebus*, is larger: males weigh 22 lbs. (10 kg). Two of the three colobines are also essentially the same size. Male Upper Guinea red colobus and king colobus (also called black-and-white colobus) weigh 19 lbs. (8.6 kg) and 22 lbs. (10 kg), respectively, while male olive colobus are about half their weight.

Generally, in primates, leaf-eating species are larger than their frugivorous relatives. At Tai, two of the folivorous *Colobus* species are larger than several frugivorous *Cercopithecus*. The olive colobus is an exception. It is the most folivorous of the three colobines, yet it is also the smallest. The olive colobus tends to use the understory, where smaller size is advantageous, and they are the most terrestrial *Colobus* species. The frugivorous sooty mangabey is as big as the larger *Colobus* species, but it is a very unusual type of frugivore. It is intensely terrestrial. The sooty mangabey forages by rummaging through leaf litter on the forest floor, searching for fallen fruits and nuts that have not yet decomposed because they are hard-shelled. The mangabeys also forage for insects on the forest floor. Insects comprise about 25% of their diet and fruit makes up nearly 70%. That proportion of fruit intake is similar to arboreal fruit-eaters.

Six of the seven Tai monkey species eat appreciable amounts of fruit. Olive colobus does not. More than 90% of its diet consists of leaves. The colobines, with their specialized guts, exhibit a combined leaf-and-fruit diet, eating more leaves than cercopithecines, whereas most of the simple-stomached cercopithecines eat insects instead as a protein source.

Insect-eating among frugivorous monkeys as large as 7–24 lbs. (~3–11 kg) places them in a body-mass category where leaves would appear more favorable as the alternative source of protein. Yet, insectivory represents 27% of the diets of four species of *Cercopithecus* and one species of *Cercocebus*. The cercopithecins are large animals that because of their enhanced manual dexterity and keen vision. are able to focus on insects that are exceptionally small.

The sympatric forms of *Cercopithecus* exemplify niche differentiation among congeneric species that are fundamentally similar in many biological needs and adaptations. Campbell's monkey, which frequents the understory, is clearly distinguished by its

Africa: lorises, galagos, Old World monkeys, and great apes 127

Figure 5.10 Examples of Tai Forest monkeys. Cercopithecines: (a) grey-cheeked mangabey, *Cercopithecus atys*; (b) swamp monkey, *Allenopithecus nigroviridis*; (c) putty-nosed monkey, *Cercopithecus nictitans*. Colobine: (d) Upper Guinea red colobus monkey, *Colobus (Piliocolobus) badius*. (The parenthesis identifies an additional taxonomic category between genus and species, the subgenus, often used in classifying *Colobus*.)

Photo credits: (a) *Cercocebus atys* is an ape of Africa, it's a baby, by Giulio Russo Photography (CC-BY-NC-ND_4.0); (b) Allen's Swamp Monkey (35039622760).jpg, by Rennett Stowe (CC-BY 2.0); (c) Putty-nosed monkey (*Cercopithecus nictitans*).jpg, by LaetitiaC (CC-BY-SA 3.0); (d) Courtesy of Dawn Starin.

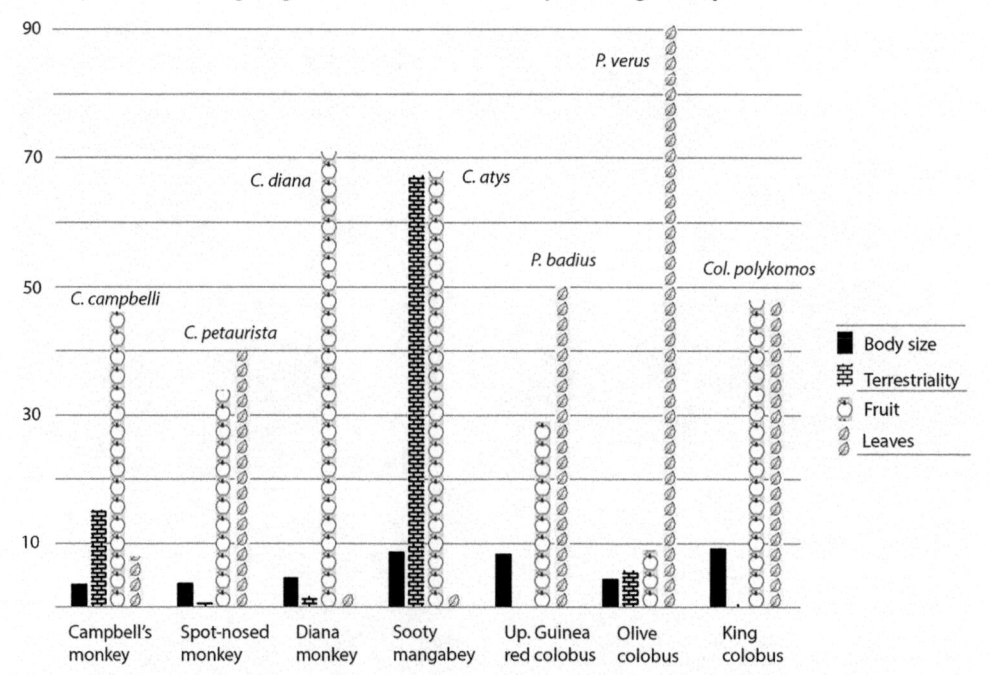

Figure 5.11 Major ecological features contributing to niche differentiation among seven sympatric Taï Forest monkey species. The vertical axis represents the average annual percentage of time devoted to ground use, fruit- and leaf-eating, and in the case of body mass, species weights measured in kilograms. Abbreviations: *C.*, *Cercopithecus*; *Col.*, *Colobus*; *P.*, *Colobus (Procolobus)*.

Adapted from McGraw, W. S., & Zuberbühler, K. (2007). The monkeys of the Taï forest: An introduction. In K. Zuberbühler, R. Noë, & W. S. McGraw (Eds.), *Monkeys of the Taï Forest: An African Primate Community* (pp. 1–48). Cambridge, Cambridge University Press.

terrestrial behavior, whereas the spot-nosed and Diana monkeys are each separated by a different balance of leaf- and insect-eating.

The sooty mangabey, the fourth species in the cercopithecine group, is ecologically most distinctive as a terrestrial frugivore. With its specialized, thick-enameled cheek teeth and powerful jaws, sooty mangabeys are able to eat much harder fruits and nuts that are inaccessible to the other monkeys. The sooty mangabey's focus on eating nuts is highlighted by their alertness to the sounds of Taï chimpanzees smashing nuts apart with rocks. After the chimpanzees leave their nut-cracking sites, the mangabeys go to the piles of discarded shells to pick through them and eat any remaining bits of nut.

Resource-switching is another factor that enables the Taï monkeys to coexist sympatrically. This refers to an animal, or a species-specific foraging routine, that shifts from eating one type of food to another during a feeding bout. It may happen because target foods are temporarily unavailable, or because a more dominant animal displaces an individual while feeding. This phenomenon has been investigated in a study of four sympatric colobine monkeys, mangabeys, and guenons belonging to the same genera as the Taï monkeys but living in another African forest, in Kibale National Park in Uganda. The four monkeys ate from most of the same plant species and devoted similar amounts of time to them. But during every 30-minute feeding bout, the more omnivorous mangabeys and guenons switched food resources more often than the leaf-eating specialist, *Colobus*. For the more flexible feeders, resource-switching provided a buffer against feeding-niche overlap.

- **Social systems and mating.** Socially, the three species of Tai colobines are organized either in unimale-multifemale or multimale-multifemale systems. Social group sizes range from about five to twenty-nine adult individuals. With a diet comprised of high proportions of hard-to-digest mature leaves that require prolonged digestion, colobine monkeys tend to be lethargic, resting for half of each day, or more. Their home ranges can be small because they are able to subsist on mature leaves without traveling far. Males defend territories in these small patches of forest, though they generally avoid one another. They often produce deep roars that can be heard from long distances by neighboring troops and may serve as a spacing mechanism. Males will fight off intruding, non-resident males.

Red colobus monkeys live in multimale-multifemale systems of 40–90 individuals that split up into subgroups when foods are scarce. Their mating system has been well studied in drier west African forests outside the Ivory Coast, in Gambia. There, the mating and rainy seasons coincide during a four-month period of intense activity, when copulations are most frequent. Females advertise when they enter estrous by showing off their just-developed, highly conspicuous swollen genitalia, **sexual swelling**, which remain puffy for about five days each month. The females run through the trees, screaming, breaking branches, and adopting display postures to make their sexual swellings obvious. Dominant males are generally the preferred mating partners, but some females may choose a familiar non-dominant mate with whom she has a strong affiliation, selecting a male that has been attentive to her over a long period of time, meaning over several breeding seasons. When this happens, the two usually pair off together quietly, remaining out of sight for several days.

These encounters with familiar males can be advantageous to a female because males tend to hold their dominant social position and role as chief copulator for only a single breeding season. The mating season stimulates intense male-male competition, leading to leadership turnovers. So, the protection provided by a dominant male to the offspring of his multiple partners because of his social status will only last a short while. On the other hand, the enduring attention of a familiar but less dominant male can offer longer-term protection as an infant grows.

The dynamic of overt sexual advertisement and mate-choice exhibited by red colobus females extends to other aspects of their social systems. In some ways, females counterbalance the strong part played by male-male dominance relationships. Females can be aggressive, and they are not subordinate to males. Adolescent females will temporarily leave their natal group to experience life outside it. When they disperse, females select which neighboring troop they prefer to join, presumably one with which they are already familiar. Afterward, they play an important role in deciding whether or not other dispersing females may be permitted to join the troop. They will chase away strangers and have been known to initiate lethal attacks against unwanted males as well.

The red colobus at Tai cooperate when confronted by their fiercest predator, chimpanzees. Red colobus monkeys are the chimpanzees' favorite prey. When a colobus troop is confronted by preying chimpanzees, the males frequently band together to counterattack. The more colobus males recruited to the effort, the more able they are to avert predation. In other situations, the red colobus retreat. For safety, they also sometimes forage with troops of Diana monkeys, a species that is very vocal and alert to the presence of chimpanzees. The Diana monkeys' alarm calls, which the red colobus recognize, warn them of the danger.

Olive colobus live in small unimale or bi-male groups averaging about five adult individuals. A dominant alpha male has priority access to females, but more than one male can breed. Olive colobus males are not able to monopolize females. This is because the reproductive strategy of olive colobus females involves leaving the home range temporarily when in estrous to visit lone males and copulate with them. Lone olive colobus males seeking breeding partners can join an existing group if the resident male is unable to deter them from doing so, or by instigating a group takeover.

The fluid group affiliations and small group sizes of olive colobus are connected with another aspect of their community ecology. Olive colobus form almost permanent associations with sympatric Diana monkeys, reducing the risk of predation for solitary males and females that have left their own groups by a variation of the **selfish herd** effect. In its classical form, this refers to the situation in which the risk of predation of any one individual is lessened by their affiliation with other conspecifics, which makes them all potential victims. When olive or red colobus are being actively hunted by chimpanzees, being in a larger group, including individuals of another species, reduces the chances of any one monkey being caught.

The three Tai *Cercopithecus* species live in unimale-multifemale systems. Group sizes vary, averaging 11–24 individuals, with a strong bias toward adult females. In Diana monkeys, females outnumber males by more than ten to one. For all the species, this means that about one-third to one-half of the animals in a troop consist of juveniles and subadults requiring management and protection. It also means that in the forest there are lone adult males and bands of males that can affiliate and occasionally cross paths with an established social group, which is most likely to occur at a feeding tree, where the meeting can elicit aggressive calling. Males of all three *Cercopithecus* species defend territories by vocalizing loudly when a troop encounters a conspecific group less than about 150 ft (45 m) away.

Other cercopithecids – the long-faced papionins

The mangabeys at Tai are part of a larger taxonomic group, the papionins, that includes six genera of African monkeys that live elsewhere. The ecological success of the widespread papionins is largely due to their extensive array of adaptations as terrestrial primates, though most feed, shelter, rest, and sleep in available trees when they wish. Some species are able to tolerate and flourish in dry, relatively treeless habitats, including semi-deserts and mountainous terrains, where, devoid of vegetation, they take shelter on rocky cliffs.

One of the papionins is the Barbary macaque, *Macaca sylvanus*. It is the only African primate species that inhabits a small northwestern area of the mainland, as noted above. Here we focus on the Sub-Saharan papionins belonging to three genera, *Papio*, *Mandrillus*, and *Theropithecus*. Phylogenetically, all of them comprise varieties of a "baboon" subgroup, though the *Mandrillus* species are commonly called mandrills and drills, and *Theropithecus* is known as gelada, which leaves the term baboon for the genus *Papio*.

General features

Papionins are the largest African monkeys. Male weight ranges from about 20 lbs. to more than 70 lbs. (9–32 kg), within the weight range of great apes. They are sexually dimorphic in size, males weighing 1.5 to 2.8 times more than the mass of females. Gender differences in pelage and other outward features are often quite marked in several genera. These are secondary sexual characteristics that are associated with the production of chemical hormones that also regulate physiology and behavior connected to chronological age, social status, and reproductive state.

Female papionins exhibit cyclic changes in primary sexual characteristics, sex-specific features that are present at birth. Females present cyclic red **sexual swellings** of the perineum (skin between and around the genitals and anus) that become quite prominent during estrous, and attract the attention of males. This type of trait is especially pronounced in papionins and widespread among female catarrhines. Female geladas, for example, have a patch of skin on the chest and throat which also serves to draw male attention as it changes with phases of the reproductive cycle. When a female is sexually receptive and ovulating, the patch becomes accentuated by a string of fluid-filled red vesicles, or blisters, some of which may become pendulous, that develop along its border and around the perineal sexual skin of the rump and groin. Adult male geladas exhibit a comparable feature in addition to their large, fur shoulder capes, which resemble the pelage of adult male hamadryas baboons, and distinctive head and facial fur. The gelada males have an inverted, heart-shaped patch of red skin on the chest that extends onto the throat and is especially noticeable when a male is in his prime. This is why geladas have been called the "bleeding heart baboons" (Figure 5.12).

Figure 5.12 An adult male (left) and much smaller female gelada in the rugged mountainous terrain of the Ethiopian highlands.

Courtesy of Jacinta Beehner.

Mandrills are the most colorful papionins, with strategically located dimorphic patterns. The males develop a striking long face that is accented by long ribs of blue, red, and white skin on a dark background. The degree of redness is correlated with social dominance among males and it elicits mating preferences among females. The colors of their faces are conspicuous against the green background foliage of the forest-dwelling mandrill. Their rumps are also multi-colored, and easily visible in the quadrupedal mandrill because they have a short, stubby tail that is carried upright as they walk.

Diet and feeding-foraging specializations

Species of *Papio* are the most ecologically flexible African monkeys. They live in a wide variety of habitats from the tropical rainforests to semi-deserts. Papionins are exceptionally omnivorous, but some also exhibit specific dietary specializations. For example, the mandrill's large body size, muscular strength, and powerful jaws are central to its terrestrial foraging in closed forest habitats. They feed on fallen fruit, hard nuts, and small prey. Using their great strength, they search for food by ripping away tree bark and rolling fallen logs aside to uncover what may be underneath. Their size and strength are advantages when foraging for nuts and seeds on the ground. As a result, the mandrills occupy a niche that resembles the *Cercocebus* mangabeys at Tai.

Geladas, *Theropithecus gelada*, exhibit the most distinctive departure from the generalized papionin morphology. It is a thoroughly terrestrial genus with a cascade of morphological and behavioral adaptations to a unique lifestyle in an extreme environment where selective pressure is likely to be intense. Geladas are endemic to the grassy high plateaus of the rugged Ethiopian Highlands, in a temperate climate where there are few food trees and no arboreal shelters. More than 75% of the gelada diet consists of grass and grass-like vegetation, meaning the majority of gelada foods are structurally comparable to leaves. Fitting the typical folivore pattern, gelada cheek teeth are more crested than the frugivorous-omnivorous savanna baboons and their incisors are relatively small. Molar cusps and crowns are also quite tall, a feature that increases the functional lifetime of a tooth when a diet is highly abrasive, such as low-growing vegetation, and rapidly wears down the enamel. Geladas inadvertently consume lots of abrasive grit on a daily basis, especially during the dry season when they dig shoots and bulbs out of the ground.

Gelada foraging methods are unique. They spend nearly all day sitting on their enlarged ischial callosities and sitting pads, and shuffling about upright in a crouch-walk, hardly lifting their bottoms. They move from one spot to another gathering handfuls of grass, sometimes picking up seeds with their fingers while also clutching a bunch of grass blades in the same hand. By comparison to other baboons, the gelada thumb is relatively long and the index finger is relatively short, a combination that enables dexterous plucking actions.

Geladas live in a complex, highly regulated social system dominated by fierce, aggressive adult males. The ability to make rapid visual assessments about gender, rank, and reproductive status is beneficial, and geladas have important signaling features that work well as they sit about and shuffle while foraging. The red chest patches of geladas are positioned high up on the trunk where they can easily be seen. Estrous females can signal their readiness to copulate, and males can signal their social and reproductive standing, because the redness varies quantifiably with age and status. In the most basic gelada social unit, a unimale-multifemale group, adult males with the reddest patches may monopolize more than six females (Figure 1.8).

Social systems and mating

Social structure varies among papionin species but they share several features in common. They are organized around dominance hierarchies enforced by aggressive behavior, which is associated with a high degree of sexual dimorphism favoring males in body mass, canine size, and secondary sexual characteristics linked with social status. Females exhibit same-sex dominance hierarchies based on maternal kinship and age-related reproductive potential. Some papionin species live in unimale-multifemale groups and multimale-multifemale groups. In different species, each male may monopolize five, six, or more females and their offspring.

Geladas, hamadryas baboons, and Guinea baboons form complex **multi-level societies** that may have three to four tiers or components. Above the basic social and breeding group are several tiers organized as social and ecological units reflecting kinship, foraging habits, and the availability of sleeping sites.

- Unimale-multifemale group. This is the core social and mating unit. It is coordinated and defended by the dominant leader male. Subordinate follower males may be attached to the unimale group but they are dissuaded from mating with the females by the leader male, who may be aggressive toward them. In hamadryas, the leader male controls the behavior of females by threatening and biting.
- Bands. These are aggregates of several breeding groups that regularly associate with one another when foraging and share sleeping sites on rocky cliffs.
- Herds. Several bands of geladas may temporarily join to form a herd at sleeping sites, or while foraging. A gelada herd may consist of many hundreds of individuals.
- Bachelor males. As a consequence of the overriding unimale power structure at the base of the baboon social system, a number of non-dominant males are always marginalized, waiting for a turn to commandeer their own females. In some species they join together to form another social unit. These bachelor males pose a particular threat of group takeover and infanticide to the unimale breeding unit.

Counts of social group sizes in papionin baboons can be complicated by their tiered structure as well as species-specific variations, but tallies ranging from 30–40 and 80–100 individuals are common. Large group size is not necessarily correlated with openness of habitat. A video study of mandrills in Gabon that live in gallery forest in a savanna recorded herds as they walked across an opening in the forest. The average group size in 20 recording sessions was 620 individuals, with a maximum count of 845.

Hominids – semi-terrestrial frugivore-folivores

The African apes, classified in the hominid family, are the chimpanzees, bonobos, and gorillas. The chimpanzee, *Pan troglodytes*, sometimes called the common chimpanzee, is the ape that occurs in the Tai Forest. It is also found in rainforests of central Africa, like its sister-species, the bonobo (Box 5.1). Gorillas are also found in lowland western African forests and in mountainous eastern habitats.

Pan troglodytes, the common chimpanzee, lives in a variety of forested environments and woodland savanna, with the widest geographical distribution and most varied ecology of any living African ape. It is highly successful. Part of its success is due to high

Box 5.1 Bonobos

Bonobos, *Pan paniscus*, are the chimpanzee's sister-species, meaning they both stem from an exclusive common ancestor. Bonobos are found in a circumscribed region in central Africa where there are no other sympatric African apes. Table 5.3 summarizes some of the features that distinguish common chimpanzees from bonobos.

Bonobos live in a generally co-dominant group structure. Females form strong relationships. Female-female sexual interactions play a large role in maintaining these relationships and limiting conflict within the social group. This differs from the nature of chimpanzee groups where the relationships among individuals and group cohesion are strongly influenced by a balance of male-on-female aggression and male-male bonding.

Table 5.3 Features that distinguish bonobos and common chimpanzees

	Bonobo, *Pan paniscus*	Chimpanzee, *Pan troglodytes*
Habitat and microhabitat	Rainforest; arbo-terrestrial	Rainforest to woodland savanna; arbo-terrestrial
Body mass	Males, 99 lbs. (45 kg); females, 73 lbs. (33 kg)	Males, 86–132 lbs. (39-60 kg); females, 68–101 lbs. (31-46 kg)
Weight sexual dimorphism	Yes	Yes
Skull	More gracile	More robust
Canine size	Small, non-projecting	Large, projecting
Canine sexual dimorphism	Some	More
Skeleton	More gracile build	More robust build, proportionately longer heavier legs and forelimbs
Diet	Frugivorous-folivorous, hunts occasionally	Frugivorous-omnivorous, hunts regularly
	Fruit, 55%; leaves, 14%; herbs, 25%	Fruit, 64%; leaves, 16%; herbs, 7%
Hunting	Rare, duiker antelopes preferred	Often, monkeys preferred
Food sharing	Yes, fruit	Yes, meat
Locomotion	Knuckle-walking quadrupedalism, frequent suspension	Knuckle-walking quadrupedalism, frequent climbing
Social organization	Multimale-multifemale, fission-fusion	Yes
Party sizes (individuals)	Large (6–15)	Small (8)
Mixed male-female parties (% seen)	75%	32–52%
Male-female relations	Strong	Weak
Social bonding	Female-female	Male-male
Grooming preference	Male-female	Male-male
Dominance relationships	Females dominant or co-dominant	Males dominant
Male-female aggression	Rare	Common
Female-male aggression	Common	Rare

	Bonobo, *Pan paniscus*	Chimpanzee, *Pan troglodytes*
Conflict resolution and appeasement	Via sex	Via aggression and submission
Sex during non-estrus phase	Common	Very rare
Female-female sexual behavior	Yes, genital-genital rubbing	Very rare
Mating systems	Females mate with one or more males, males tolerate multiple partners	Females mate with one or more males, dominance-based, mate-guarding
Infanticide	None reported	Yes but rare
Territoriality	Tolerant of neighbors	Hostile toward neighbors
Boundary patrols	None reported	Yes
Tool use	Rare	Habitual

intelligence and the behavioral flexibility it affords. This is especially evident in the wild by differences in learned social behaviors, which includes tool use, that have developed in geographically distinct populations. In captivity, chimpanzees have learned to communicate with researchers using symbols, and their well-known intelligence has long been exploited in many ways for entertainment purposes. Chimpanzees' wide-ranging ecological capacity is a function of a generalized ape anatomy supporting an omnivorous diet and an arboreal-terrestrial lifestyle.

Chimpanzees range in weight from about 74–132 lbs. (34–60 kg). They are sexually dimorphic in body mass (males weigh about 30% more than females), canine size and shape, and in features of the skull that reflect size, including bony ridges marking the attachment of muscles. Their heads are distinctive in combining a small snout with a pronounced brow ridge and flattened forehead, a shape that emphasizes the deep-set eyes. Their appearance is also distinguished by their quadrupedal stance and gait, the notably raised head and sloping back, a result of fully extended long arms combined with the vertically aligned knuckle-walking hand posture and the bent-knee attitude of the hindlimb, which keeps the rump low (Figure 5.7).

The chimpanzee is larger, stronger, smarter, more communicative, more intimidating than any other Tai primate. The result is that they have a distinct ecological separation from all the cercopithecids. Chimpanzees live a very different lifestyle from the other members of the community. Some of the foods they eat are physically inaccessible to monkeys and can only be accessed by using tools. Chimpanzees use both terrestrial and arboreal microhabitats, and they can sleep anywhere in large trees by building nests.

- **Diet and locomotion.** Chimpanzees are essentially ripe-fruit specialists, but they are also carnivorous – they eat meat. At Tai, the chimpanzees habitually prey on monkeys as well as the potto, a strepsirhine. Chimpanzees prefer to eat the relatively large and lethargic leaf-eating monkeys, particularly the red colobus. More than 70% of their diet consists of fruit and they eat vegetation from over 300 plant species. Chimpanzees also eat a variety of insects.

Chimpanzee dentitions exhibit a typically frugivorous pattern. Their incisors are relatively large and robust. The molars have blunt cusps and large, shallow basins that are functionally designed for crushing and grinding food. They lack flesh-slicing cheek teeth but that does not constrain them. They tear away their prey's soft tissue and crack its bones with physical strength, robust teeth and jaws, and strong, dexterous hands. They have been seen cracking open the skulls of monkeys they prey on to eat their brains.

Chimpanzee limb anatomy and proportions are effective for a large semi-terrestrial primate, with approximately equal forelimb and hindlimb lengths, and large hands and feet. Their main locomotor mode is knuckle-walking quadrupedalism. Chimpanzees are especially adept at climbing large trunks with their long reach, muscular upper limbs and shoulders, and strong legs. In the canopy, they combine climbing with quadrupedalism and arm-swinging to travel short distances, and use suspensory postures to feed.

Apart from meat-eating, the foraging patterns and diet of chimpanzees is also unusual among primates to the extent that they use tools to secure embedded foods. They are extractive foragers. The first instance of chimpanzee tool use reported in the literature involved "fishing rods" made of long slender sticks fashioned from branches by stripping away lateral twigs (Box 5.2). The sticks are used to extract termites from the terrestrial mounds in which the insects live. The chimpanzee pushes the probe into the mound and, when removed, it has a bunch of termites clinging to it, and the animal eats them off the stick. This is an efficient method for a large primate to obtain a worthwhile supply of protein-rich food without expending much energy. To maximize the opportunity for termites to cling to the stick, some populations of chimpanzees have devised a way of feathering one end to form a brush.

The most unique aspect of chimpanzee feeding is hunting. Hunting behaviors and frequencies differ at the research sites where it has been studied in detail, but it is evident that all chimpanzee populations hunt. Hunting mammals is common in chimpanzees and involves preying on more than 30 species. But this only accounts for a small fraction of the chimpanzee diet, 1%–3%.

Box 5.2   Jane Goodall

Jane Goodall, born in England in 1934, is most famously known for her pioneering, long-term field studies of chimpanzees in Tanzania, Africa, beginning in 1960. Since 1968, the Gombe Stream National Park, established through her award-winning work, has continued to be a center for research, training, and conservation.

Goodall was the first scientist to observe and report that chimpanzees use tools that they make, such as twigs stripped to extract termites from mounds on the ground. Up until that time, it was believed that no animals, only humans, used tools. She was also the first to discover that chimpanzees hunt, and eat meat.

Because of her many years of living in the forest and getting close to the chimpanzees she studied, Goodall came to know and describe their individual personalities, naming the animals, and documenting their family relationships for their entire lives. Her work revolutionized the study of animal behavior and the threats they face, including poaching and habitat loss.

Goodall, the author of 15 books, has won numerous awards for her work. In 2002, she was named by the United Nations as a United Nations Messenger of Peace. In 2019, she was recognized as one of the 100 most influential people in the world.

However, the role of hunting in the chimpanzee lifestyle is profound. It is a risky and complex endeavor. Hunting in groups requires that individuals coordinate and collaborate. Afterward, a hunt generates intense social interactions in the demands and pleas for sharing the meat among the hunters and others. The chimpanzees' favored target, red colobus monkeys, are not easy prey. The pursuit takes place high in the canopy where the monkeys forage and feed. Red colobus are large primates and excellent arborealists that are protected by living in social groups. They use anti-predator alarm calls and behaviors, and they often form polyspecific associations that provide another layer of alertness and protection via the selfish herd effect. On the run, the colobus can more easily leap between trees via small branches and twigs than the heavier chimpanzees. They also have slashing canines and will at times mount counterattacks when severely threatened.

- **Social systems and mating.** Chimpanzees live in multimale-multifemale groups that are characterized by flexible fission-fusion subgrouping patterns. Troop size varies among the populations that have been studied but they are all large, some involving 150 individuals. Groups of 29–82 have been documented at Tai, where foraging party subgroups average about eight individuals. Party size varies seasonally. It can also vary throughout the day in size and composition, by gender and the age of individuals. All-male parties are slightly more common than all-female parties, but more than 50% of the subgroups include both males and females. Smaller parties occur when food is scarce and larger parties form when it is abundant. There are also socio-sexual factors that influence party size. Parties tend to be larger when estrous females are involved.

Chimpanzee social organization revolves around the dominance of males and strong bonds that form between adult males, though bonding partnerships are fluid and can change quickly. Consequently, males are more often in the company of other males and groom one another frequently. Closely allied males support one another to attain and hold higher ranks in the dominance hierarchy. They also collaborate in mate-guarding, which means staying close to a female to ward off other potential rivals when she is in estrous. Males cooperate when chimpanzee troops meet one another and aggressively defend their territories. Females also form alliances and dominance hierarchies with one another that may be beneficial in securing food for themselves and their offspring.

Behavior within chimpanzee troops tends to be regulated by dominance displays and the physical aggression of males. Displays are bluffs that avert physical altercations. While females are subordinate to males, male-female aggression is less common than male-male aggression, because aggression against other males is a mechanism that males use to discern and maintain their status. This is because in the context of the flexible fission-fusion social system in which the composition of parties is fluid, a male has limited opportunities to monitor his social relationships, and many males may be present at one time. Displaying an aggressive bluff or having an encounter is a way for a male chimpanzee to test his own strength and the strength of his affiliations, to see who will collaborate with him or intercede on his behalf, or come to the rescue of his opponent.

Both males and females initiate mating in chimpanzees, though it is mostly the males. Females are in estrous for 10–15 days in the course of a 36-day menstrual cycle, during which time the perineal region becomes conspicuously red and swollen, which attracts male attention. Males and females pair up in different ways. Females may have one mating partner or many in succession, without eliciting aggression by other males. A male-female pair may leave the group for a few days or weeks. While they are residents, females rarely mate with males from another social group. They appear to show preferences toward certain partners, mating more frequently with higher-ranking males, even when they copulate with several individuals.

Chimpanzees often appear to be mild-mannered but they can be lethally aggressive toward neighboring troops. They form scouting parties to patrol the boundaries of their territories, intentionally invade the territories of neighboring troops by penetrating well beyond the other's territorial boundaries, kill individuals belonging to other social groups, and commit infanticide. Patrolling squads are quiet before an intertroop encounter, practicing the same skills they employ when hunting monkeys but under much more dangerous circumstances, because now they are opposed to an enemy that is likely to be equally powerful, intelligent, and experienced in conflict. Parties of allied males are also known to kill members of their own troop, including adult males.

Other African apes – gorillas, the largest living primates

Gorillas, the other African genus of the hominid family, which do not live in the Tai Forest, are exceptionally large primates but not the largest that has ever lived, as discussed in Chapter 8. That record is held by a fossil, *Gigantopithecus*. The gorilla's huge size and all of its dietary, locomotor, and behavioral correlates widely separate them ecologically from all sympatric primates.

There are two gorilla species, each one geographically located on either side of Africa's equatorial rainforest; the eastern gorilla, *Gorilla beringei*, and the western gorilla, *Gorilla gorilla*. Adult males range widely in body mass, roughly 320–420 lbs. (145–190 kg). Adult females weigh roughly 120–160 lbs. (54–74 kg). Reflecting their extreme size differences, male skeletons are much more robust than females'. Sexual dimorphism is also expressed in canine size and cranial morphology.

As males age, the deep brown or black fur of the back turns silvery white, which is why they are called silverbacks. On the chest, a hairless broad patch of black skin develops which males use in an audiovisual display, chest-beating, to project their presence and competitively "size up" one another. Alternately slapping the massive barrel-shaped chest rapidly with both hands produces a drumming sound that can be heard for about a half-mile. Body size is correlated with the sound's pitch, and larger individuals with bigger chests produce a deeper sound, which conveys their greater size and strength, and also signifies an elevated social status.

- **Diet and locomotion.** Gorillas are frugivorous, folivorous, and they also eat ground-level herbs, which are also a type of leafy vegetation. The fruit component varies greatly seasonally and according to the elevation of the habitats in which they live. The mountain gorillas, a subspecies of eastern *G. beringei*, live in the high forest slopes where little fruit is available due to the altitude, while *G. gorilla* living in lowland western Africa experience fruit scarcity during long dry seasons. Herbs are important food resources for all gorillas as is the pith of bamboo for mountain gorillas, and swamp plants where they are available. The gorillas' large body size, capacious guts, large teeth, and powerful jaws allow them to eat a bulky diet of leaves and herbs, and crack open hard bamboo shoots to get at the pith inside.

  Gorillas are semi-terrestrial. Given their enormous size and diet of herbs and other ground-based plants in addition to fruits and leaves, most of their time is spent on the forest floor, using knuckle-walking quadrupedalism to move about.

Silverback males usually sleep on the ground in nests they build each night. At times, other males may sleep on the ground, but they mostly sleep in tree nests like females and young.

The amount of gorillas' arboreal activity varies regionally depending on the forest structure and fruit availability. In the trees, gorillas use climbing and quadrupedal locomotion to reach feeding sites capable of sustaining their weight on the largest branches of the canopy. But to travel between trees gorillas typically have to come to the ground. Mountain gorillas in the east, living in habitats with few trees of substantial size, spend little time in the canopy. In the western lowlands, trees are larger and rich with leaves and fruits. There, gorillas spend more of their time feeding and foraging in the canopy, even using small-diameter branches as they enlist a wide variety of locomotor patterns.

- **Social systems and mating.** Gorillas live in unimale-multifemale groups averaging eight to ten individuals, typically led by a dominant silverback male in his prime. Most, but not all of the young in a group are sired by the silverback. Since gorillas grow slowly and multiple females are reproductively active at one time, groups may periodically have resident subadult males and even adult males that are not dominant, but multimale breeding groups are not common. These groups typically revert to a unimale-multifemale structure when younger males age out and leave the troop, or are expelled. Groups composed of bachelor males that are unattached to an established socio-sexual troop also form in the forest.

Protection from predators, and the protection of offspring from infanticidal outsider males, are vital roles played by a silverback male. Male-female relationships are central to gorilla social life, and females spend much of their time feeding and resting in close proximity to males, who are dominant over females. Grooming between adult males and females is an important activity. Male-female antagonism is mild, usually consisting of low-level growls or chest-beating. Gorilla home ranges overlap to a great extent, so there are frequent opportunities for groups to meet. During these intertroop encounters, males of opposing social groups may chest-beat but the levels of aggression are typically non-threatening. The members of groups that meet may peacefully interact with one another. They may even rest at night a few meters apart.

Despite their imposing size, gorillas are not immune to natural dangers. Unaffiliated males living in the forest pose a serious threat to infants when a silverback male is not present. Predation by leopards is a threat to western gorillas, and sympatric chimpanzees are a source of danger as well (Box 5.3).

Box 5.3   Dian Fossey

Dian Fossey (1932–1985) was the first scientist to conduct a long-term study of gorillas in the wild, observing them daily for decades. In 1966, Fossey set up a research program in a remote spot in the Virunga Mountains of Rwanda, where, at the time, gorillas were being illegally captured, hunted, and killed. She worked there continuously for nearly 20 years, documenting the lifestyles of gorilla families – which were hardly known before then. Tragically, she was murdered in her camp. Her assailants were never identified.

Fossey's efforts helped to establish a research and conservation center that continues to thrive, the Karisoke Research Center in Volcanoes National Park, in Rwanda. It focuses on monitoring the resident population of mountain gorillas, providing opportunities to study them, educating the local community to encourage their participation, and supporting community health and development projects as well as employment opportunities that engage local people. The Virunga region is now home to a slowly growing gorilla population that comprises about two-thirds of all wild mountain gorillas, about 1,000 individuals as of 2022.

Paleocommunities

A fossil assemblage discovered in Egypt at a site known as the Fayum, consists of a regional paleocommunity spanning over eight million years, in the late Eocene and early Oligocene, from about 38 to 30 million years ago, as discussed further in Chapter 8. The fossils show that there was a taxonomic shift in the composition of the Fayum community over time. Strepsirhines were abundant and lived alongside anthropoids during the earlier phases, but none have been found in the younger fossil beds where anthropoid remains are common. Dietary adaptations of the strepsirhines, almost all of which were small-sized, weighing less than 1 lb. (453 g), varied. There were insectivores, frugivores, folivores, and tree-gougers.

Changes in body size are evident among the anthropoids. Body mass increased with time. The anthropoid families that are found in the younger geological stages were large-sized primates, more than 10 lbs. (4.5 kg) in weight, whereas the older anthropoids overlapped with some of the strepsirhines in body mass and were smaller, weighing between 8 oz. (242 g) and 1.8 lbs. (800 g). Dietarily, the fossil anthropoids, even the smallest species, probably relied on fruits, and gums to an extent, as their molars tend to have blunt rather than pointed cusps. The largest anthropoid genus, about 15 lbs. (6.7 kg), was a frugivore that probably also ate leaves.

One fossil site at the Fayum (known as L-41) may approximate a sympatric assemblage. It has produced ten species and genera, including six anthropoids and four strepsirhines. That collection suggests ecological segregation of species by activity rhythm, size, and diet, as in today's primate communities.

Fossils from Rusinga Island in Kenya reveal a much younger paleocommunity, from 18 million years ago. Ten species belonging to five or six genera of primates have been discovered there, four or five early ape species and five strepsirhine species. The two strepsirhine genera are ecologically distinct, from one another and from the catarrhines. Both are nocturnal lorisoids, like Africa's extant strepsirhines: one is a lorisid and the other is a galagid. The galagid shows evidence of leaping locomotion, and the lorisid was a slow-climbing quadruped much like the extant genera. They were both small-sized primates that probably ate fruits and insects. The catarrhines were considerably larger and diurnal. They were large and extra-large primates, fruit-eaters that used quadrupedal locomotion.

Fossils of extinct primates are abundant at Rusinga, but the composition of the primate community differs from modern African communities in several ways, including:

- The co-occurrence of 4–5 ape species contrasts markedly with the diversity and geographical distribution patterns of extant apes. Today, there are only two extant genera

Africa: lorises, galagos, Old World monkeys, and great apes 141

in all of Africa, and no more than one species of *Pan* and one species of *Gorilla* live together in sympatry.
- Ecologically, the Rusinga fossil apes more closely resemble extant Old World monkeys – of which there are none at Rusinga – than the living African apes, in body mass, diet, and locomotion.

These fossils indicate major changes in the structure of the primate fauna in Africa. During this period, about 18 million years ago, apes were the dominant catarrhines in Africa. Then, about 10 million years ago, monkeys replaced them. In general terms, the evidence from Rusinga suggests that the rise of cercopithecoid monkeys could have presented a level of competition with the monkey-like early apes that ultimately contributed to the decline of their taxonomic diversity, and to their ecological shift toward semi-terrestriality.

The conservation status of African primates

Six African primates are currently among the 25 most imperiled species according to the IUCN SSC Primate Specialist Group (Figure 1.11). They include all the major taxonomic groups in Africa: a galago (lorisid); one leaf monkey (colobine); three cheek-pouched monkeys (cercopithecines); and a chimpanzee (hominid). Apes are the most severely threatened primates (Figure 5.13). Today, all ape species are either Endangered or Critically Endangered. There are several galagos that are Endangered but none are threatened to the level that makes them Critically Endangered. The high levels of threat faced by the

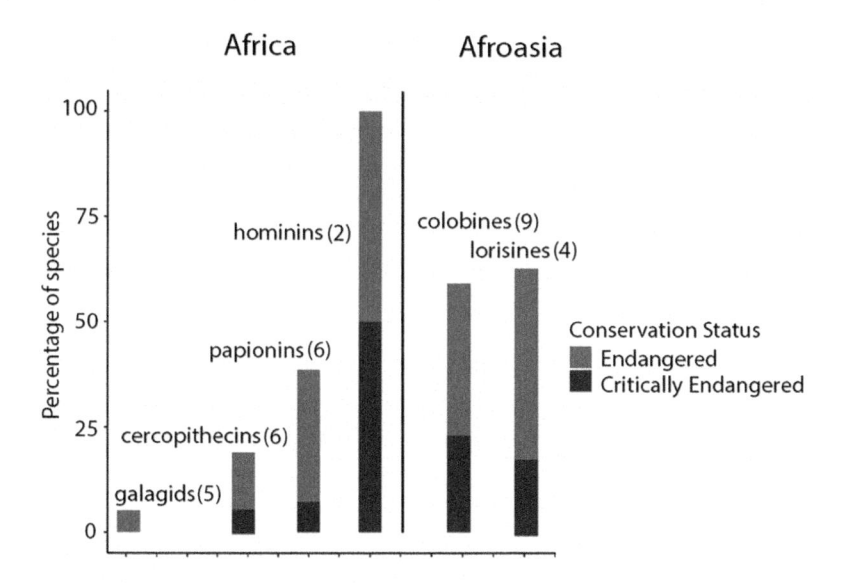

Figure 5.13 The threat levels to six families, subfamilies, and tribes (a category below the subfamily and above the genus) of primates found in Africa and jointly Afroasia (occur both in Africa and Asia). Numbers in parentheses refer to the number of genera in each group.

Adapted from Fernández, D., Kerhoas, D., Dempsey, A., Billany, J., McCabe, G., & Argirova, E. (2021). The current status of the world's primates: Mapping threats to understand priorities for primate conservation. *International Journal of Primatology*. https://doi.org/10.1007/s10764-021-00242-2.

nocturnal lorisines and by the colobines is noteworthy. The nocturnal lorises are small, exclusively arboreal, and restricted to rainforest habitats, a lifestyle that would seem to offer them protection. The forest-loving, arboreal colobines are even more at risk than the open country, terrestrial papionins that may live in ecosystems harboring crocodiles, lions, and other predatory large cats. Why are both taxa gravely at risk?

More than a third of Africa's primates are threatened and nearly half of the others are declining in numbers. The chief threats include habitat loss and hunting. For the African apes, large-scale agricultural expansion to cultivate palm trees to produce palm oil poses a particular threat, especially in regions occupied by chimpanzees and gorillas that fall outside protected areas. Mining in Central and Western African forests also exacts a large toll.

Conservation programs have been effective in protecting some populations of African primates. One example is a legacy of Dian Fossey's heroic efforts to highlight the plight of mountain gorillas in Rwanda (Box 5.3). Another conservation program in the Democratic Republic of the Congo, the Virunga National Park, a UNESCO World Heritage Site, has established a variety of projects to sustain the gorillas that live there and preserve the area's flora and fauna. There is a facility that houses and is dedicated to caring for orphaned gorillas. A team of nearly 800 rangers routinely patrol the park. The stability afforded by conservation initiatives has led to a doubling of the park's gorilla population in 15 years, to about 350 individuals.

Jane Goodall's research on chimpanzees in Tanzania led to the establishment of the Gombe Stream National Park in 1968 (Box 5.2). Chimpanzees have been studied there continuously for more than 60 years, the longest-running animal behavior study in the world. Goodall's scientific work developed into a life-long effort to preserve chimpanzee habitats in Tanzania and elsewhere. Additionally, the Jane Goodall Institute collaborates with agencies focusing on health and wellness in villages in the Democratic Republic of the Congo that are near chimpanzee populations. The Institute is a global leader in the intergenerational conservation movement. It partners with organizations that are developing methods to manage the life-threatening consequences of climate change in parts of Africa.

See Chapter 9 for more on threats to primates globally and conservation efforts.

Key concepts

Contest competition
Behavioral displays
Multi-level societies
Carnivory
Tool use

Quizlet

1 What geographic features describe the terrestrial habitats of baboons?
2 What are some features that chimpanzees and bonobos have in common?
3 How does the potto use its protective shield?
4 Why are colobines called leaf monkeys?
5 Describe two of the features that contrast catarrhines and platyrrhines.
6 What is the one visible trait that distinguishes apes from monkeys?
7 Name three characteristics of the ape's skeletal morphology.

Bibliography

Abernethy, K., White, L., & Wickings, E. (2006). Hordes of mandrills (*Mandrillus sphinx*): Extreme group size and seasonal male presence. *Journal of Zoology*, 258, 131–137. https://doi.org/10.1017/S0952836902001267

Bergman, T., Ho, L., & Beehner, J. (2009). Chest color and social status in male geladas (*Theropithecus gelada*). *International Journal of Primatology*, 30, 791–806. https://doi.org/10.1007/s10764-009-9374-x

Goné Bi, Z. B., & Wittig, R. (2019). Long-term diet of the chimpanzees (*Pan troglodytes verus*) in Taï National Park: Interannual variations in consumption. In C. Boesch & R. Wittig (Eds.), *The Chimpanzees of the Taï Forest: 40 Years of Research* (pp. 242–260). Cambridge, Cambridge University Press.

Kalan, A. K., Carmignani, E., Kronland-Martinet, R., Ystad, S., Chatron, J., & Aramaki, M. (2019). Chimpanzees use tree species with a resonant timbre for accumulative stone throwing. *Biology Letters*, 15(12), 20190747. https://doi.org/10.1098/rsbl.2019.0747

Lambert, J. E. (2002). Resource switching and species coexistence in guenons: A community analysis of dietary flexibility. In M. E. Glenn & M. Cords (Eds.), *The Guenons: Diversity and Adaptation in African Monkeys* (pp. 309–323). New York, Springer.

McGraw, W. S., & Zuberbühler, K. (2007). The monkeys of the Taï forest: An introduction. In K. Zuberbühler, R. Noë, & W. S. McGraw (Eds.), *Monkeys of the Taï Forest: An African Primate Community* (pp. 1–48). Cambridge, Cambridge University Press.

Nowack, J., Mzilikazi, N., & Dausmann, K. H. (2010). Torpor on demand: Heterothermy in the non-lemur primate *Galago moholi*. *PloS One*, 5(5), e10797. https://doi.org/10.1371/journal.pone.0010797

Southern, L. M., Deschner, T., & Pika, S. (2021). Lethal coalitionary attacks of chimpanzees (*Pan troglodytes troglodytes*) on gorillas (*Gorilla gorilla gorilla*) in the wild. *Scientific Reports*, 11(1), Article 1. https://doi.org/10.1038/s41598-021-93829-x

Starin, E. D. (1994). Philopatry and affiliation among red colobus. *Behaviour*, 130(3/4), 253–270.

Swedell, L. (2015). *Strategies of Sex and Survival in Hamadryas Baboons: Through a Female Lens*. Oxfordshire, Routledge.

Watts, D. P., & Mitani, J. C. (2015). Hunting and prey switching by chimpanzees (*Pan troglodytes schweinfurthii*) at Ngogo. *International Journal of Primatology*, 36(4), 728–748. https://doi.org/10.1007/s10764-015-9851-3

6 Asia
Lorises, tarsiers, Old World monkeys, and apes

Chapter Contents

Southeast Asia's geography and climate	145
The Asian radiation: lorises, tarsiers, Old World monkeys, and apes	146
Asian lorises – nocturnal stalking predators	147
Tarsiers – nocturnal ambush predators	149
Hylobatids – brachiating lesser apes	153
Southeast Asian primates living in communities: diversity, ecology, and adaptation	155
The slow loris – a nocturnal gumivorous frugivore-faunivore	155
Cercopithecids – frugivores and folivores	157
Hylobatids – brachiating frugivore-folivores	159
Other Southeast Asian primates	159
Macaques and Hanuman langurs	159
The odd-nosed monkeys	160
Orangutans – the largest arboreal mammal	161
Paleocommunities	163
The conservation status of Asian primates	164
Key concepts	165
Quizlet	165
Bibliography	165

Superfamily Lorisoidea (lorisoids) – lorises and galagos
    Family Lorisidae (lorisids) – lorises
Superfamily Tarsioidea – tarsiers
    Family Tarsiidae (tarsiids) – tarsiers
Superfamily Cercopithecoidea (cercopithecoids) – Old World monkeys
    Family Cercopithecidae (cercopithecids) – Old World monkeys
        Subfamily Cercopithecinae (cercopithecines) – cheek-pouched monkeys
        Subfamily Colobinae (colobines) – leaf monkeys
Superfamily Hominoidea (hominoids) – apes
    Family Hominidae (hominids) – great apes
        Subfamily Ponginae (pongines) – orangutans
    Family Hylobatidae (hylobatids) – lesser apes

Southeast Asia's geography and climate

Asia is divided into five geographic regions. Primates occur in three of them, South Asia, Southeast Asia, and East Asia. Most of this chapter is devoted to Southeast Asia, where the majority of today's primates live.

The geography of tropical Southeast Asia is unlike any other region where primates are found (Figure 6.1). It spreads across a continental landscape, a long peninsula, and an extensive archipelago comprised of thousands of islands with a range of sizes, from extremely large to very small. Most of the Indonesian Peninsula and several major islands, including Borneo and Sumatra – the third and sixth largest islands in the world, respectively – and Java form a distinct biogeographic region.

The main primate habitats in Southeast Asia are located in the extensive closed-canopy tropical rainforests of the Malaysian Peninsula and the large islands of Sumatra, Java, and Borneo. Together, these areas encompass the third-largest primate-inhabited rainforest ecosystem in the world, following the Amazon and Congo basins. Dozens of primate species representing six major primate families and subfamilies are found in the region: lorisids, tarsiers, cercopithecine and colobine monkeys, lesser apes, and two, or possibly three, great ape species.

These primates are found mostly in the tropical and subtropical regions of southern Asia, which stretches from the Indian subcontinent in the west to China in the north to the Indonesian archipelago in the east. Our focus is on primates living on the mainland

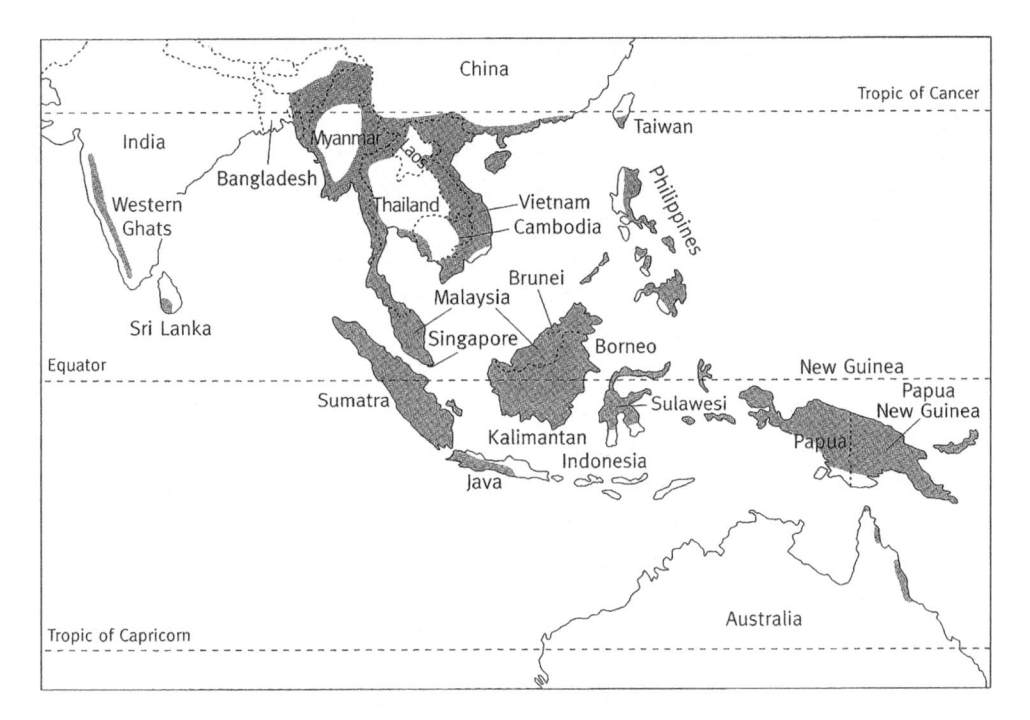

Figure 6.1 Geographical location of rainforests, shaded, in South and Southeast Asia.
From Corlett, R., & Primack, R. (2011). *Tropical Rain Forests: An Ecological and Biogeographical Comparison* (2nd edition). New York, Wiley-Blackwell.

and island rainforests of Southeast Asia, a distinct biogeographical zone (Figure 6.1). A few species also live farther north where temperatures are cooler, winters with snowfall occur, and forested habitats are temperate rather than tropical or subtropical, producing far less floral and faunal diversity. The temperate forests are dominated by deciduous broad-leaf trees that lose their leaves seasonally and evergreen **coniferous** trees that have needle-like leaves. Of particular significance to primates is the manner in which seeds grow there. Most coniferous trees' seeds are contained in woody cones rather than inside fleshy fruits that are the core of many primate diets.

Southern Asia's climate, and its forest habitats, are strongly influenced by seasonal rains originating over the Indian and Pacific Oceans, in addition to the moisture-trapping hills in India and the Himalayan mountains, which contain some of the highest peaks on Earth. The weather of tropical Asian rainforests is rainy, hot, and humid. Some areas have a dry season lasting four to eight months. The equator cuts through the archipelago, making it perpetually warm, moist, lush, and biologically diverse. These forests and others in Southeast Asia comprise more than a half-dozen of the world's terrestrial biodiversity hotspots.

Elsewhere in Southern Asia, fewer than 20 primate species live in deciduous woodlands and montane forests, including the semi-terrestrial macaques, the most versatile and flexible primate. One, the Japanese macaque, *Macaca fuscata*, also known as the snow monkey, is wide-ranging. They are found in deciduous subtropical forests and mountainous forests in Japan that are characterized as a subarctic biome. During harsh winters when little edible vegetation is available, these monkeys forage for fallback foods by reaching into a river, catching and eating fish, aquatic plants, and aquatic insects and mollusks. On the Indian subcontinent, the rhesus macaque, *M. mulatta*, is found in urban areas as well as natural habitats. Tens of thousands of rhesus macaques are estimated to live in New Delhi, India. The Hanuman or gray langurs, *Semnopithecus entellus*, are habitat generalists that are also part of an urban landscape and ecology.

The Asian radiation: lorises, tarsiers, Old World monkeys, and apes

Four of the seven major groups of primates occur in Asia. Three of these are found both in Africa and Asia, lorises, Old World Monkeys, and apes. The fourth, tarsiers, are endemic to Asia.

The Asian fauna constitutes a separate biogeographic assemblage, though it shares family- and subfamily-level taxa with the primate radiation of mainland Africa. The Asian primate fauna, compared to Africa, has:

- A complex geographical distribution across continental landscapes and many islands, large and small
- Less available open country within the tropical and semi-tropical zones
- A distinctly different taxonomic composition of Old World monkeys and apes:
  - Five Asian colobine genera and one cercopithecine genus – a higher proportion of leaf-eaters and fewer omnivorous frugivores – compared with two or three African colobines and about 12 cercopithecine genera
  - One genus in common with Africa, the macaque, *Macaca*
  - Lesser apes, gibbons and siamang, not found in Africa
  - Great apes represented only by the endemic arboreal orangutan genus

Asia: lorises, tarsiers, Old World monkeys, and apes 147

- Fewer small and medium-size species (weighing less than 5 lbs., 2.3 kg) and more larger species (weighing roughly 10–20 lbs., or 5–10 kg, and more)
- Fewer terrestrial or semi-terrestrial monkeys and no semi-terrestrial apes
- Communities with fewer sympatric monkey species
- Tarsiers and hylobatids, two major extant endemic taxonomic groups that occur only in Asia
- A haplorhine genus, the tarsier, and two lorisid genera, strepsirhines – two distantly related nocturnal lineages – that are the most predatory faunivorous primates, verging on carnivory
- Three major taxa that have radically distinct locomotor morphologies and behaviors
  - Extreme vertical-clinging-and-leaping in the small tarsiers
  - Brachiation in the large gibbons and siamang
  - Quadrumanous climbing in super-large orangutans

Table 6.1 summarizes some of the main features describing the groups of primates that inhabit both Asia and Africa. More detail is given in Chapter 5 in the profiles describing lorises and galagos, Old World monkeys, and apes.

Ecologically, folivorous primates predominate in Asia. Most of the Old World monkeys are leaf- and seed-eating colobines. Adding to the high representation of Asian leaf-eaters is an ape, the siamang, *Symphalangus*, also a folivore. Arborealists are far more common in Asia. There are only two genera that are ground-dwellers. One is the only Asian cercopithecine genus, *Macaca*, that includes several terrestrially inclined species. Another is a colobine, the semi-terrestrial langur genus, *Semnopithecus*.

Asian lorises – nocturnal stalking predators

Two endemic Asian genera, *Loris*, the slender loris, and *Nycticebus*, the slow loris, comprise the lorisine subfamily, informally called lorises, represented by about nine species.

Table 6.1 General features describing the three main primate groups that appear in both Asia and Africa

General features	Lorises	Old World monkeys	Apes
Size	Small to medium	Medium to extra-large	Super-large
Activity rhythm	Nocturnal	Diurnal	Diurnal
Domain	Arboreal	Arboreal, semi-terrestrial, terrestrial	Semi-terrestrial, arboreal
Diet	Fruit, insects, gums	Fruit, leaves	Fruit, leaves
Locomotion	Quadrumanous climbing, vertical-clinging-and-leaping	Quadrupedalism	Knuckle-walking quadrupedalism, quadrumanous climbing, brachiation
Sexual dimorphism	Monomorphic	Dimorphic	Dimorphic
Communication	Primarily by scent	Visual and vocal signals	Visual and vocal signals
Social organization	Semi-solitary	Unimale-multifemale, multimale-multifemale	Unimale-multifemale, multimale-multifemale, pair-bonded
Reproduction	Singletons and twins	Singletons	Singletons

They are the only members of the lorisid family that live in Asia. Other lorisids, the pottos and angwantibos, are African.

Asian lorises have a distinctive outward appearance. They are small-to-medium in size, weighing about 3 oz. to 4.5 lbs. (~85 g –2 kg). Their heads have a rounded shape because their eyes are very large, close together, and forward-facing. Their faces seem very short because they are overshadowed by the huge eyes, and the muzzle ends in a small, pointy nose. Their ears are also small (Figure 6.2d). The slender loris has thin forelimbs and slim hindlimbs. The slow loris has a more robust appearance.

The genus *Loris* has the largest eyes of any primate other than tarsiers, which are Asian haplorhines. Their eyes bulge out from the eye sockets and are supported underneath and on the side by a very wide, ribbon-like postorbital bar (see Figure 4.3) that extends outward from the cheek to wrap around the eyeball. The extra bony support for the super-enlarged eye is similar to that of tarsiers, which also have an eye socket with extra support to accommodate their enormous eyes.

The Asian lorises are wholly arboreal and live in dense forest habitats. *Loris* is highly faunivorous, while species of *Nycticebus* variously rely on fruits, gums, nectar, and prey. They both tend to have fur that is dull in color, browns and grays on the back and lighter shades on the underside. In several species, a jagged-looking dark stripe runs along the back. The facial markings of lorises, with light-colored patches and/or stripes, draw attention to the eyes even in darkness.

Through an evolutionary phenomenon called mimicry, slow lorises developed a remarkable specialization, a venomous bite. No other primates have a venomous bite. It complements one of the most unusual primate adaptive complexes.

Mimicry occurs when one species evolves imitations of another species' anatomy or behavior because it provides a selective advantage. In this case, slow lorises have evolved a form of mimicry, called Müllerian mimicry. It simulates the anti-predator warning signals of the carnivorous, venomous spitting cobra. These cobras live in the same region as lorises, are nocturnal, and also hunt in trees. Essentially, the signals mean "stay away" or risk a toxic, potentially lethal bite. Both cobras and lorises benefit from sharing the same pattern. So, if a predator learns to avoid one, it will stay away from the other (Figure 6.2).

Cobras have a pair of markings that are visible on the back of the head, called eyespots, and a threatening pattern of morphology and behavior visible from the front, when the snake rears up to strike (Figure 6.2). It then produces a fearsome hood, flaring the skin of its neck and arching its head forward into attack position. Seen from above, the conspicuous eye patches of the slow loris and its long midline stripe of fur on the back that runs from the neck to the rump present striking resemblances to the appearance and shape of an attacking or locomoting cobra. As the loris ambles along quadrupedally by twisting its way forward, its winding stripe and head resemble the snake – the bent body shape in attack mode or the serpentine locomotion that propels a set of eyespots. Head-on, the large dark eyes of the slow loris face, framed by light-colored fur, resemble the hood of the spectacled Javan cobra: a pair of large black spots bordered by a field of white. This protective system appears to be effective, and may be the reason why slow lorises can be seen walking near predatory animals, even when carrying an infant.

The slow lorises are one of roughly ten species of venomous mammals. Observations and experimental research show that the slow loris produces its venom by mixing saliva with secretions licked from a gland near its elbow. The poison is delivered by biting. The venom is sufficiently potent to cause extreme pain in a human who is bitten and a severe

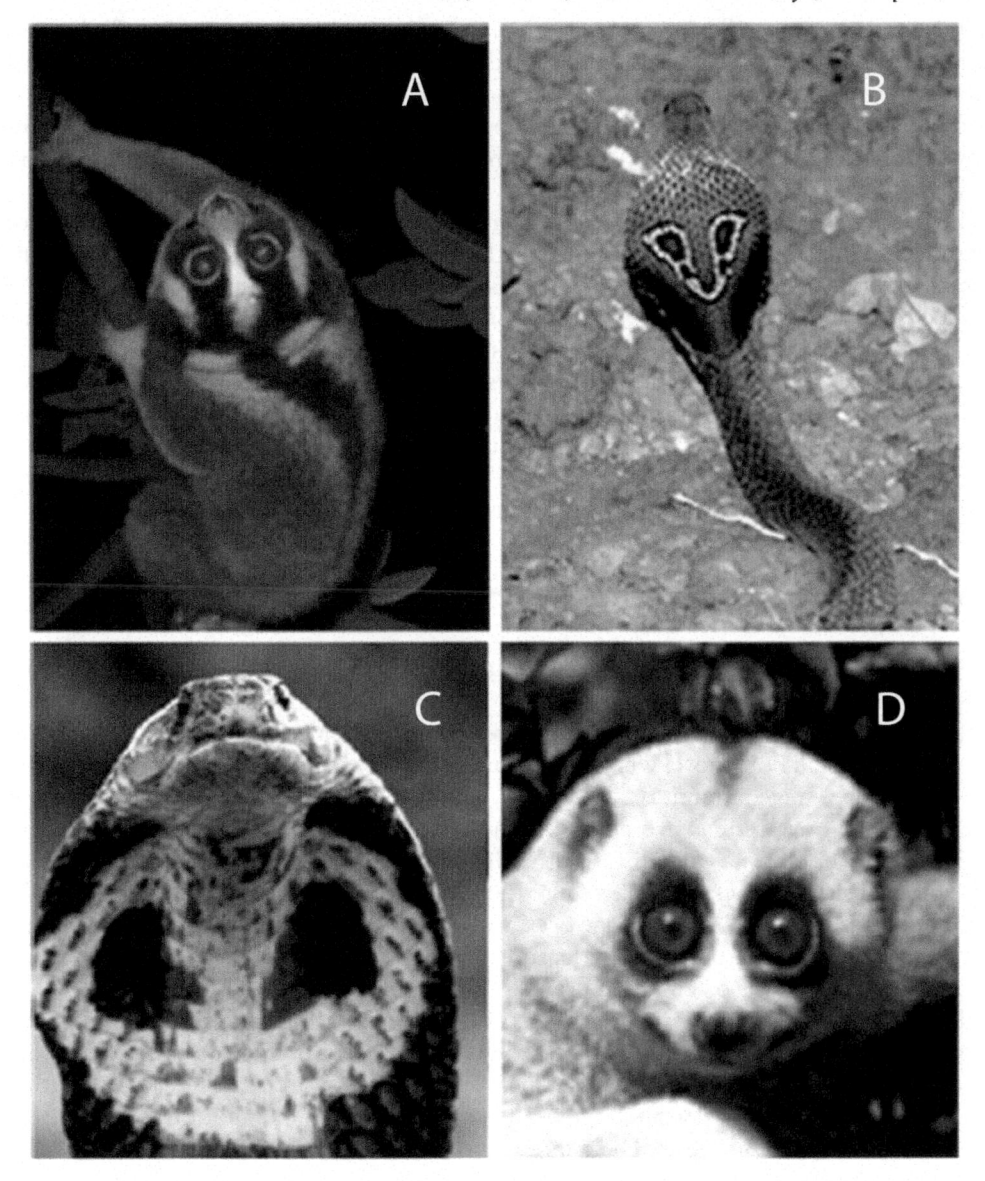

Figure 6.2 Mimicry of the Javan slow loris (a, d) and the spectacled cobra (b, c) seen from above (a, b) and in a head-on view (c, d).

Courtesy of K. A. I. Nekaris.

allergic reaction (anaphylactic shock) that can be deadly. Lorises that fight with one another also inflict severe, sometimes fatal, bites.

Tarsiers – nocturnal ambush predators

Tarsiers are the most isolated primates, geographically, phylogenetically, and historically, and the most adaptively divergent. They are a group found on many Southeast Asian islands, on larger ones like Sumatra, Borneo, Sulawesi, and Mindanao in the Philippines, and on tiny islands as well. The taxonomy advocated here recognizes only a single genus, *Tarsius*, the only living members of the tarsiid family. Estimates of the number of *Tarsius* species vary from four to over a dozen. They constitute a small radiation of highly uniform primates that are all small, weighing roughly 2–5 oz. (~60–140 g), and they only exist in tropical rainforests.

Tarsiers are unique. Few aspects of tarsier biology resemble any other primates, from head to tail, fingers to toes, legs, ankles, teeth, body size, growth patterns, diet, locomotion, sociality, and communication. Their extensive set of divergent adaptations are all connected with the tarsier's ecological specialization as a nocturnal vertical-clinging-and-leaping, hyper-faunivorous primate that is essentially carnivorous.

It is important to emphasize that tarsiers are haplorhines (Figure 1.10). In spite of their unusual, extreme adaptations, tarsiers share several phylogenetically and adaptively important features with anthropoids, including:

- The dry haplorhine nose with a continuous upper lip and associated facial musculature
- A "daytime retina" with a dense concentration of photoreceptor cells of a certain type (cones) located in a pit (fovea) where the eye functions in highest resolution
- A particular kind of placenta

The tarsiers' position is on the visual side of the olfactory-vision tradeoff that distinguishes strepsirhines from haplorhines even though tarsiers are nocturnal. A functional reduction in the sense of smell is indicated by external nasal morphology and the structure of the internal bony nose, which is anatomically reduced, making room for the orbits which take up a huge amount of space in the snout (Figure 6.3). Tarsiers are thoroughly nocturnal, yet their retinas are not like other nocturnal primates. The retina, like the retina of anthropoids, has a high-resolution foveal pit that works best in bright light, and it is color-sensitive. It lacks the reflecting tapetum lucidum of nocturnal strepsirhines that amplifies dim light. Instead, the huge eyes are an adaptation to maximize the amount of light that is cast on the retina. These features indicate that the nocturnal tarsiers evolved from a species that was originally diurnal.

Tarsiers have the largest eyes, relative to body size, of all primates and all mammals (Figures 1.6 and 6.3). Each eye is larger than the size of the tarsier's brain. They also have the longest foot of any primates. The tarsier foot is more than 80% the length of the trunk. The tarsier's vertical-clinging-and-leaping style of locomotion is powered by hyper-long feet and long legs. Their large feet and large hands are also important for grasping supports when landing after a jump, and the big hands are beneficial in capturing prey.

Tarsier eyes are permanently fixed in the face-forward position. To change the direction of its gaze, the tarsier must turn its head because their eye muscles cannot move the eyeballs. Tarsiers can turn the head around nearly 180° to set up their line of sight. This simultaneously orients the ears toward the same focal point, enhancing the coordination of the visual and auditory systems when finding prey. Swiveling the head to such a degree is made possible by a specialized morphology of the cervical vertebrae. It nearly doubles the roughly 90° right-left range of motion that the primate head is typically capable of by allowing the neck to twist as well.

Being able to rotate the head around is an important adjunct to the vertical-clinging-and-leaping locomotor system. It enables a tarsier to scan the surroundings behind its

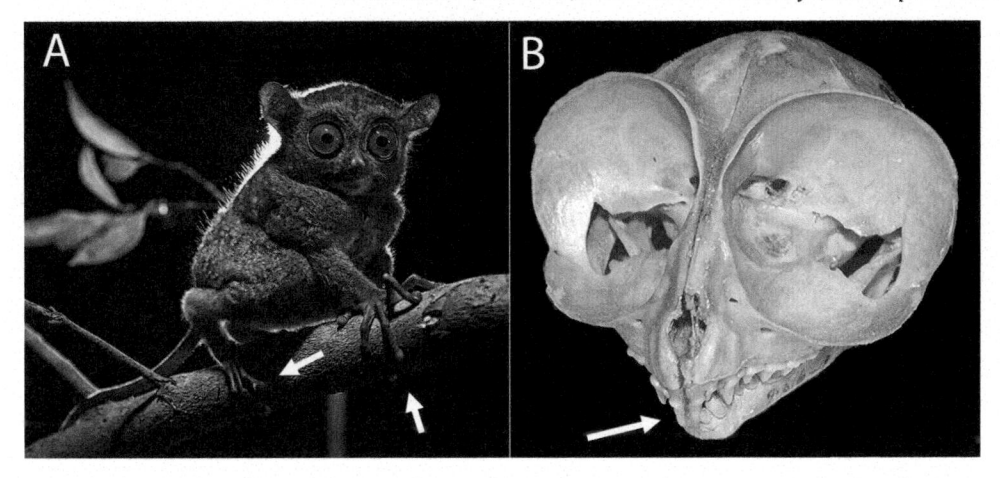

Figure 6.3 (a) Horsfield's Tarsier in Indonesia (arrows point to large hands and feet). (b) skull of a tarsier showing the enormous eye sockets that support the eyes, and large, pointy upper central incisors (arrow).

Photo credit: (a) *Tarsius* Fransiskus Simbolon.jpg., by Fransiskus Simbolon (CC-BY-SA 4.0); (b) Tarsier skull.jpg, by Andrew Bardwell (CC-BY-SA-2).

back. With eyes and ears forming a sensitive audiovisual tracking system, the tarsier can face in the opposite direction from its prey while waiting in ambush. After a backward leap, the tarsier spins around midflight to catch an insect or pounce on a grounded frog.

- **Phylogeny.** Tarsiers are the haplorhine sister-group of anthropoids.
- **Body mass and shape.** The small-bodied tarsiers present a compact shape, with a short trunk and a large, round head. The eyes bulge from their bony sockets and take up nearly all of the face, but there is a long row of large cheek teeth underneath the huge orbits and a prominent pointy snout. Tarsier ears are large and leathery, capable of twisting and folding into various shapes to capture sound. The tail is also proportionately long and hairless near the rump. While sitting in the vertical-clinging position, tarsiers often bend that part of the tail against vertical substrates to form a three-point support system together with the clutching feet. Toward the tail's tip, some species develop a tuft or vane of fur that has aerodynamic qualities.
- **Activity rhythm.** Tarsiers are strictly nocturnal and active throughout the night.
- **Ecological domain.** Tarsiers are arboreal, and prefer the lower levels of the canopy, especially the understory where relatively narrow-diameter, vertical supports are abundant, and where they can easily land – then take off – when pouncing on prey.
- **Diet.** Tarsiers are exclusively predaceous, and more carnivorous than any other primates. They do not eat vegetation of any kind. They are excellent hunters. Some of the insects that tarsiers eat are almost half their body length, excluding the tail. The vertebrates they eat include birds, bats, frogs, lizards, snakes, and mice.
- **Locomotion.** Tarsiers are the most specialized vertical-clingers-and-leapers. The tarsier's extreme version can be summarized by a set of three numbers: (1) the intermembral index, 52–58 (Box 3.1), showing the extraordinarily long hindlimbs relative to the forelimbs; (2) the ratio of foot length to trunk length, 83%; (3) to strengthen the lower leg, the lower part of the fibula is fused to the tibia for about 2/3 of its length.

- Tarsiers are the only living primates in which a long length of the lower leg bones is fused. Tarsiers have long toes (and fingers) with large disc-like touch pads which provide non-slip grips that hold fast while the leg and hindfoot extend fully to generate take-off thrust.
- **Craniodental morphology.** The cranium of tarsiers accommodates specializations connected with their vision, hearing, posture and locomotion, and feeding. Enormous orbits are the dominant feature. The housing for the middle ear, the petrosal bulla on the underside of the cranium, discussed in Chapter 2, is huge, which contributes to sensitive hearing. The connection between vertebral column and cranium is located in the middle of the head, in a biomechanically efficient location for balancing the head when the trunk is orthograde while the animal clings. The skull is very thin-boned, light in weight. The cheek teeth are very large, with tall, pointed cusps and well-developed shearing crests. The canines are low-crowned. The anterior teeth are unique in shape, arrangement, function, and number. The dental formula is 2.1.3.3/1.1.3.3. It lacks one of the lower incisors. The incisor teeth are vertical, tall, and pointy. By way of natural selection, tarsiers pack more specialized features into the head than any other primates.
- **Pelage.** Tarsiers are drably colored in greys and browns, with few distinguishing, species-specific patterns.
- **Sexual monomorphism.** Tarsiers are sexually monomorphic in all respects. There are no differences between males and females other than in the reproductive organs.
- **Communication.** Vocalization and olfaction play important roles in communication, to different degrees among different species. Tarsiers mark surfaces with urine and scent mark by rubbing trees using the ano-genital region and a chest gland. Their high-pitched vocalizations reach the ultrasonic range, above the frequency that human ears can hear.
- **Social systems and mating.** Group sizes are small and vary in composition among the species. Both unimale-unifemale and multimale-multifemale groups have been observed, based largely by identifying the sex and relative ages of individuals found sleeping together. Social groups tend to use a small number of sleeping trees in primary forests. In some species, mated pairs foraging apart use vocal duets to find one another. When together, they may duet to announce their joint presence to neighboring groups.
- **Reproductive pattern.** Tarsiers give birth to singletons. Wild adult females are reported to have interbirth intervals of 12–19 months.
- **Parenting styles.** Females are the sole care providers. Infants are carried by their necks in their mothers' mouths for the first three months. Afterward, they are parked in a nearby tree as she forages. Males do not transport or physically interact with young. Neither males nor females share food with offspring after mothers stop nursing them.

With so many aspects to their biology uniquely divergent, the tarsier niche has its own place in the ecology of Southeast Asian primates, which are mostly comprised of anthropoids. Nocturnal, small, active mostly in the understory, eating only insects and vertebrates and no vegetation at all, the needs of tarsiers are vastly different from nearly all co-occurring primate species. So, while they are found on many islands that are also inhabited by other primates, *Tarsius* has only a small interactive role in today's primate communities.

There are a few non-anthropoids in Southeast Asia that are somewhat similar to tarsiers ecologically, like slow lorises. Comparing how the Bornean tarsier, *T. bancanus*, and the Bornean slow loris, *Nycticebus menagensis*, another faunivore, live in sympatry provides additional insight into the tarsier niche.

The key to their ecological separation is stratification, meaning in what part of the forest are the animals most active. Lorises are found in taller microhabitats, higher up in the forest layers, roughly between 30 ft. and 100 ft. (9–30 m). Tarsiers live in areas where trees are not as tall, and they are most active less than 10 ft. (3 m) above the forest floor. These patterns reflect differences in locomotion and prey-capture methods. The slow-climbing lorises quietly stalk prey and require a continuity of substrates. They do not leap between branches. The dense canopy provides the lacework of supports they need to move about, including relatively thick branches on which they can move quietly by grasping quadrumanously with their specialized hands and feet.

Tarsiers pounce on their prey from a distance, grabbing them and then leaping back to a tree to eat. Thirty-five percent of the insects that a tarsier eats are snatched in mid-air, and nine percent are caught on the ground. *Tarsius* requires low, small-diameter vertical supports to perch and the openness of the understory as a hunting ground where prey is revealed. In this way, vertical separation provides an important dimension by which these faunivorous primates maintain ecologically differentiated niches.

Hylobatids – brachiating lesser apes

Hylobatids are the only catarrhine family endemic to Southeast Asia. They are a family of hominoids, apes, that includes gibbons and siamang. As hominoids, their detailed profile and cladistic relationships among the catarrhines were discussed in Chapter 5 (Figure 5.3). Hylobatids are known as the lesser apes because they are smaller in size than the other ape family, the hominids, composed of Asian orangutans, and African chimpanzees, bonobos, and gorillas. The three gibbon genera, *Hylobates*, *Hoolock*, and *Nomascus* are smaller than the siamang, *Symphalangus*. The three gibbons weigh 12–15 lbs. (5.4–6.8 kg), whereas siamang weighs roughly 24 lbs. (11 kg).

In hylobatids, the principal form of traveling locomotion depends almost entirely on the forelimbs – hand-over-hand brachiation. Their style of brachiation is the most acrobatic form of arboreal locomotion of any mammal. It is fast and highly efficient biomechanically. At times it involves a free-floating phase during which there is no contact between the body and any support. Using powerful arms, hylobatids launch themselves into the air when moving from one tree to another across a large gap, unlike other primates that leap across openings propelled by the hindlimbs. Hylobatids move bipedally when walking on branches. Due to the disproportionately long forelimbs and hands, they are not able to be quadrupedal. When walking bipedally, hylobatids balance by raising up the arms with elbows flexed, as if walking a tightrope – it also keeps the forelimbs from hindering locomotion. Their legs are also long relative to the trunk, which makes them excellent climbers, but hindlimb proportions tend to be overshadowed by the extreme length of the forelimb. With the longest arms and forearms of any primate, the range of intermembral indices for hylobatids is 126–147; the forelimb is roughly 25%–50% longer than the hindlimb. Hylobatids are the only primates in which the order's hindlimb-dominated locomotor paradigm is reversed (Figure 5.7).

Gibbons are frugivores but leaves are also important in their diet. Ripe fruits make up roughly 60% of the diet, followed by leaves (30%), other vegetation, and insects. Siamang have roughly the opposite pattern: 50% is composed of leaves and 45% is from fruit.

The specialized skeletons of these exclusively arboreal primates have far-reaching locomotor and postural significance for the biology of hylobatids. They enable access to all parts of the canopy for multiple purposes, using brachiation, climbing, and suspension. The animals eat and rest in suspended postures, copulate while hanging by the arms, and vigorously brachiate during behavioral displays.

As a traveling adaptation, brachiation is speedy and energetically efficient, advantageous to hylobatids feeding on ripe fruits that are widely dispersed in the forest and may be attractive to other, competing primates. Their suspensory locomotor and postural behaviors are supported by a large ensemble of anatomical features that are extensions of the hominoid pattern, relating to shoulder mobility and muscular power to operate the arms in overhead positions, ability to extend the elbows fully, and wrists that have enhanced maneuverability. The long arms are a fundamental advantage, by extending an animal's reach when collecting food. Hand anatomy, with its long, curved fingers, is distinctive in facilitating different grasping modes, like a four-fingered hook-like clasp over a branch while brachiating. The thumb is also offset by a deep cleft in the soft tissue that increases the spread between it and the lateral digits, thus maximizing the efficacy of a full-hand grasping action when applied to a wide support, with the thumb facing the palm.

A suspensory feeding posture is an advantage in a larger-bodied primate that moves from the sturdy inner branches of a tree outward into the thin branches to feed. There, for a large above-branch quadruped, such as a macaque, the slender peripheral branches tend to bend under the animal's weight, risking a fall even while it sits, and requiring active countermeasures to prevent it. The further out on a limb the macaque goes to reach a fruit, the harder it becomes to maintain balance. The hylobatid's ability to hang from or drop to a lower branch, even one that bends, offers a safer and more energetically efficient method. By combining above-branch sitting with below-branch hanging, hylobatids double the feeding sphere at a single feeding location, thus saving the energy required to move to a different branch, as the quadrupedal macaque must do when reaching its tipping point.

Hylobatids have a rounded skull, with a distinctly short face, shallow lower jaw, and large circular orbits that are widely spaced and prominently rimmed by bone. Consistent with a frugivorous morphology, incisor teeth are robust and wide, and cheek teeth are bluntly cusped with large crushing basins. Hylobatids are monomorphic in most features. Both sexes have a tall, dagger-like upper canine and a sectorial p3 that reciprocates the upper canine. Male and female gibbons are often distinguished by pelage color, or markings on the face and head. The pelage of siamang is monomorphic, always black.

Many hylobatid species are characterized by one of the primates' most distinctive vocalization patterns, the whooping daily dawn call, frequently called song, that reverberates through the forest. It is often performed at territorial boundaries by a mated pair as an extensive duet that lasts as long as 15 minutes, tightly coordinated as they harmonize and also sing their own parts in alternating three-minute solos. Both males and females have enlarged throat sacs that are filled with air and then expelled to produce the calls.

Hylobatids form pair-bonded groups composed of a single adult male, a female, and offspring, although other arrangements, possibly involving groups that are in transition, have been observed. Functionally, these small social groups operate as restricted foraging and feeding parties that also defend a territory that contains an adequate supply of ripe fruits. Small cohesive social groups are more manageable than large diffuse ones in species prone to cover long distances rapidly. The pair-bonds are not necessarily exclusive, long-term sociosexual relationships. Females may affiliate with non-resident males and extra-pair copulations, matings outside the pair-bond, do occur. Over several years, new adult pairings also form due to death, departure of an adult, or displacement by an outsider. To recognize these dynamics, their system has been called serially monogamous. It does not always rely on sexual or reproductive fidelity year-after-year, but individuals nevertheless maintain pair-bonded relationships.

With adults inclined to acrobatic locomotion, unlike most primates that carry offspring on their backs, hylobatids' newborns and infants are carried ventrally by the mother, on her abdomen. When the young are about two years old, they are weaned. Then, among siamang, the group's male and resident subadults may also carry a youngster occasionally. Generally, male hylobatids are supportive in playing with, grooming, and protecting offspring.

Southeast Asian primates living in communities: diversity, ecology, and adaptation

The primate community at Kuala Lompat in the Krau Wildlife Reserve, Malaysia, our case study of niche differentiation in this chapter, is comparable in composition to primate communities that are found across Southeast Asia. The Krau is peninsular Malaysia's largest reserve and home to a diverse tropical flora and fauna, including a rich ensemble of primates. Kuala Lompat has been an important biological research station for decades.

The primate community is composed of seven sympatric species belonging to six genera (Table 6.2). The two cheek-pouched macaque species are the only congeners. Two genera of colobines are recognized, *Presbytis* and *Trachypithecus*. Together with the genus *Semnopithecus*, these Asian colobines are commonly called langurs. The gibbon and siamang, *Hylobates* and *Symphalangus*, are the other catarrhine genera at Kuala Lompat. Other Malaysian primates found there and/or in comparable communities include the slow loris, *Nycticebus*, the only strepsirhine; the orangutan *Pongo*; and the tarsier, *Tarsius*.

As we found in Madagascar, South America, and mainland Africa, the major ecological features that distinguish these Asian primates within their communities – activity cycle, body mass and sexual dimorphism, microhabitat preferences as reflected by locomotor pattern, and diet – correspond with their phylogenetic positions (Table 6.2). The ecological niche partitioning that evolved both within and between clades in each of the regions also follows the same patterns.

The slow loris – a nocturnal gumivorous frugivore-faunivore

The non-overlapping, ecological separation of *Nycticebus coucang*, the slow loris, from all other sympatric primates at Kuala Lompat is determined by activity rhythm, body mass, diet, and locomotion. They are nocturnal primates, by far the smallest species. They rely on a distinctive combination of foods, and employ a singular pattern of locomotion reflecting their preference for a particular microhabitat in the canopy.

- **Diet and locomotion.** Fruit, gum, and animal prey are important parts of the slow loris diet, but it relies on other vegetation as well, including nectar, leaves, flowers, tree gum, and tree sap. Sap is a type of exudate found beneath bark, deeper into the tree than where gum is located. Their locomotor style, slow-climbing

Table 6.2 Ecological characteristics of Kuala Lompat primates

Family/subfamily	Common name	Genus and/or species	Body mass Male	Body mass Female	Locomotion	Microhabitat	Diet
Lorisinae	Slow loris	*Nycticebus coucang*	Small	Small	Slow-climbing quadrupedalism	Arboreal	Faunivorous
Hylobatidae	**Lar gibbon**	*Hylobates lar*	Large	Large	Brachiation	Arboreal	Frugivorous
Hylobatidae	**Siamang**	*Symphalangus syndactylus*	Extra-large	Extra-large	Brachiation	Arboreal	Folivorous
Cercopithecinae	**Long-tailed macaque**	*Macaca fascicularis*	Extra-large	Large	Quadrupedalism	Arboreal	Frugivorous
Cercopithecinae	**Pig-tailed macaque**	*Macaca nesmestrina*	Extra-large	Large	Quadrupedalism	Terrestrial	Frugivorous
Colobinae	**Dusky langur**	*Trachypithecus obscurus*	Large	Large	Quadrupedaism	Arboreal	Folivorous
Colobinae	**Sumatran langur**	*Presbytis melalophus*	Large	Large	Quadrupedalism	Arboreal	Folivorous

Note: The activity cycle of all species is diurnal, with the exception of the nocturnal slow loris.

quadrumanous quadrupedalism, serves both as a stealthy means to surprise prey and a method to avoid being detected by predators. At a small body size, this method of locomotion is only suited to the peripheral branches of the canopy where a virtually continuous, dense set of supports is available. None of the other co-located primates use twigs as a major avenue for dynamic locomotion though hylobatids efficiently feed on fruits and leaves in the terminal branches using suspensory postures.

- **Social systems and mating.** Little is known about the social systems and mating habits of wild slow lorises. They are semi-solitary so their small-group lifestyle does not significantly impact the ecology or behavior of the more social primates living in larger groups. Adult males and females with offspring travel separately but maintain some level of contact with one another via olfactory and vocal signals. Adults are also known to sleep together in pairs.

Cercopithecids – frugivores and folivores

There is a sharp ecological distinction between the two genera of cercopithecines and colobines at Kuala Lompat (Figure 6.4) based on diet and preferred microhabitat,

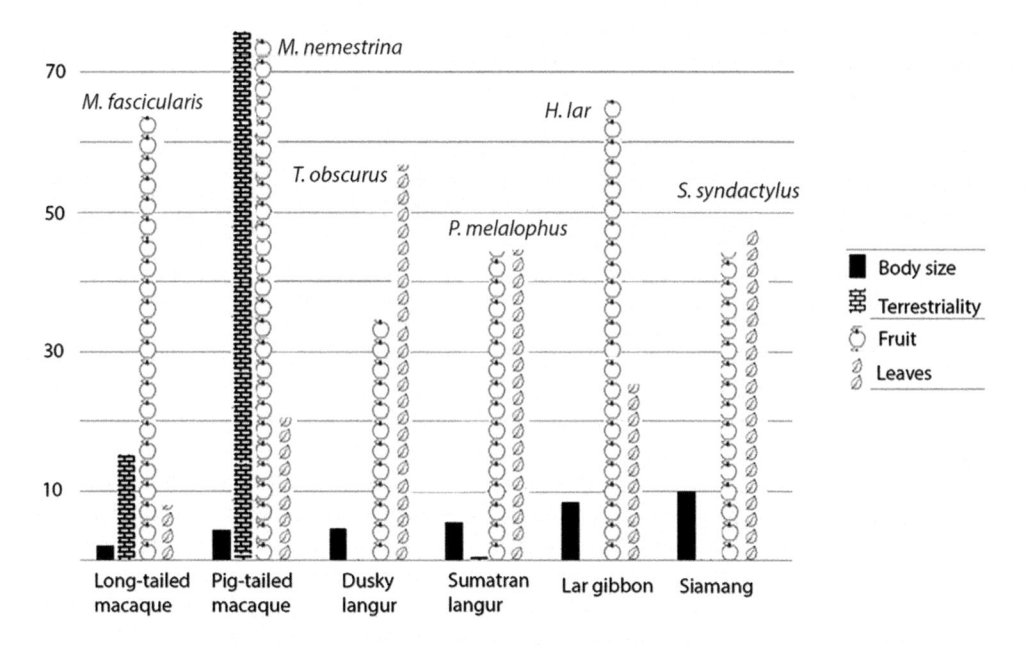

Figure 6.4 Ecological characteristics of six sympatric monkeys and apes at Kuala Lompat. The vertical axis represents the average annual percentages of time devoted to ground use, fruit- and leaf-eating, and in the case of body mass, species weights measured in kilograms. Abbreviations: H, *Hylobates*; M, *Macaca*; P, *Presbytis*; S, *Symphalangus*; T, *Trachypithecus*.

Adapted from Chivers, D. J. (2013). Southeast Asian primates: Socio-ecology and conservation. *Raffles Bulletin of Zoology*, Suppl. 29, 177–185 with additions from other sites: Bartlett, T. Q. (2015). *Seasonal Variation in Behavior and Ecology, CourseSmart eTextbook*. London, Routledge. https://doi.org/10.4324/9781315664088).

whether it tends toward terrestriality or is exclusively arboreal. It is important to note that ground use here refers to the forest floor, not to open-country habitats.

- **Diet and locomotion.** The macaques are frugivorous, and their niches are well separated ecologically. Roughly 60%–75% of their feeding time is devoted to fruits, while 20% or less is spent on eating leaves. The remainder involves feeding on animals. However, the two species differ in size, degree of terrestriality, and habitat usage. The pig-tailed macaque, *Macaca nemestrina*, is roughly twice the size of the long-tailed macaque, *M. fascicularis*, and spends much more time on the ground. It focuses its activities in the forest, as opposed to the edges of forest preferred by *M. fascicularis*. It travels farther each day, and occupies home ranges that are about three times larger. The long-tailed macaques are more arboreal and they spend the majority of their time in the upper reaches of the canopy, which is where the colobines *Presbytis* and *Trachypithecus* devote most of their feeding and traveling time. Arboreal competition is minimized because the colobines are folivores while *M. fascicularis* is a frugivore with omnivorous tendencies.

  The two langur species spend most of their time feeding on leaves. The leaf-to-fruit balance of their diets differs, as does their preference for younger or older leaves, certain tree species, and emphasis on seed-eating. What separates the two most markedly is microhabitat. The Sumatran langur, *Presbytis melalophus*, is a forest-edge species whereas *Trachypithecus obscurus*, the dusky langur, favors intact forest habitats. When examined quantitatively, their postural and locomotor patterns also differ, in ways that correlate with osteological and muscular differences. The dusky langur tends to feed in the continuous upper canopy and travels mostly via large branches in quadrupedal mode. The Sumatran langur favors less continuous parts of the forest, including the understory, uses leaping more often, and also engages in more climbing, arm-swinging and suspensory feeding. The anatomy of the Sumatran langur's hips and knees tend to resemble primate leapers, and they have larger hindlimb muscles to provide more leaping power.
- **Social systems and mating.** There is another basic distinction between the macaques and langurs that is both ecological and phylogenetic. As is generally the case with cercopithecines, the macaques live in large multimale-multifemale groups, range widely in the forest, feed independently while troops are spread out, and are more omnivorous as a correlate. The size of feeding groups is only loosely limited by food. It varies according to the immediate distribution, availability, and variety of foods that the macaques eat opportunistically. That is why the larger macaque groupings at Kuala Lompat, which average more than 30 individuals in the terrestrial *Macaca nemestrina*, are more flexibly organized in a fission-fusion manner, while also being shaped by the typical cercopithecine form of social regulation, dominance and hierarchy. In macaques generally, males and females mate with multiple partners. But high-ranking males do not try to maintain exclusive access to ovulating females.

  The langurs, like many Asian colobines, live in smaller unimale-multifemale groups. At Kuala Lompat, groups are composed of 15–17 individuals. The animals have relatively small daily ranges, and individuals can feed in close proximity when harvesting relatively abundant leaves. Troops are also territorial, and use loud calls as a spacing mechanism, like other colobines.

  The structure of Asian colobine social groups revolves around females. Females do most of the grooming and they initiate sex. It is common for langur mothers to share parenting with other adult females, who also hold, groom, and carry infants for substantial periods. Other juveniles may provide assistance, too. Males do little, other than providing protection when necessary. Adult females are flexibly attached to individual adult males. Females transfer between social groups and also interact with lone males.

Hylobatids – brachiating frugivore-folivores

There are two sympatric lesser apes living at Kuala Lompat, the lar gibbon, *Hylobates lar*, and the siamang, *Symphalangus syndactylus*. Their niches are differentiated from the other primates in their community by their brachiating body plan, which enables them to make efficient use of the high canopy and its terminal branches for travel and feeding, without much competitive interference. Among the Old World monkeys living in this primate community, only the dusky langurs show a preference for the high canopy while the other species tend to use the lower crown and/or forest floor.

- **Diet and locomotion.** *Hylobates lar* and *Symphalangus syndactylus* are both frugivore-folivores (Figure 6.4). What differentiates their niches at Kuala Lompat and elsewhere is the ratio of leaves and fruits that compose their diet. *Hylobates lar* is a ripe-fruit frugivore and *Symphalangus syndactylus* is a folivore (Figure 6.4). The gibbon eats many more fruits than leaves and prefers ripe fruit. The siamang eats about equal amounts of fruits as they do leaves. Siamang are able to subsist on a leafier diet than the gibbons, partly because of their larger size and proportionately larger guts. The dentition of the more folivorous siamang is essentially that of a frugivore.
- **Social systems and mating.** The male-female nucleus of the pair-bonded hylobatid social group is relatively stable and remains intact over the course of several birthing seasons, which may come 2–3 years apart. Newborn gibbons rely heavily on their mothers for food and assistance during the first two years of life. But in siamang, males take over caregiving duties after the first year. As noted, males and females jointly defend territories within their home range. The territory is usually located in a highly productive area within a larger home range. Over the long term, this means that a family-group's offspring come to broadly know the area that sustains them, which often enables dispersing females to pair with a male in or near their formative home range when establishing a new social group.

Other Southeast Asian primates

Macaques and Hanuman langurs

Representing two different cercopithecid subfamilies, the cheek-pouched macaques and leaf-eating Hanuman langurs that live outside the Krau Reserve at Kuala Lompat are noteworthy for their ecological flexibility.

The macaques are the world's most widespread primate genus, found in Asia, north Africa, and Gibraltar, in Europe. They are an extremely successful genus, with more than 20 species ranging from the Indian subcontinent in the west, to the whole of Southeast Asia and Japan in the east, and to China in the north. Ecologically versatile, macaque habitats vary from tropical to semi-desert to cold temperate zones, from flatland to the foothills of the Himalayas. Species can be arboreal, semi-terrestrial, or fully terrestrial.

Macaques are omnivorous but focus on fruits, except for species that are habitat specialists and favor leaves when fruits are in short supply. Insects are also important foods, and the more terrestrial species may eat roots and herbs.

Asian macaques are medium or large in size and can be quite sexually dimorphic. Those that inhabit urban areas sometimes live in large troops nearing 100 individuals. In natural habitats, social group sizes and systems vary considerably among the species. They maintain complex multimale-multifemale groups. Larger groups may split up into smaller foraging parties. Male dominance hierarchies play a large role in structuring relationships, in conjunction with strong bonds among females that are based on kinship. These **matrilines** form dominance hierarchies among the females in a social group.

The Hanuman langur, *Semnopithecus entellus*, also known as the gray langur, is one of about a half-dozen species of *Semnopithecus*. It is an extra-large monkey that is considerably dimorphic in size: males can be nearly 40% larger than females. They are semi-terrestrial, flexible feeders.

Their behavior has been intensively studied since the 1970s, when the species became a prominent early example of the phenomenon of infanticide, which has since been reported in many primate species and other mammals where males or females are known to kill conspecific infants. These langurs often live in unimale-multifemale groups, and resident males disperse to find mating partners when sexually mature. The dispersing male, or a band of males, typically force their way into another sociosexual group by overpowering the resident alpha male.

The takeover may involve killing one or more infants that are part of that troop. Without an infant to feed, a nursing mother's physiological status changes and her ovulatory cycle automatically restarts, which means she soon becomes a potential reproductive partner for a new dominant male.

The odd-nosed monkeys

There is a group of Asian arboreal colobines that do not occur at Kuala Lompat. They are known as the odd-nosed monkeys. They are geographically widespread and comprised of four genera: snub-nosed monkey, *Rhinopithecus*; Douc langur, *Pygathrix*; proboscis monkey, *Nasalis*; and the pig-tailed langur, *Simias*. The common names used for the first two genera and the group as a whole reference the appearance they all share, the shapes of their noses. Their noses range from being short and upturned, as in *Rhinopithecus*, to being extremely large and sexually dimorphic in *Nasalis* – bulky and drooping downward in males, smaller and prominently jutting forward in females (Figure 6.5). The odd-nosed monkeys tend to be large and extra-large in size, robustly built, good leapers, and many species are quite dimorphic. The proboscis monkeys favor habitats near rivers and swamps, and they are good swimmers. They are highly selective feeders and prefer young leaves, unripe fruit, and seeds.

Rhinopithecus, the snub-nosed monkey, includes species that live in mountainous, temperate coniferous forests in China and elsewhere where winters are severe. Temperatures may drop well below freezing, and it snows. They forage in the trees and on the ground and have an unusual diet. In the trees, they eat buds, flowers, young leaves, and seeds, as well as tree bark and resins. Tree resin is a viscous liquid that is present in bark. On the ground during winter, they eat tree roots, fallen nuts, dry grass, and invertebrates that they find under rocks and in decaying wood. Arboreal lichens, which may be rich

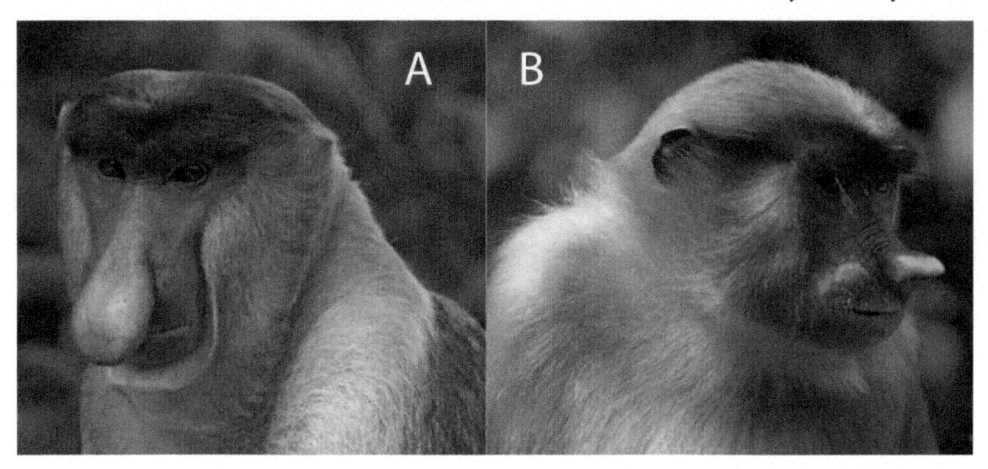

Figure 6.5 Male (a) and female (b) proboscis monkeys illustrating sexual dimorphism in nose size and shape.

Photo credit: (a) Proboscis monkey (*Nasalis larvatus*) male Labuk Bay 2.jpg, by Charles J. Sharp (CC-BY-SA-4/0); (b) Proboscis monkey (*Nasalis larvatus*) female Labuk Bay.jpg, by Charles J. Sharp (CC-BY-SA-4.0).

in carbohydrates, also make up a substantial part of their diet year-round, and particularly as a **fallback food** during the winter when other items are scarce and the ground is covered in snow. Lichens are curious organisms that may have a leaf-like, bushy or sponge-like morphology. They develop through a biological interaction between fungus and algae. Since they are not plants, lichens can grow under harsh conditions, with limited sunlight and warmth. Other mammals that subsist on lichen during the cold and snowy winter include reindeer.

Orangutans – the largest arboreal mammal

A number of the distinctive features of orangutans were reviewed as part of the hominoid profile in Chapter 5. With a typical estimated female-male weight range of about 65–190 lbs. (~30–86 kg), the long-haired red orangutans are the largest arboreal mammals in the world. Two species are usually recognized, one in Borneo, *Pongo pygmaeus*, and the other in Sumatra, *P. abeli*, though some scientists maintain there is a third species that also lives in Sumatra, *Pongo tapanuliensis*. These primates are slow-growing and long-lived – their lifespans can exceed 50 years.

Orangutans are extremely sexually dimorphic in weight, canine size, skeletal features, and outward appearance. Adult males are well over twice the weight of females. Males also increase in weight as they age, some reaching 220 lbs. (100 kg). There are two categories of adult males, and they play different social roles as discussed below. As they mature, they exhibit distinctive features which set them apart, a phenomenon called **bimaturism**. Bimaturism involves differences that develop among individuals of the same sex, unlike sexual dimorphism, which refers to differences that develop between the male and female sexes. The expression of bimaturism in orangutans is pronounced in their facial appearance. In addition to having an enlarged throat sac, some dominant adult males develop a large, flabby flange of skin that frames the face. They are called flanged males. The flanged males also develop longer, shaggy hair, distinctive beards, and are prone to

Figure 6.6 Flanged (a) and unflanged (b) adult male orangutans.

Photo credit: (a) Orang Utan in Tanjung Puting National Park, Kalimantan.jpg, by George Arif Kurniawan (CC-BY-SA-4.0); (b) Orang Utan lives in Tanjung Puting National Park.jpg, by Charlsmile (CC-BY-SA-4.0).

emitting loud long calls. The others are called unflanged males. They are fully adult, subordinate to flanged males, and lack the embellishments (Figure 6.6).

The preferred habitats of orangutans are lowland rainforests, inland swamps, and areas near rivers and lakes. They are highly intelligent; they score very high in laboratory tests of their cognitive ability when compared with other primates. Like chimpanzees, orangutans make tools to probe for insects. They use sticks to break off sections of pineapple-like, large durian fruits; to pry seeds from hard fruit or extract honey from a beehive; and to scratch themselves. And, they repurpose leaves for a variety of functions, like self-cleaning and protection from the rain. Different orangutan populations exhibit their own variations in tool-use behaviors.

Adult orangutans build tree nests nightly for sleeping. Nest-building involves a complicated process that begins with site selection, a choice of branches that may be transported from elsewhere, and a particular construction method. Sometimes, several relatively small trees are bent and fastened together to form a foundation. In the final stages of assembly, an orangutan braids leafy branches together to form a mattress, stands on it to test its strength, and then arranges a variety of enhancements that produce a pillow, blanket, roof or an extra platform above the original one. The entire process may take 7–9 minutes. Nests may also be put together for a midday rest, taking less time to construct. Daytime nests are reused or rebuilt more often than nighttime nests.

- **Diet and locomotion**. Orangutans are frugivores, but they eat other vegetation, including tough materials like the inner bark of trees, leaves, and seeds, and small invertebrates. They prefer fruits with soft pulp that grow in large crops. When available, such fruits can comprise 100% of the orangutan diet. As large arboreal mammals that require a large volume of food, their diet is strongly influenced by the region's rainforest phenology. In Southeast Asian forests, a very large percentage of the major tree families have long periods when they do not bear fruit, followed by a massive synchronous production, called **mast fruiting**. These events occur every 2–10 years, depending on the area's annual climate pattern. This means that orangutans must at times contend with prolonged food shortages, when less nutritious items become important fallback foods. Some populations of orangutans then switch to feeding on bark, stripping it directly from tree trunks and twigs with their powerful jaws and well-developed anterior dentition. In spite of periodic food scarcity, the massive size attained by some orangutans is actually an advantage. Their ability to gain body weight, to store fat that is convertible into energy, buffers them against food shortages when fruits are rare and they must rely on lower-quality staples.

Orangutans exhibit a distinctive craniodental and skeletal anatomy. The orbits are circular and closely spaced. Incisors are wide and the cheek teeth are covered in thick enamel. Powerful jaws that can open in a large gape, large incisors, and reinforced molars are advantageous in accessing and eating fallback foods like hard fruit and bark, and ingesting large chunks of tough durian fruit. The craniodental adaptations are essentially a hard fruit, frugivorous design.

Postcranially, the forelimb is very long, about 39% longer than the hindlimb, the hands and feet are large, and the shoulder and hip sockets are very shallow, which maximizes limb mobility. This pattern facilitates quadrumanous arboreal locomotion. The ability to reach out at all angles to grab supports with hands and feet to maintain balance while moving is crucial to such a large, heavy tree-dwelling primate. In spite of these features, locomoting male Sumatran orangutans have been observed to accidentally break branches and suffer falls about twice a day. Orangutans also use the ground for traveling, especially the large flanged males, and when feeding on termites, drinking water, and eating soil – not an uncommon practice among primates – which is a source of minerals and may have pharmaceutical properties.

- **Social systems and mating.** Orangutans are the only semi-solitary anthropoids. They may be the least social anthropoid primates, and can spend weeks without interacting with a conspecific, although individuals appear to keep track of one another in the forest. This pattern is a social adaptation influenced by their very large size, extreme dimorphism, and the potential for intra-specific food competition, especially when resources are scarce for prolonged periods. Males live alone and females live with their young until they age-out of the natal group. The home ranges of males overlap with the smaller ranges of several females simultaneously. Females actively maintain connections with flanged males, often by approaching them after hearing their loud calls. Aware of one another, in this way a flanged male is in a position to protect a female from roaming unflanged males.

Paleocommunities

A large fossil ape, *Sivapithecus*, has been a focal point of primate paleontology in Asia. *Sivapithecus* was an important part of mammalian paleocommunities in India, Pakistan, and Nepal during the 8–12-million-year interval of the Miocene Epoch. The Miocene Epoch is a period when other primates, including strepsirhines, other apes and Old World monkeys, were present in the region. The genus is part of the orangutan clade and very similar anatomically to the living orangutan.

The habitats of mammals that comprised paleocommunities associated with *Sivapithecus* in Pakistan, including species from more than ten families of mammals, were investigated through chemical analyses of elements absorbed in dental enamel. This showed how they were ecologically differentiated on the basis of food and microhabitat. Additionally, the fossil samples were selected from two intervals that were about a million years apart, bracketing the extinction of *Sivapithecus* in the fauna at 8.4 million

164 Asia: lorises, tarsiers, Old World monkeys, and apes

years ago. During both periods, both forested and open habitats were utilized by different mammals. Between 9.2 and 9.3 million years ago, when *Sivapithecus* was most abundant, it may have fed in the upper canopy of a closed, moist forest. By 8.1 million years ago, rainfall decreased, seasonality increased, the forests had become fragmented, an open-country grassland ecology dominated, and the mammals that preferred the wetter environment became extinct, including *Sivapithecus*. This ecological turnover in the ecology of the mammal paleocommunity in Pakistan corresponds with the time when now extinct Old World monkeys became prolific. The phylogenetic connection and similarity of *Sivapithecus* to living orangutans indicates that these apes have been part of Asian primate communities for at least 12 million years.

The conservation status of Asian primates

Asia ranks a close second as the region with the highest proportion of threatened primate species, following Madagascar (Figure 6.7). Species living on Madagascar are five times more at risk, and in Asia they are three times more at risk, than primates living in mainland Africa or the neotropics. In Madagascar, 71% of species are Critically Endangered or Endangered. Overall, according to the IUCN SSC Primate Specialist Group, 77% of all Asian primate species are now threatened, and nearly 60% are Critically Endangered or Endangered. By far, the orangutans are most at risk, with all three species classified as Critically Endangered.

Habitat loss contributes most to the precarious conservation status of Asian primates. Deforestation to support soybean cultivation, palm oil production, and mining are the

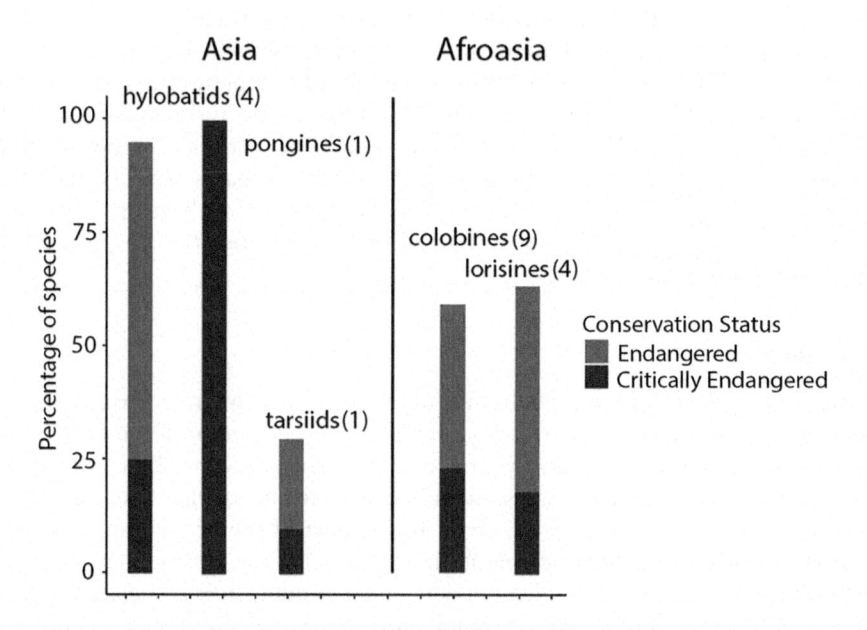

Figure 6.7 The threat levels to five families and subfamilies of Asian and Afroasian primate species. Numbers in parentheses refer to the number of genera in each group.

Adapted from Fernández, D., Kerhoas, D., Dempsey, A., Billany, J., McCabe, G., & Argirova, E. (2021). The current status of the world's primates: Mapping threats to understand priorities for primate conservation. *International Journal of Primatology*. https://doi.org/10.1007/s10764-021-00242-2.

most severe threats. Deforestation to accommodate small-scale agricultural needs accounts for roughly one-third of the total loss.

The conservation status of Asia's primate populations is being monitored, and vigorous efforts to mitigate the threats are ongoing by primatologists and conservationists, and by international partnerships. The Gunung Palung Orangutan Project is one model of an international, multi-faceted, science-driven program to protect and preserve orangutans and their habitat in western Borneo. About 5,000 orangutans live in and around the Gunung Palung National Park where the project is located. Using innovative technologies like aerial drones, as well as traditional field methods, research conducted since 1992 has examined the phenology of the forest, the ecological needs and behaviors of the orangutans, their nutritional status, health, stress hormones, and more. The population densities of orangutans living in primary and degraded forest habitats are determined by surveys of nests that the animals build in trees and by using drones to investigate areas that are difficult to reach on foot. The aerial drone surveys also help monitor habitat loss. The Gunung Palung Orangutan Project places a strong emphasis on local community involvement and education.

See Chapter 9 for more on threats to primates globally and conservation efforts.

Key concepts

Müllerian mimicry
Mast fruiting
Matrilineal hierarchies
Bimaturism
Ecological turnover

Quizlet

1 What are two of the habitats in Southern Asia where macaques are found?
2 Why are hylobatids, the only catarrhine family endemic to Southeast Asia, known as lesser apes?
3 Hylobatids have the longest forelimbs of any primate. How are the forelimbs used in their principal form of traveling locomotion?
4 Describe the one nocturnal strepsirhine genus and one nocturnal haplorhine genus that live in Asia.
5 Compare the phenomenon of bimaturism with sexual dimorphism.
6 Describe how an adult orangutan builds a nest for sleeping.
7 Orangutans are the only semi-solitary anthropoids. What are the features of this social adaptation?

Bibliography

Bartlett, T. Q. (2015). *Seasonal Variation in Behavior and Ecology, CourseSmart eTextbook*. London, Routledge. https://doi.org/10.4324/9781315664088
Chivers, D. J. (Ed.). (1980). *Malayan Forest Primates*. New York, Plenum.
Chivers, D. J. (2013). Southeast Asian primates: Socio-ecology and conservation. *Raffles Bulletin of Zoology*, Suppl. 29, 177–185.
Corlett, R., & Primack, R. B. (2005). Dipterocarps: Trees that dominate the Asian rainforest. *Arnoldia*, 63(3), 2–7.

Delgado, R. A., Jr., & Van Schaik, C. P. (2000). The behavioral ecology and conservation of the orangutan (*Pongo pygmaeus*): A tale of two islands. *Evolutionary Anthropology: Issues, News, and Reviews, 9*(5), 201–218. https://doi.org/10.1002/1520-6505(2000)9:5<201::AID-EVAN2>3.0.CO;2-Y

Gursky, S. L. (2007). *The Spectral Tarsier*. Upper Saddle River, Pearson.

Knott, C. D., Kane, E. E., Achmad, M., Barrow, E. J., Bastian, M. L., Beck, J., Blackburn, A., Breeden, T. L., Brittain, N. L. C., Brousseau, J. J., Brown, E. R., Brown, M., Brubaker-Wittman, L. A., Campbell-Smith, G. A., de Sousa, A., DiGiorgio, A. L., Freund, C. A., Gehrke, V. I., Granados, A., ... Susanto, T. W. (2021). The Gunung Palung Orangutan Project: Twenty-five years at the intersection of research and conservation in a critical landscape in Indonesia. *Biological Conservation, 255*, 108856. https://doi.org/10.1016/j.biocon.2020.108856

Munds, R. A., Ali, R., Nijman, V., Nekaris, K. A. I., & Goossens, B. (2013). Living together in the night: Abundance and habitat use of sympatric and allopatric populations of slow lorises and tarsiers. *Endangered Species Research, 22*(3), Article 3. https://doi.org/10.3354/esr00556

Nekaris, K. A. I., Moore, R. S., Rode, E. J., & Fry, B. G. (2013). Mad, bad and dangerous. *Animal and Toxins Including Tropical Diseases, 19*(1), 21. https://doi.org/10.1186/1678-9199-19-21

Nelson, S. V. (2007). Isotopic reconstructions of habitat change surrounding the extinction of *Sivapithecus*, a Miocene hominoid, in the Siwalik Group of Pakistan. *Palaeogeography, Palaeoclimatology, Palaeoecology, 243*(1), 204–222. https://doi.org/10.1016/j.palaeo.2006.07.017

Ruf, T., Streicher, U., Stalder, G. L., Nadler, T., & Walzer, C. (2015). Hibernation in the pygmy slow loris (*Nycticebus pygmaeus*): Multiday torpor in primates is not restricted to Madagascar. *Scientific Reports, 5*(1), Article 1. https://doi.org/10.1038/srep17392

Takenaka, M., Hayashi, K., Yamada, G., Ogura, T., Ito, M., Milner, A. M., & Tojo, K. (2022). Behavior of snow monkeys hunting fish to survive winter. *Scientific Reports, 12*(1), Article 1. https://doi.org/10.1038/s41598-022-23799-1

Xiang, Z.-F., Huo, S., Xiao, W., Quan, R.-C., & Grueter, C. C. (2007). Diet and feeding behavior of *Rhinopithecus bieti* at Xiaochangdu, Tibet: Adaptations to a marginal environment. *American Journal of Primatology, 69*(10), 1141–1158. https://doi.org/10.1002/ajp.20412

7 Primate communities compared
Ecology, morphology, and behavior

Chapter Contents

Arboreal frugivorous primates in ecological context	168
The rainforest biome	170
Tropical rainforests are not all the same	172
Madagascar	172
Amazonia	172
Africa	175
Southeast Asia	176
The arboreal, frugivorous epicenter of primate life	177
Convergent evolution of morphology	178
Phylogeny is the starting point	178
Body mass, size, and diet	180
Size and diet	180
Giants and dwarfs	181
Nocturnal, diurnal, or cathemeral	182
Diet, dentition, and gut	183
Tree-gougers	183
Enamel caps and bilophodont molars – hard-fruit eaters	183
Food-handling	184
Fat storage	184
Locomotion and posture	184
Leapers	184
Hangers, climbers and clamberers, and brachiators	185
Primates without tails	186
Cranial morphology	186
Parallel evolution in platyrrhines – grasping tails	187
Social behavior	188
Morphology and behavior	189
Primate breeding systems	192
Categories and subcategories of breeding systems	192
Convergent evolution of primate societies	194
Social dispersal	197
Key concepts	198
Quizlet	198
Bibliography	198

DOI: 10.4324/9781003257257-7

In each of the four regions where they are found today, primates and their communities exhibit a significant amount of convergence in morphology and behavior, based on a fundamental body plan and genetic structure that developed in the arboreal frugivorous adaptive zone. In this chapter we discuss the ecological synergy between primates and the rainforest. We compare and contrast the rainforest environments and how the primates have adapted to them.

The lemurs in Madagascar and the platyrrhines in South America are adaptive radiations that developed in isolation. Madagascar is an island and South America had been an island for most of platyrrhine history until tectonic changes connected it to Central America several million years ago. The Old World catarrhines also result from a single adaptive radiation, but it grew and spread among three continents, Africa, Asia, and Europe, as the fossil record shows (see Chapter 8). The primate radiations of Madagascar, the neotropics, and Afro-Eurasia each evolved from a unique ancestral species or a distinct clade, meaning they each had a separate phylogenetic beginning. Each began with species characterized by a particular set of adaptations pertaining to a certain ecological niche.

In the two regions where primates were isolated, Madagascar and South America, and in the two where they were not, Africa and Asia, we find that in the adaptive radiations the extant species partition their niches along similar patterns in the ranges of diet, locomotion, body size, social organization, and the other factors that describe a primate profile. How they divide available food resources leads to resemblances in their dentitions and digestive processes. How they use the strata of the three-dimensional forest habitat, or the ground, influences their similarities in foraging systems and locomotor styles. There are similarities even among species that are adaptive outliers.

These same patterns, similar ranges, and many specific traits that evolved independently in each region are examples of convergent evolution, the evolution of the same features in different species that are not closely related. An example of convergent evolution is the locomotor style of vertical-clinging-and-leaping that evolved independently in all four regions in a number of genera and families in four of the seven major groups of primates.

Arboreal frugivorous primates in ecological context

Primates are an ecological cornerstone of tropical and subtropical rainforest systems. Their existence has important consequences for the structure and function of these ecosystems. As premier arboreal frugivores, primates perform critical functions that benefit the reproduction of countless rainforest trees and species. They do this by excreting the seeds of the fruits that they eat. Viable seeds are dispersed through the forest as the animals travel and forage. It is also common for primates to defecate near their sleeping trees before moving out in search of food. Critical interactions that connect frugivorous primates to the ecosystems in which they live result in, or are the consequences of, reciprocal adaptations of both the animals and plants that are linked through **coevolution** (Box 7.1).

Primates account for a very large percentage of fruits that are directly eaten from trees by rainforest vertebrates. Their value as seed dispersers has been proven by experimentally testing the viability of seeds to germinate after passing through a primate's gut, and the distance a primate travels after feeding on a particular tree species before defecating its seeds. As a consequence, primate ecology is tightly integrated within the rainforest

> **Box 7.1 Evolutionary principle: coevolution**
>
> Coevolution refers to the link between two (or more) interacting species that adapt to one another by the force of selective pressures that drive correlated changes in both. One form of coevolution has been likened to an evolutionary contest, an interlocking feedback loop in which each side periodically evolves adaptations to elude or counteract an advantage gained by the other side. This has been a feature of predator-prey relationships that is documented in the fossil record of many animals.
>
> A simple action-and-reaction evolutionary model applies to the interactions of animals and plants as well. For example, by eating leaves folivores reduce the ability of a plant's energy-producing, photosynthesis machinery, thus exerting selective pressure on the plant to avoid being damaged. A generalized counter-adaptation to limit the harm done by leaf-eaters (which include insects and many vertebrates) happened more that 125 million years ago. Trees began to evolve leaves that incorporate a diverse array of chemical compounds – which are unimportant in other functions – to deter folivores by producing a bad taste, a toxic reaction, or limiting digestibility. To get around these chemical defenses, primate folivores and other herbivores evolved gut adaptations to neutralize the chemicals so those leaves can be eaten.
>
> Another form of coevolution is mutually beneficial to the parties that are ecologically linked. Each performs a service that is advantageous to the other, so the process becomes fixed by natural selection. Seed dispersal of angiosperm plants by animals in the rainforests is a primary example.
>
> Plants use a variety of methods to deposit or spread their seeds. Seeds can simply drop to the ground, be blown away by the wind, or get eaten and swallowed by animals, and then excreted. Seeds that are dispersed by animals have a nutritious built-in attractant, food, the material that surrounds the seeds, which is often fleshy fruit pulp. The animals become vehicles that scatter seeds when they defecate during their daily rounds. That is beneficial to a parent tree because seeds that are moved away from the tree's crown have a better chance of germinating than those landing beneath it.

ecosystem through the mechanisms of seed dispersal and coevolution. Given the relatively large body-sizes of many species compared to other arboreal seed dispersers like birds and bats, primates are effective dispersers of large-seeded fruits in particular.

A study in French Guiana estimated that a single social group of six semi-folivorous, large-size howler monkeys disperses more than 1,000,000 seeds per year, many of which are very small, comparable to the size of strawberry seeds. This figure was calculated by observing the troop for two years, counting the average number of seeds (223) present in howler monkey feces of a standard weight (3.5 oz.; 100 g), and measuring the collective weight of the group's feces that is excreted each day (3.3 lbs.; 1.5 kg).

Only a fraction of the seeds dispersed by these howler monkeys were likely to develop into trees, but scale is important here. Many angiosperm trees naturally generate a fantastically large number of seeds as an adaptation to maximize their reproductive potential. It is also important to note that not all frugivorous primates, and not all fruit-eating bouts, offer dispersal possibilities. There are arboreal and semi-terrestrial primates that habitually chew seeds, eating them for their protein content. Some, like the South American saki and uacari monkeys, have specialized seed-crushing molars, an adaptation that can diminish the reproductive potential of individual trees. However, this does not appear to damage the overall productivity of the species that the monkeys feed on because of the enormous volume of seeds that the trees produce. Other frugivorous monkeys, and other fruit-eating vertebrates, eat from the same tree species but do not chew the seeds.

Why are there fewer primate folivores than frugivores?

Arboreal frugivory is the epicenter of primate life, but leaves, the most abundant forest vegetation, would appear to be the most convenient source of food to which arboreal primates have access. So, we may ask, why don't all primates subsist on leaves, to save them the energetic costs of traveling to locate edible fruits and other food? There are several reasons why specialized folivory works for some primates but not others, including:

- For any mammal, a balanced diet normally involves eating a varied menu to obtain all the necessary nutrients. It is very rare for a mammal to subsist entirely on the nutrients provided by leaves alone, or any single category of food.
- Since leaves do not provide large amounts of nutrients they must be consumed in bulk, which makes it incompatible with a small primate's body mass and gut.
- As food, not all leaves are equal because plants exhibit a wide range of adaptations in structure and chemistry that deter animals from eating them. Folivores require specialized adaptations in the gut to process or neutralize some chemical components.
- The pliable physical structure of leaves requires blade-like cutting or shearing edges on the crowns of the animal's cheek teeth, and larger teeth can have more of them. This involves: the growth of large teeth; the time needed for offspring to develop and set in place enough cheek teeth to sustain them while becoming dietarily independent; and the investment of growing and maintaining bone and muscle that can endure a lifetime of extensive chewing.
- A predominantly leafy diet does not eliminate the need to forage. The production and growth of leaves in the rainforest is not synchronized and nutritional quality differs among plant species and even individual trees, so choice leaves are not available everywhere at all times.
- In the life cycle of a leaf, it may take only a few weeks to mature but adult leaves may last on trees for years, while their texture changes. So, the majority of leaves present in the rainforest at any one time are older, tougher ones that typically require more specialized dental and digestive adaptations.

The rainforest biome

Tropical rainforests constitute the world's richest non-marine ecosystems, which is a principal reason why primates have diversified and flourished. Though they cover only 6% of the planet's land surface, tropical rainforests include two-thirds of a certain plant type that give them shape and generate vegetation that primates eat. They are called vascular plants because they have tube-like channels (vessels) that transport water and nutrients for growth and maintenance. They are built with woody tissues that house the channels and develop into tree trunks and branches, the skeletal structure of the rainforest, and they produce leaves, fruit, and flowers. The tropical rainforests are also characterized by

Primate communities compared: ecology, morphology, and behavior 171

a high density of trees that form a closed canopy of interconnected branches above the forest floor.

Angiosperms, the most prolific of all such plants, with many families, are dominant in tropical rainforests. They produce buds, flowers, fleshy fruits, nectar, and seeds in addition to leaves, pith, and gums, all of which are eaten by primates and other animals. The angiosperms generate and attract an immense biomass of potential vegetable and animal food for primates, essentially year-round. They also produce an environment that attracts many predators and organisms that carry disease.

Tropical rainforest trees vary in many ways, including: how high above the ground they grow; whether crowns form low down or high up on the trunk; how wide and interconnected are the crowns of adjacent trees; how densely packed are species and individuals locally and across the landscape (Figure 7.1). The proportion of trees that are

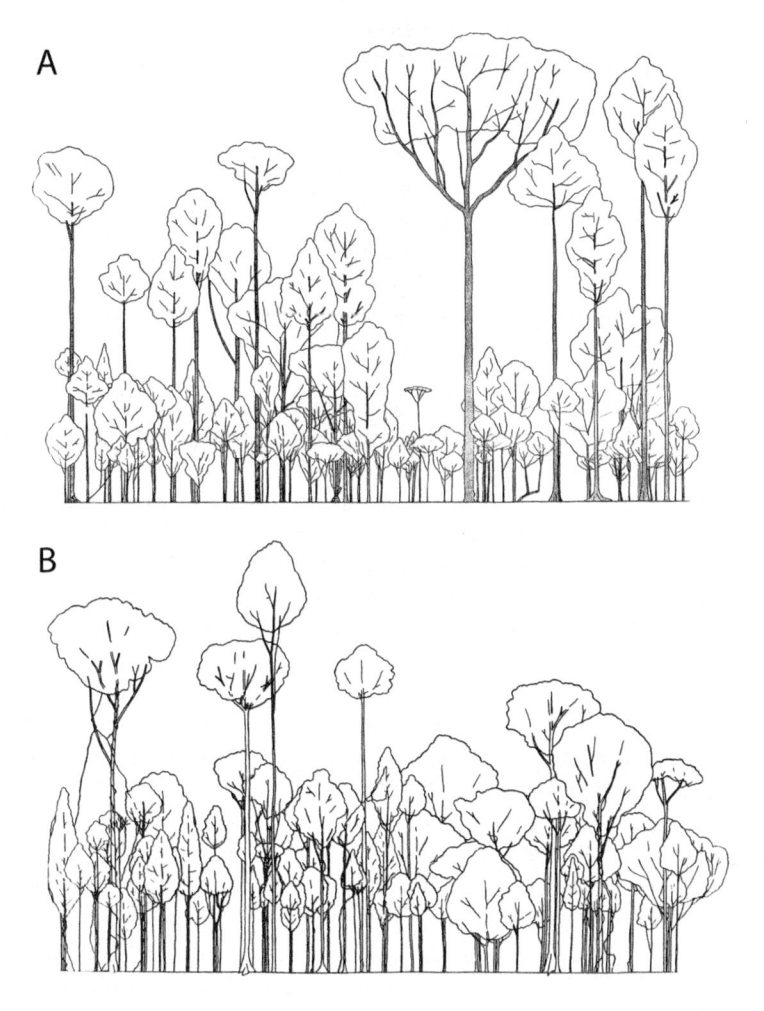

Figure 7.1 Structural profiles of tropical rainforests dominated by angiosperms in Nigeria (a) and by dipterocarps in Borneo (b). The understory and trees less than 15 ft. tall are not shown.

Adapted from Richards, P. W. (1966). *The Tropical Rain Forest: An Ecological Study* (1st edition). Cambridge, Cambridge University Press.

deciduous, that lose their leaves, is small. The tropical rainforests are mostly evergreen, flush with leaves throughout the year.

Tropical rainforests are not all the same

There are profound differences in the rainforests of the regions where primates live that have bearing on the composition of the primate communities, diversity, evolution, and adaptations. For a variety of reasons – paleontological history, biogeography, soils, topography, climate, and more – tropical rainforests are not all the same botanically. In general, the neotropics have the lushest flora and the most distinctive combination of plant families, while African and Asian forests are more similar to one another in composition. This reflects the long isolation of South America when it was an island continent, and the proximity between Africa and Asia that has influenced the intercontinental dispersal of plant species and ecosystems' spread.

Madagascar

Madagascar's rainforest ecosystems are distinctive in a number of ways. The rainforests cover the smallest geographical area by far. Relative to the other forests, there are few large-diameter trees, canopy heights are lower, and edible fruits are not plentiful. There are few plants that produce fleshy fruits, they are highly seasonal when present, and individual trees do not produce large fruit crops. A fallback food that many frugivorous primates in other regions resort to when preferred fruits are scarce, figs, are not well represented in Madagascar. Additionally, the protein content of Madagascar's fruits is lower than in those in the neotropics.

These factors explain why few lemurs, Madagascar's endemic primates, rely on fruits year-round, and why a large proportion of lemur genera have evolved several radical feeding and lifestyle adaptations. Madagascar has more ecological outliers than any of the other regions where primates live, both in the absolute number of genera and in their relative contribution to the adaptive radiation. These include the extractive insect-foraging aye-aye, the gum-eating fork-marked lemur, the bamboo-eating lemurs, the hibernating mouse lemur, and the semi-terrestrial ring-tailed lemur (Figure 7.2). These five taxa alone represent about one-third of the island's primate genera.

Amazonia

The Amazon in South America is a primary example of an equatorial tropical rainforest. It is particularly rich in towering tree species, and also many low-growing plants. Angiosperm plants predominate in the neotropics, where there are more than 25 endemic plant families and nearly 100,000 plant species. In Africa there is much less species variety and tree density, while Southeast Asia presents a fundamentally different botanical pattern. In Southeast Asian rainforests the entire ecosystem is dominated by one family of angiosperms, the dipterocarps.

The vertical stratification of the neotropical rainforest is another aspect of regional distinction that has consequences for primate ecology. There is abundant feeding-niche space that small primates exploit in the neotropics by using the canopy and the sub-canopy level, the understory. The understory is filled with flower-and fruit-producing shrubs and small fruiting trees, including a great variety of small-stature palms. In Southeast Asia,

Figure 7.2 A young bamboo lemur eating a bamboo stalk in Madagascar.
Courtesy of Jukka Jernvall.

stratification is quite different, in part because the structurally distinctive dipterocarps, as discussed below, are by far the most dominant family. There, the understory is essentially a nursery where dipterocarp saplings grow.

For primates, the enormous diversity of angiosperm trees in the neotropics makes South America a fruit-eaters paradise – angiosperm fruits are plentiful, edible, and they come in all sizes. This abundance and diversity also sustain a huge fauna of non-primate animals, invertebrates and vertebrates, that constitute another food resource for a platyrrhine, particularly among the predaceous cebids. Additonally, the Amazon supports a dense population of angiosperms that do not produce fleshy fruit but their seeds are eaten by specialized seed eaters, the sakis and uakaris. They occupy a feeding niche that has no counterpart among Old World anthropoids or strepsirhines (Figure 7.3).

The richness of South America's fruit crops is a factor that contributes to other distinctions of platyrrhine ecology. For example, there are no neotropical monkeys dedicated to folivory on the scale of the colobines, an entire radiation that relies on large, chambered fermenting stomachs to store and metabolize leaves in bulk, in addition to a leaf-eating dentition. Folivory is rare among New World monkeys. *Alouatta*, the howler monkey, is the only platyrrhine genus that has dental and gut adaptations for leaf-eating. They ferment in the hindgut, which is less efficient than the foregut system found in colobines.

The platyrrhine radiation lacks any ground-dwelling species. There are no terrestrial or semi-terrestrial South American primates, either inside the rainforest or outside in savanna-like terrains. This is a marked contrast with the radiation of African primates where there are semi-terrestrial monkeys and apes living in forests and fully terrestrial

Figure 7.3 A South American seed-eating uakari holding a fruit.
Photo credit: *Cacajao calvus novaesi*.jpg, by Paul Schlarman (CC-BY 2.0).

monkeys inhabiting open country, including deserts and mountainous plains. In this respect, the radiation of neotropical monkeys is more similar to Madagascar's lemurs, all arborealists, with the exception of the semi-terrestrial ring-tailed lemur.

These ecological factors help explain why New World monkeys are the most diverse living anthropoids. There are several reasons related to environment and habitat that make their diversity remarkable in comparison to the catarrhine Old World monkeys and apes:

- New World monkeys' adaptations are responses to a single exclusive biome, the tropical rainforest. Catarrhine diversity, on the other hand, reflects adaptations to a range of ecosystems, from tropical rainforest to near-desert environments, to wintry, snowy forests outside the tropical zone.
- Platyrrhine adaptive diversity results from selective pressures encountered exclusively in the tree canopy or the understory of the tropical rainforest, and does not include forest ground-dwelling adaptations in any individual genus or clade.
- The platyrrhine radiation developed on a single, geographically confining continent. South America is much smaller than and lacks the historical biogeographical connections that have linked mainland Africa and Asia (and in the past Europe, where primates lived for millions of years). Intercontinental connections in the Old World provided openings for geographical dispersals over an enormous spatial scale that presented new ecological opportunities and mingled flora and fauna.

Africa

The floral diversity of African rainforests is relatively small in comparison to the neotropics and Southeast Asia, and the forests are warmer, drier, more seasonal, more open at the ground level, and there are fewer tree species overall. This is an environment where the ability to forage in places beyond the canopy, on the ground, are advantageous. With a smaller variety of fruits and a smaller diversity of insect life attracted to fruits, there is limited potential for arboreal primates to exploit small-bodied, frugivore-insectivore niches.

The relatively limited botanical variety in Africa relates to the evolution of semi-terrestrial and fully terrestrial anthropoids, the array of papionin Old World monkeys and African apes (Figure 7.4). Competition for a limited range and supply of fruits may have imposed selective pressure among medium-to-large size catarrhines to exploit the ground where fallen fruits and nuts can be found, as temperatures began cooling about 14 million years ago and grasslands became widespread 10–6 million years ago at the expense of forests. The growing prevalence of more open spaces within the rainforests that enable many individuals to move and forage together may have influenced many new adaptations, including those relating to social structure. African Old World monkeys are noteworthy for having large social groups

Figure 7.4 An African, knuckle-walking, male silver-backed gorilla.
Photo credit: Goryl (14923279426).jpg, by Lukas Plewnia (CC-BY-SA 2.0).

Southeast Asia

Dipterocarps are the sole dominant angiosperm family in Southeast Asia. Several aspects of dipterocarp biology influence primate ecology and diversity. The trees grow tall and straight, with few lateral branches below the crown. They typically have a buttress root system that solidly anchors trees to the ground. So, when a tree dies, its canopy breaks down but the trunks tend to remain in place. This produces a rather homogeneous forest structure of living and standing dead trees, and a dark understory, because only a limited amount of light can penetrate to the ground far below a tree's crown. As a result, the subcanopy zone of the forest is not botanically diverse or biologically rich. In contrast, the more stratified Amazonian rainforests are taxonomically and structurally diversified, and treefalls are common. There, downed trees produce large open spaces of light that help generate localized patches of dense, low-growth vegetation, and an abundance of insects and small vertebrates that frequent this microhabitat.

The forest structure of Southeast Asia is less conducive to the lifestyle of small primates that are inclined to use sub-canopy microhabitats for feeding and locomotion, in comparison to South America, for example. This is one of the reasons why there are only three genera occupying small-body-size niches in Southeast Asia, none of which are anthropoids. The two lorises, *Loris* and *Nycticebus*, are fruit- and gum-eaters or highly predaceous faunivores, with typical lorisid musculoskeletal specializations designed particularly for canopy locomotion. With limited leaping ability, they are virtually shut out of the lowest stratum of the forest where leaping locomotion is required to navigate between narrow, vertical saplings that are spaced well apart. The other small-sized primate in Southeast Asia is *Tarsius*, a genus that eats no vegetation and is extremely specialized for living in the dim-lit understory as a nocturnal, vertical-clinging-and-leaping predator. Gaps in locomotor substrate matter less to a leaping tarsier than to a quadrumanous loris that hunts in the lacy twigs and branches of the canopy.

The tall, high-branching structure of dipterocarp trees, with emergents that can rise another 65–98 ft. (20–30 m) above the continuous canopy, is conducive to postural and locomotor styles of species at the large end of the primate body-size spectrum, with long limbs that benefit climbing. This physical setting may have influenced the evolution of the endemic hylobatids and their novel locomotor style, brachiation, which is a derivative of climbing. The huge orangutans are also specialized as quadrumanous climbers.

Most of the canopy fruits that are produced by dipterocarps are not attractive to primates because they lack the nutritious fleshy fruit pulp that primates eat. The basic structure of a dipterocarp fruit consists of a small, dry nut, which is the seed, with a wing-like structure attached to it. These are wind-dispersed seeds.

This means that Southeast Asian primates live in a region where fleshy fruits are limited. They are the most desirable and plentiful arboreal food types enjoyed by other anthropoids. The leaves of dipterocarps are also relatively unpalatable and hard to digest due to a high concentration of a bitter, acidic group of secondary chemical compounds called tannins. As a consequence, the anthropoid Asian radiation consists of few frugivores and many folivores (Figure 7.5). It is taxonomically dominated by the large-bodied, specialized, leaf-eating colobines. There is only one resident cercopithecine genus, the macaque, and several of its species are semi-terrestrial, frugivorous-omnivores.

The distinctive phenology of Southeast Asian dipterocarps also has a major influence on the feeding behavior of primates that makes these ecosystems more favorable

Figure 7.5 A leaf-eating silvered leaf monkey from Asia.

Photo credit: Silvered Leaf Monkey (*Trachypithecus cristatus*) (8127485334).jpg, by Bernard Dupont (CC-BY-SA 2.0).

to well-adapted folivores than to frugivores. Flowers and fruits are rare in a long-term sense because most dipterocarps do not flower and fruit annually like many other angiosperms. That means they are scarce during the lifetime of any long-lived anthropoid species. Large crops of fruits only become available in a 2–7-year interval, a phenomenon known as **mast fruiting**. During the intervening years, tropical Asian forests may seem like fruit deserts where only a few individual trees yield fruits at the same time. The combination of dipterocarp dominance and mast fruiting is an ecological disincentive to frugivory, which explains why the primate fauna is heavily biased in favor of folivores.

The arboreal, frugivorous epicenter of primate life

These regional ecological comparisons reinforce the idea that arboreal frugivory is the epicenter of primate life. In the ideal rainforest ecosystem, where fruits are varied, plentiful, and seasonally available, primates are essentially omnivorous frugivores balancing their diets with either leaves or insects and small vertebrates. When the ecosystem's fruit yield is less than ideal because it is neither diverse nor abundant, relatively low in protein, or scarce for extensive periods, arboreal primates have alternative feeding strategies, a core diet of leaves or seeds or gums or fauna, all of which are more challenging to access

178 *Primate communities compared: ecology, morphology, and behavior*

and process than fruit. These types of food have rarely become the dietary foundations of major primate clades. They usually occur only in a limited number of species in various taxonomic groups.

In light of the radical but successful ecological shift of the papionin Old World monkeys, the primates' only ground-dwelling adaptive radiation, scientists have long been puzzled why a similar non-arboreal transition did not also occur among New World anthropoids that appear to have comparable adaptive potential. One simple explanation for this is that there was no need – no selective pressure in neotropical communities for a genus or lineage to shift toward terrestriality – because there is little to be gained relative to what is available in the fruit-eaters paradise of the neotropics. There, the prolific fruit supply of the rainforest system, coupled with its physical structure, is the foundation that supports extensive niche differentiation in the arboreal realm, in the canopy and understory.

Primates in Madagascar and Southeast Asia are also not disposed to a thoroughly terrestrial lifestyle as a means to differentiate niches within their communities. In Madagascar, arboreal ecological separations are accomplished in other ways, frequently by eating radical diets, like toxic bamboo or wood-boring insect larvae. Only the ring-tailed lemur, which lives in a parched environment, is semi-terrestrial. In Asia, a few folivorous species have non-arboreal feeding habits but this only occurs seasonally, as a last resort, because the environments are more temperate than tropical and edible fruits are scarce for prolonged periods.

In Africa, the terrestrial adaptive radiation among the cercopithecines is a one-off evolutionary phenomenon. It emerged under a combination of novel baseline conditions – ecological, temporal, phylogenetic, and genetic – that are not repeated elsewhere, in other regions. Cercopithecine-like terrestriality is a model ecological adaptive strategy that cannot reasonably be applied to other primate adaptive radiations except in rare instances, like the case of subfossil lemurs in resource-poor Madagascar, discussed in Chapter 8. Arboreal frugivory remains the epicenter of primate life.

Convergent evolution of morphology

In this second part of our comparison of geographical regions and their primate communities, we look at the convergence of morphological traits that contributes to the adaptive resemblances among primates. Table 7.1 highlights examples that are organized largely according to the factors and functional-adaptive complexes addressed in previous chapters that define the adaptive profiles of taxa. Several of the categories are further discussed below.

Phylogeny is the starting point

Phylogeny is the starting point that sets the ecological, anatomical, and behavioral framework for the primate adaptive radiations. The adaptive varieties of primate communities and regional faunas are fundamentally influenced by phylogeny and the genes associated with relatedness.

Madagascar was colonized only by lemurs. They all belong to the olfactory branch of the olfaction-vision divide, which effectively is a **phylogenetic constraint** that strongly influences activity rhythm. So, nocturnal and cathemeral taxa dominate. The opposite is true of the neotropics, colonized by diurnal anthropoids. Only the owl monkeys exhibit

Table 7.1 *Examples of the convergent evolution of adaptations exhibited in a family or genus living in primate communities in the four geographical regions*

Traits and systems	Madagascar	South America	Africa	Southeast Asia
Body mass				
Small to medium	√			
Small to large		√		
Medium to super-large			√	
Large to super-large				√
Dwarfism	√	√	√	√
Gigantism		√	√	√
Activity rhythm				
Diurnal	√	√	√	√
Cathemeral	√			
Nocturnal	√	√	√	√
Ecological domain				
Arboreal	√	√	√	√
Semi-terrestrial	√		√	√
Terrestrial			√	√
Diet				
Frugivory	√	√	√	√
Folivory	√	√	√	√
Faunivory	√	√	√	√
Specialized seed-eating		√		
Gum-eating	√	√	√	√
Tree-gouging	√	√		√
Extractive foragers	√	√	√	
Bilophodont molars	√		√	√
Specialized fermentation	√	√	√	√
Locomotion				
Quadrupedalism	√	√	√	√
Quadrumanous climbing	√		√	√
Suspensory locomotion		√	√	√
Leaping locomotion	√	√	√	√
Foot-powered leaping			√	√
Claw- or nail-clinging	√	√	√	
Opposable thumbs		√	√	√
Thumbless hook-grips		√	√	
Shortened external tails		√	√	√
Absent tails	√		√	√
Pehensile tails		√		
Other				
Sexual monomorphism	√	√	√	√
Sexual dimorphism		√	√	√
High encephalization	√	√	√	√
De-encephalization	√	√	√	√
Tool use		√	√	√
Periodic fat reserves	√			√

a nocturnal (rarely cathemeral) lifestyle. Diurnality is a phylogenetically inherited trait that was universal among New World monkeys until the owl monkey lineage shifted its activity rhythm.

Fruit-eating in lemurs is rare in contrast to anthropoids, where it is widespread. The rarity of frugivory in lemurs is a function of phylogeny and to a limitation of Madagascar's fruit supply, as discussed above. Lemur dietary adaptations are phylogenetically constrained to some degree by their dentition and metabolism. The toothcomb morphology is not conducive to varied and intensely frugivorous diets that may require biting into tough fruit husks. A low basal metabolic rate, which is typical of lemurs, is associated with folivory rather than frugivory.

Phylogeny is not only a constraint. It is also a foundation that influences the direction that adaptations take. Madagascar's lemurs evolved from an ancestor that was prone to leaping, and then vertical-clinging-and-leaping became a prominent form of locomotion as the lemurs diversified. Long-legged skeletons are not an ancestral trait among anthropoids. New World monkeys and Old World monkeys are mostly generalized quadrupeds. Few of those have relatively long legs, and few platyrrhines and no catarrhines have evolved vertical-clinging-and-leaping locomotion.

Body mass, size, and diet

Primates exhibit a very large range of body sizes. At the species and genus levels, body size is highly significant ecologically. Body mass is always subject to selective pressure as a fundamental, genetically variable attribute that correlates closely with the functions of many biological systems. Diet is one of several factors that drives the direction of body-size shifts, or stabilization. This is one reason why convergence in body-size range is a common, non-random pattern among primates occupying different regions.

Size and diet

Of the three main dietary categories, fruit-eating generally does not impose size-specific selective pressures. Because the rainforest typically produces a great variety of fruits in a large range of sizes, both smaller and larger primate species in a given community usually have access to fruits that are appropriate for their body size, meaning the dimensions of their mouths and hands, and their muscularity. These features are correlates of body mass. Processing leaves in bulk requires a large gut, meaning a relatively large body mass, which is why folivores are consistently larger than their frugivorous relatives. A reliance on insects is closely associated with smaller body sizes that enable efficient foraging, manipulative handling, and capture methods, like stalking or ambushing. With respect to food alone, it is often the protein dietary complement, leaves or insects, and other animal matter, that exerts a strong influence on the evolution of body mass.

While each of the regions exhibits a range of body sizes, there are differences in how the size classes are distributed among them. The differences are taxonomic and also relate to ecology. The largest genera, the super-large great apes, like the gorilla and orangutan, are found in Africa and Southeast Asia. Semi-terrestriality is an ecological factor associated with large size in African apes. The orangutans are exceptional in being the largest arboreal mammals. The less massive categories, extra-large and large, are found among Old World monkeys. In Africa, the largest ones are cercopithecines, cheek-pouched monkeys, particularly the semi-terrestrial and terrestrial forms like the baboons. In Asia, the majority of Old World monkeys are arboreal colobines, the extra-large leaf-and seed-eating monkeys. Their body mass reflects selection for size in connection with folivory.

Madagascar's lemur species are smaller than African, Asian, and neotropical primates. The range of lemur body weights is most similar to the platyrrhines, mostly small- and medium-sized species. Neither the rainforests nor the more open, dry habitats in Madagascar support the larger body sizes seen among mainland African or Asian anthropoids.

These regional size-scale patterns reflect botanical differences of the ecosystems. Because Madagascar offers few nutritionally valuable fleshy fruits and the climate is hypervariable year-after-year, the animals rely more on insects, gums, and flowers, all foods that are associated with smaller body sizes.

Even the more folivorous lemurs are relatively small in comparison with anthropoid leaf-eaters. Lemurs may be size-constrained, because they are unable to combine leaves with sufficiently rich fruits, like the semi-folivorous, larger-bodied howler monkeys in South America. Lemurs also cannot process large volumes of leaves that might sustain a large body mass because stomach size is proportional to body mass. As hindgut fermenters, they cannot bulk-process leaves and seeds like colobines that are large-bodied, with capacious stomachs where fermentation takes place.

In South America, the rich supply of varied fruits in the canopy and understory supports many small-and medium-sized species that are predatory frugivorous omnivores, the cebid family, as well as dedicated frugivores, the pitheciids. It is significant that while many primates feed on seeds at times, only in South America is it a specialized year-round feeding niche supported by many anatomical adaptations, in the pitheciid radiation of sakis and uakaris.

Giants and dwarfs

There are individual genera and monophyletic groups that have evolved a considerably larger size than their closest relatives or ancestral stocks; they are called giants. Where size reduction has occurred, the animals are called dwarfs (Tables 7.1 and 7.2). It is important to note that these are relative terms that have meaning in a phylogenetic context and do not refer to actual measurements of body mass or the size classes used throughout

Table 7.2 Weights of the smallest and largest primate species in five major taxonomic groups

	Strepsirhines	*Platyrrhines*	*Old World monkeys*	*Great apes*	*Lesser apes*
Smallest	Madame berthe's mouse lemur *Microcebus berthae* 1.1 oz. (30 g)	Pygmy marmoset *Cebuella pygmaea* 4.3 oz. (122 g)	Talapoin *Miopithecus talapoin* 2.4 lbs. (1.1 kg)	Bonobo *Pan paniscus* 71 lbs. (32.2 kg)	Pileated gibbon *Hylobates pileatus* 12 lbs. (5.4 kg)
Largest	Indri *Indri indri* 15 lbs. (6.8 kg)	Muriqui *Brachyteles arachnoides* 22 lbs. (10 kg)	Chacma baboon *Papio ursinus* 59 lbs. (27 kg)	Gorilla *Gorilla berengei* 386 lbs. (175 kg)	Siamang *Symphalangus syndactylus* 26 lbs. (11.9 kg)

this book. Giants and dwarfs occur in all the regions where primates live as a result of convergent evolution.

Some scientists have proposed that the cheirogaleid family of Madagascar is a dwarfed clade. There are five genera and many species and all of them weigh less than a pound (460 g). The other three lemur families are much larger. Among the cheirogaleids, Madame Berthe's mouse lemur is the smallest, at 1 oz. (30 g). It is 25% smaller than the next largest cheirogaleid species, suggesting that Madame Berthe's mouse lemur is a dwarf species. Some extinct subfossil lemurs were super-large giants. The extinct subfossil genus *Archaeoindris* was the size of a gorilla.

South American callitrichines are another dwarf clade, in which the largest species are about half the mass of medium-sized platyrrhines. The smallest callitrichine genus is the pygmy marmoset, *Cebuella*, roughly 3.5 oz. (100 g), a dwarf that is roughly a third the size of its closest relative, the marmoset, *Callithrix*. The largest callitrichine genus, the lion marmoset, is a giant, about five times larger than the pygmy marmoset. Ecological adaptations have driven these examples of dwarfism and gigantism, the derived gum-eating diet in the pygmy marmoset and the derived form of predation in the lion marmoset. In Asia, the smallest tarsier, *Tarsius pumilus*, is a dwarf weighing less than 2 oz. (50 g), about half the mass of the second smallest tarsier species and about one-third the size of the largest one.

In Africa, among catarrhines the smallest monkey is a talapoin. It is a dwarf that weighs less than three pounds, little more than a kilogram, a quarter the size of the smallest closely related species. The great apes represent two distinct lineages of gigantism, the African apes and the Asian orangutan. Among the African apes, compared to chimpanzees the gorilla is also a giant. In Asia, there was an extinct relative of the orangutan, called *Gigantopithecus*, that was even larger than any living ape. More on this in Chapter 8.

Nocturnal, diurnal, or cathemeral

There are three primate taxa that are nocturnal: strepsirhines in Madagascar, Africa, and Asia; owl monkeys in South America; and tarsiers in Asia. Each taxon has eyes that independently evolved adaptations to bolster vision in dim light. In strepsirhines, the tapetum lucidum is the critical adaptation. Situated behind the transparent retina, it increases the amount of light that stimulates the retina's photoreceptor cells by reflecting light back onto the retina (Figure 3.4). In owl monkeys and tarsiers, the main nocturnal adaptation of the eye is a larger eye size, relative to body mass, which increases the amount of light that enters the eye (Figures 1.6 and 6.3). Enhanced vision in dim light improves the nocturnal animals' ability to use both the canopy and subcanopy levels of the forest; the darkest areas of the rainforest are close to the ground.

The primates with the largest eyes are all predaceous. This includes tarsiers, a haplorhine, and lorises and galagos, which are strepsirhines. Huge eyes are particularly beneficial when hunting for prey at night. Tarsiers in Asia and a number of galagos in Africa also evolved independently as microhabitat specialists that use the darkest part of the forest, the understory, as their hunting grounds. Other convergent adaptations present in tarsiers and galagos are associated with their methods of detecting prey, including very large, mobile, external ears that work like dish antennae to localize prey by sound. Tarsiers and galagos use similar capture techniques, pouncing on prey using vertical-clinging-and-leaping locomotion, an understory locomotor behavior.

The shift from a nocturnal (or cathemeral) activity pattern to a diurnal one happened several times among primates, notably in Madagascar's strepsirhine lemurid family and in the ancestral stock of haplorhines. The shift from diurnal to nocturnal happened twice among haplorhines, in the tarsier lineage and in owl monkeys. Tarsiers and owl monkeys evolved the same anatomical method of adjusting to nocturnal vision by developing an enlarged eyeball size, which compensates for the haplorhine retina's lack of a light-reflecting tapetum lucidum.

Diet, dentition, and gut

Primate communities in each of the four regions where they live are composed of species that are basically frugivores, folivores, and insectivores or faunivores, as well as omnivores. Table 7.1 uses several other terms for descriptive purposes. The diversification of primate feeding niches resulted in convergent evolution of specific dental morphologies and gut adaptations. For example, the pointed, piercing cusp design of the molars of the squirrel monkey and tarsier, one of the most insectivorous platyrrhines and the most predacious primate, evolved convergently. So did the long shearing blades of the large molars found in distantly related leaf-eaters, like colobines, howler monkeys, and woolly lemurs. Among leaf-eaters, the foregut and the hindgut evolved as alternative storehouses of microbes that assist digestion by fermentation. In colobines it is the foregut; in strepsirhines and howler monkeys it is the hindgut.

Tree-gougers

The fork-marked lemur in Madagascar is a gum-eating, tree-gouging species. They use their upper incisors to anchor the jaws in the bark of a tree while the toothcomb is used to scrape an opening where the gum accumulates in response to the damage. This strategy is similar to the technique employed by tree-gouging, gum-eating marmosets in South America that have differently shaped, anthropoid anterior teeth. It also resembles the approach used by Madagascar's aye-ayes that gash holes into trees to capture wood-boring larvae. All these patterns evolved convergently to meet the physical challenge of gnawing wood.

Tree-gougers are faced with postural challenges while they prepare a gum-eating site. Africa's needle-clawed galago, Madagascar's fork-marked lemur, and South America's marmosets and pygmy marmosets have various specializations that help them to hang on to a tree trunk while they scratch and scrape away the bark. They include wide hands and feet to keep them from slipping, or sharp nails or claws to help them stay pinned to the bark.

Enamel caps and bilophodont molars – hard-fruit eaters

Old World monkeys, New World monkeys, and apes, including species such as mangabeys, capuchins, and orangutans that feed on very tough fruits, all have thick enamel caps on their molar crowns to resist tooth breakage and natural wear. In contrast, leaf-eating primates tend to have thinner molar enamel, which can wear down in a way that exposes cutting edges on the sides of cusps and maintains their sharpness by rubbing off miniscule particles of enamel that form at the edges.

Bilophodont molars (Figure 5.4), a hallmark of the cercopithecid family of Old World monkeys, that evolved in an omnivorous-frugivorous species, is an adaptation to reinforce cusps against breakage and to crush hard food particles in mortar-and-pestle fashion by confining them to the basins between the ridges. Bilophodonty also occurs among Madagascar's indriids that are basically folivores that turn to eating bark and seeds as fallback foods when leaves are scarce.

Food-handling

South American capuchin monkeys, mangabeys and other Old World catarrhines in Africa and Asia, have precision-grip thumbs that enhance the handling of food, especially when an item is small. The morphologies of their thumbs and gripping behaviors are different. Catarrhines have opposable thumbs. They are able to rotate the thumb opposite the other digits to apply pressure precisely to grab or pinch an object between the touch pads of the fingertips. Capuchin precision grips are limited to squeezing the thumb against the index finger to control a small object. They are the only platyrrhines that use this type of grip. This means that manual precision grips evolved in one platyrrhine genus and convergently in catarrhines.

Fat storage

Small mouse lemurs in Madagascar and super-large orangutans in Southeast Asia have independently evolved fat storage systems to help them survive periods of food scarcity. This is an extreme example of convergence pertaining to habitats in two distantly related and adaptively divergent primates. The extra-small mouse lemurs, largely insectivorous animals that rely on gums and fruits also, store fat in their tails (Figure 3.2). The fat is converted to energy while they hibernate for months each year when food is scarce. Super-large male orangutans gain a lot of weight when fruits are abundant and they eat huge amounts. Stored as body fat, it converts to energy in order to supplement a fallback diet of relatively non-nutritious leaves and bark during periods when fruits are in short supply.

Locomotion and posture

While arboreal quadrupedalism is the norm for many primates, other, similar locomotor specializations have evolved independently in each of the regions. Variations in postural and locomotor adaptations appear at least as often as similar dietary preferences. Some cut across very basic ecological adaptations, appearing in species that differ greatly in body mass. The quadrumanous, arboreal orangutan in Asia, for example, is more than seventy-five times the weight of the quadrumanous, arboreal potto in Africa. Their similar locomotor abilities are made possible by a high degree of limb-joint mobility and strong hands and feet that evolved independently.

Leapers

Leaping adaptations appear among genera of strepsirhines and platyrrhines, and a few arboreal, quadrupedal colobines, like grey langurs and proboscis monkeys (Figure 2.4). As noted, vertical-clinging-and-leaping specializations are found in galagos and tarsiers. Vertical-clinging-and-leaping also occurs in a few platyrrhines, like the seed-eating saki monkey, the gum-eating pygmy marmoset, and the fungus-eating Goeldi's monkey, though

Figure 7.6 The long-legged sifaka (a) of Madagascar is a leaper while the long-armed Asian gibbon (b) is a climber and brachiator.

Photo credits: a, *Propithecus verreauxi* i.jpg, by Neil Strickland (CC-BY-2.0); b, *Hylobates agilis*.jpg, by K. Rudloff (CC-BY-SA-4.0).

these South American monkeys do not have comparable anatomical specializations of the limbs and feet of tarsiers and galagos. A common ecological factor of the galagos, tarsiers, and platyrrhines is their reliance on the subcanopy microhabitat where their preferred foods are located, and where narrow, vertical substrates for locomotion dominate.

Two specialized anatomical variants of leaping adaptations have evolved in animals of different body mass classes. Thigh-powered leaping occurs in the large-bodied sifaka, an indriid lemur (Figure 7.6). Foot-powered leaping evolved in the small galagos and tarsiers (Figure 6.3). All the leapers have evolved long hindlimbs. Their low intermembral indices fall between about 55 and 70, meaning the forelimbs are 55% to 70% shorter than the hindlimbs (Box 3.1). In galagos and tarsiers the length and leverage of the lower limbs is extended by their extremely long feet.

Hangers, climbers and clamberers, and brachiators

Different suspensory postures and suspensory locomotor styles have evolved among primates for feeding and traveling functions. Sifakas in Madagascar typically locomote as vertical-clingers-and-leapers but they also frequently use suspensory postures while feeding on leaves, hanging by their hands and feet. Head-down, foot-hanging feeding postures are employed by South American uakaris as they manipulate seed pods with two hands. While positioning themselves in the flexible end-branches of trees, the prehensile-tailed howler monkeys often hang by their tails so they can reach down to feed on a bunch of leaves from a lower branch.

Atelids are climbers and clamberers, and hylobatids are brachiators. In these two families, in which below-branch locomotor behaviors are most important, long forelimbs are the prominent adaptation. Their forelimbs are longer than their hindlimbs. No other primates match the brachiating hylobatids in relative forelimb length (Figure 7.6). In hylobatids the intermembral index ranges from a low 128 in gibbons to a high of 147 in siamang (Box 3.1).

Spider monkeys and muriqui employ arm-swinging styles of suspensory locomotion that are similar to hylobatids as they climb and clamber through the trees and swing beneath branches. Unlike hylobatids that are apes and don't have a tail, these platyrrhines use their prehensile tails and arms together in a coordinated motor behavior.

There are examples of convergent evolution in the morphology of primate hands connected with arboreal locomotion. Several Old World and New World anthropoids have curved hand bones that effect a strong, hook-like clasp while climbing and locomoting in a suspensory manner. Brachiating gibbons and siamang exhibit an additional adaptation that is unique to them. They have a deep cleft between the thumb and palm that widens the span of their grasp, so they can grab thick supports.

Primates generally have long digits, but short fingers and toes have evolved convergently in several cases. In arboreal neotropical spider monkeys and muriqui, and in African colobines, the thumbs are vestigial or missing. These primates have four-fingered hands. In lorises, the smallest primate climbers and clamberers, it is the second digit, the index finger, that is a tiny nubbin. In many strepsirhines, the second digit of the foot, which has a grooming claw, is short; in lorisids and galagids it is exceptionally short. Among South American platyrrhines, the hallux (large toe) is very short in the clawed callitrichines, and the fifth digits of hands and feet are short in titi and owl monkeys.

Primates without tails

Primates typically have relatively long tails as a feature of the arboreal locomotor complex and only a few do not. Tails play a fundamental role in locomotion as a flexible balancing appendage. This is commonly seen when a primate whips its tail from side to side to steady itself while walking on narrow branches, or when a tail is used to make mid-air adjustments when leaping or approaching a landing spot. Or, simply to be draped around a nearby branch for proprioceptive input. There are examples of convergent evolution in primates whose tails are short or missing entirely.

All apes are missing a tail. Asia's stumped-tailed macaques, Madagascar's indris, and Asia's lorises also lack a tail. The African pottos and angwantibos have a short, stubby tail. In the neotropical uacari, the tail is also distinctly short.

In apes, the absence of a tail is attributable to several factors. With their large, extra-large and super-large sizes, apes require a firm grasp to maintain balance in the trees. Large hands, feet, and muscular strength are what is needed for a heavy primate to negotiate the arboreal environment. This means that selective pressure to maintain the tail as a locomotor organ is relaxed, and that other functions can influence the morphology and evolution of the tail region. When in the trees, the heavy apes habitually adopt a relatively orthograde (erect) body posture to climb vertically and to haul themselves through the branches (Figure 5.7). This places stress on the body and internal organs that differs from the stresses encountered by quadrupedal, pronograde (horizontally disposed) primates. The selective advantage of a coccyx, the internal bony vestige of a tail, is that it plays an important role as an attachment site for soft tissues that support internal organs (including the bladder and intestines) from below, in the form of a pelvic floor.

Cranial morphology

The skull is the most complicated unit of the skeleton because it is responsible for the housing and functioning of many systems, including the brain, the eyes, nose, and middle and inner ears, teeth and chewing muscles. The skull is attached to the vertebral column and the rest of the postcranial skeleton, so its shape is also influenced by locomotion, posture, and head carriage. There are many examples of convergence in the anatomical details of primate crania.

The short faces in gibbons and marmosets are an example of convergence. The long faces of baboons and lemurs are another example. When looking at the overall construction of the cranium in a capuchin monkey, a platyrrhine, and a macaque, a catarrhine, the degree of independently evolved cranial resemblances is striking (Figure 4.4), particularly in their relatively short faces and large braincases.

The huge orbits of lorises and tarsiers evolved convergently to accommodate their large eyes. Tarsiers and lorises each have a wide, ribbon-like bony frame that wraps around the jutting eyes to provide eyeball support. These are expanded postorbital bars, derivations of the slender postorbital bars that occur in all strepsirhines and were present in the first haplorhines. In lorises, the frame is positioned on the side of the eye. In tarsiers, it is positioned behind the eye, where it meets expansions of other bones that enlarge the socket posteriorly and produces a unique bowl-shaped orbit (Figure 6.3).

The differences in position of the expanded postorbital bar, and the size of the expansions, relate to locomotor differences. Lorises are slow-moving quadrupeds. The width of their postorbital bar reflects the contour of the gigantic eyeball that presses against it. But because tarsiers are extreme leapers, and the force of their powerful leaps pushes the eye backwards in the socket, the bowl-shaped orbit provides needed support behind the eye to help keep it in place and maintain its shape.

Parallel evolution in platyrrhines – grasping tails

Parallel evolution refers to the evolution of similarities among closely related taxa. Grasping tails evolved only in New World monkeys. There are two types of grasping tails: semi-prehensile and fully prehensile tails. Each type evolved in a different clade of platyrrhines. Among cebids, only the capuchin monkey has a semi-prehensile tail. Among atelids, the howler monkeys, woolly monkeys, spider monkeys, and muriqui all have a fully prehensile tail. The two types of tail are an example of parallel – not convergent – evolution.

The behavioral capabilities of semi-prehensile and fully prehensile tails differ, as do their morphology and the ways in which the tails are used in locomotion. Capuchin tails are fully covered with fur and short relative to the animal's medium body-size mass. In quadrupedal locomotion, capuchins typically carry the tail stretched out horizontally behind the rump. Its grasping abilities are not applied during locomotion. While feeding, capuchins often use the tail in a postural adaptation, as a hook that is curled around a branch while it hangs from it, or in a tripod stance that is formed with the feet, to free the hands for handling food. Squirrel monkeys, the closest relatives of capuchins, use their long tails much like the capuchins use their shorter tails although they have not been categorized as having prehensile tails.

Atelid tails, in contrast, are long relative to body size. Their tails have a long belt of sensitive, naked skin on the underside that enhances tactile acuity and resistance to slippage (Figure 4.6). In woolly monkeys, spider monkeys, and muriqui, the tail tends to be carried aloft where it can be slipped around a tree limb above the body while walking or clambering. During arm-swinging locomotion, spiders and muriqui use the tail as a third hand during the hand-over-hand locomotor cycle. In howlers, the tail is fully prehensile and it tends to be used as a postural support for stability, not in dynamic movements.

This is similar to the capuchins that also use the tail as a postural adaptation even though it is semi-prehensile.

Tail use among other platyrrhines suggests that these prehensile abilities stem from a genetic potential that is expressed in various behaviors. Eight of the sixteen genera of living New World monkeys exhibit high degrees of musculoskeletal control of the tail. Owl and titi monkeys tail-twine for social reasons (Figure 4.9). Squirrel monkeys wrap their tails around themselves to keep warm, around a partner in a social context, or around a tree trunk while stretching away from it. Eight genera in all three major platyrrhine clades use their tails in special ways: owl, titi and squirrel monkeys, capuchins, howlers, woolly monkeys, spider monkeys, and muriqui. Comparable degrees of tail control do not exist among strepsirhines or Old World monkeys and, of course, apes have no tails at all, meaning platyrrhines are the only primates that exhibit parallel evolution of the tail in this way because they are phylogenetically predisposed to the behavior.

Social behavior

Primates' behavior consists of the actions or inactions of individuals and the varying styles by which their social groups are organized and function. These systems evolve through natural selection to benefit the individual's survival and reproductive success. Like other adaptations, varied systems of social organization have evolved, and are seen in primate communities in all major taxonomic groups living in the four regions where primates occur. How males and females interact with each other and their young, how they communicate, how they forage, and how they interact with neighboring conspecific groups – these are some of the factors that contribute to social organization.

Primates are social mammals even though some of their social systems are called semi-solitary. Individuals living in semi-solitary systems, like mouse lemurs, are also social although they do not live in cohesive groups and do not interact much. While they do not forage together, clusters of adult conspecifics occupy the same area, maintain awareness of conspecifics and their movements, sometimes sleep together, and, of course, they mate. Several other forms of social organization have evolved, including systems that are structured around adults that are pair-bonded, live in unimale-multifemale groups, and multimale-multifemale groups. Primates can benefit from being social in several ways (Box 7.2).

Box 7.2 Primates benefit from being social

Group-living comes with a basic cost: resources must be shared. It can also involve hidden costs, like the stress associated with social hierarchies that determine access to resources and may be enforced by aggression. There are several reasons why the benefits of sociality to an individual may outweigh the costs:

- Predation. There is safety in numbers. With multiple individuals around, predators can be spotted more easily and resisted, and the chances of being caught by an intruder are lowered by the presence of other targets.
- Food sharing and defense. Foraging as a group that is naturally spread out in space but cooperates by communication increases the opportunity to locate food and share the information. Feeding locations, or areas where foods are

concentrated, can be better defended by a group from other conspecifics or species that are competitors.
- Mating opportunities. Group-living provides built-in mating opportunities. Even if individuals disperse out of their natal group when they mature, they will associate with conspecifics that have the same patterns of behavior.
- Socialization and learning. Interacting in play with same-aged individuals, and interacting in other circumstances with adult relatives and individuals of the other gender, are critical for a juvenile to learn the skills and social rules required in adulthood.
- Well-being. There is a positive effect that primates can derive from affirming interactions like grooming, tail-twining, affiliative vocal signals, and social bonding, as further discussed below.

Morphology and behavior

Some patterns of social behavior are closely tied to and supported by morphology. Grooming is an example. Grooming serves both hygienic and social functions.

Self-grooming and allogrooming

One of the most distinctive and common social interactions exhibited by primates is grooming, and primates have evolved anatomical features specifically for this purpose. All primate species groom by stroking their fur to clean it of debris and organisms like parasites that live on the skin or fur. Self-grooming in many species is facilitated by a grooming claw, an upturned claw-like nail on the foot that is used to scratch at the fur. Strepsirhines all have a grooming claw on the second toe. Tarsiers have two grooming claws, on the second and third toes. The only anthropoids that have a grooming claw are two related platyrrhines, the owl and titi monkeys.

Allogrooming, also known as social grooming, refers to the grooming of one individual by another, typically of the same social group but also of conspecifics belonging to other social groups, should the individuals interact. Allogrooming plays an important role in regulating the behavior of individuals within a social group, and in building social bonds between individuals. In living strepsirhines, the toothcomb is the principal organ used in allogrooming, so the process becomes an opportunity to exchange scent, the critical way in which strepsirhines communicate. The retention of a toothcomb in all but one genus and species, the aye-aye, is an indication of the importance allogrooming has had in the evolution of the nocturnal and cathemeral radiation of lemurs. In anthropoids, operating in daylight with superior eyesight and hand-eye coordination, the fingers are used in allogrooming.

The nature of reciprocal allogrooming is related to social system structure and kinship. Females generally show a preference for female grooming partners, especially those that are related. In male-dominant, hierarchical groups like baboons, females groom males far more often than males groom females. The amount of time spent allogrooming is correlated with group size. In large groups, individuals do not groom with more members, rather, they increase the frequency of mutual grooming with established partners. In groups where males and females are co-dominant, as in some pair-bonded gibbon species, males groom females more often. Among chimpanzees, where male bonding and male alliances are very important, adult males groom each other far more often than they allogroom with females.

Sexual dimorphism and monomorphism

Sexual dimorphism, differences between males and females other than in primary sexual characteristics, is a common feature among anthropoids but not in strepsirhines. Sexual dimorphism may be exhibited by differences in body size, canine size, and the coloration of pelage. Monomorphism is the condition in which male and female primates do not exhibit such differences. Even if only one feature is different in males and females, we still call this sexual dimorphism. In South American saki monkeys, for example, males and females only differ in their pelage but not in their body size, shape or other features (Figure 4.7).

Sexual selection is a form of natural selection that drives sexual dimorphism. Sexual selection favors the development of traits in either males or females that are attractive to the opposite sex. Sexual selection may result in the evolution of traits that increase a male's competitive advantage for access to females in social systems that include multiple males. Baboons are an example. They live in multimale-multifemale groups that are hierarchically organized. Body size differences between the sexes can be extreme; males may weigh twice as much as females. Differences in canine size are also dramatic (Figure 7.7). The high degree of sexual dimorphism is correlated with the amount of competition that takes place between males for the number of females that a male can monopolize.

The distinctive, sexually dimorphic coloration of mandrills that live in large multimale-multifemale groups plays an important role in mating preferences. Mandrills have dual dominance hierarchies, one rank for males and one for females. Females prefer to mate with males that display the brightest red colors, who usually are the highest ranking males. Males prefer to mate with the highest ranking females and those that have already given birth. Among females, social rank is associated with kinship. Related

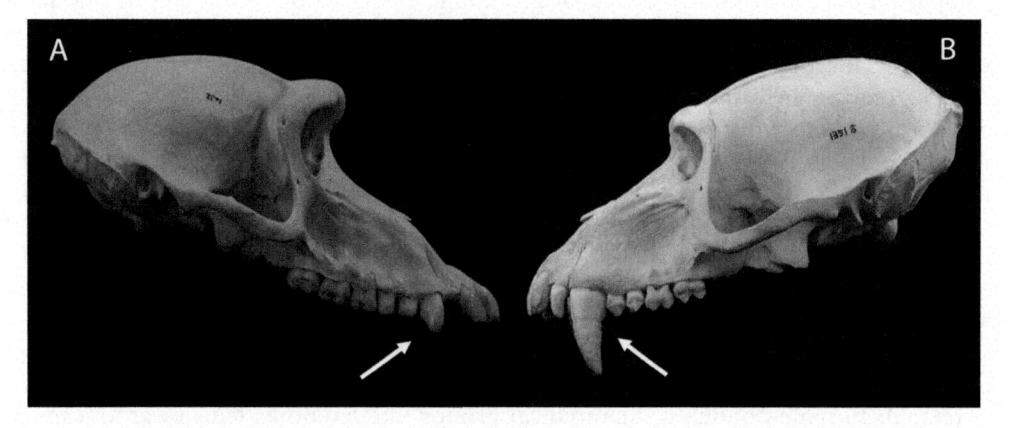

Figure 7.7 Crania of an adult female (a) and male (b) hamadryas baboon, brought to the same cranial length to highlight the size difference of the upper canine teeth (arrows).

Photo credit: Department of Anatomy, Dokkyo University School of Medicine, https://dept.dokkyomed.ac.jp/-dep-m/macro/mammal/en/order_list.html

females maintain strong social bonds with one another. Mothers are higher ranking than daughters

An unusual example of sexual dimorphism occurs among males in some squirrel monkey species that temporarily gain weight by water retention, which is part of what is called the fatted-male syndrome. This happens annually, when males grow larger in mass during the breeding season to gain a competitive edge over other males in an effort to attract estrous females.

The expression of monomorphism in canine size occurs when canines are either large or small in both males and females. In two platyrrhines, the pair-bonded titi and owl monkeys, the canines of males and females are low-crowned. In muriqui, another monomorphic platyrrhine that lives in multimale-multifemale groups, canines are also low-crowned in both sexes. In pair-bonded, monomorphic lesser apes, gibbons and siamang, both sexes have high-crowned, slashing canine teeth. Monomorphic strepsirhine males and females also have large, slashing upper canine teeth.

Long-distance vocalizations

Primates make significant use of vocalizations to communicate (Box 4.1). Many primates produce species-specific loud calls that carry long distances in the forest. Their general purpose is to announce a group's presence in the area to conspecifics within earshot. To generate these long-distance calls, various strepsirhines, platyrrhines, and catarrhines have an air sac in the throat that is associated with the vocal cords. It amplifies and helps modulate sound. Gibbons are a striking example. During loud calling, the throat sac blows up like a large balloon. Primate vocal sacs vary morphologically and are variously distributed among taxa, including some species of lemurs, platyrrhines, Old World monkeys, and in chimpanzees, which indicates that some occurrences evolved by convergent evolution.

Among primates, howler monkeys have the most unusual and loudest sound amplification system, which is also sexually dimorphic; females have a smaller vocalization system. Males roar more loudly than a lion with a sound that is nearly as low-pitched as an elephant's roar. The mechanism involves a greatly enlarged hyoid bone and larynx that act like resonating chambers (Figure 7.8). Situated in the throat and extending into the space below the cranium where the lower jaw sits, the entire skull is structured to accommodate the system. The result is that howlers have one of the primates' most distinctive skulls. The cranium is very large, with a protruding upturned snout and a massive, tall mandible.

Tail-twining

The tail-twining, twisting tails together, seen in two closely related platyrrhines that belong to the same subfamily clade, titi and owl monkeys, is another behavior that is tied to a special morphology-related characteristic (Figure 4.9). It is the ability to control the tail and wind it through several turns along its length. Tail-twining reinforces the social bonds among group members, adults and offspring. It is a tactile gesture that has functional interpersonal qualities like those associated with allogrooming. Monogamous titi monkeys engage in tail-twining for long periods when they bed down on a branch in the late afternoon to prepare for sleep.

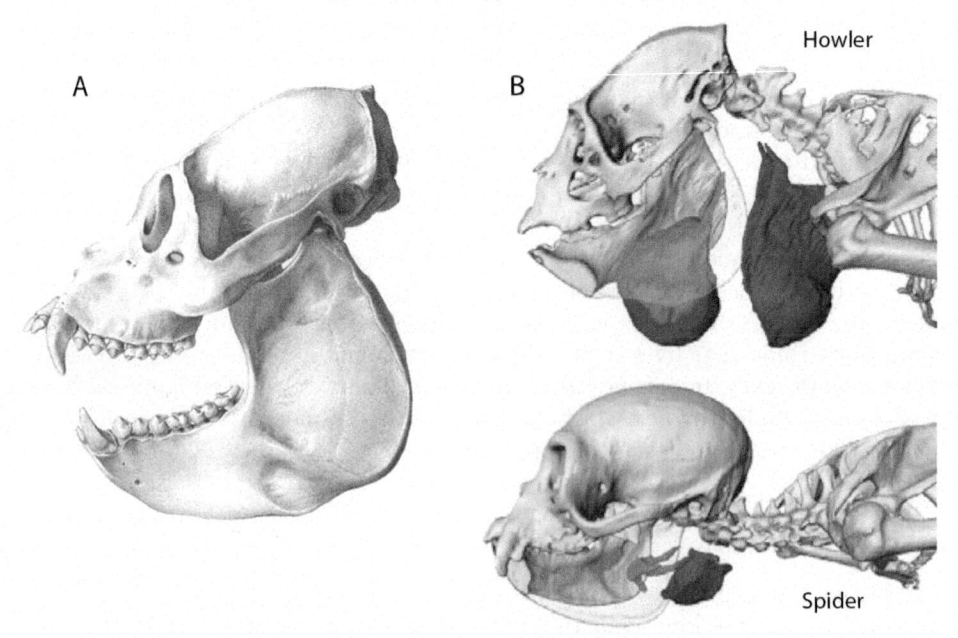

192 *Primate communities compared: ecology, morphology, and behavior*

Figure 7.8 Skull of a male howler monkey (a) and comparison (b) of the size and position in the body of the hyoid and larynx, shaded, in a howler monkey and a spider monkey.

From Rosenberger, A. L. (2020). *New World Monkeys: The Evolutionary Odyssey*. Princeton, Princeton University Press.

Primate breeding systems

The range of group sizes exhibited by primates, from adult pairs to dozens or hundreds of adults living together, reflects different kinds of breeding and mating configurations. There are social systems where only one female is a breeder in the group and others where two or more females breed. A single-female breeding arrangement may involve one or more males as partners if the social system is based on pair-bonds or is semi-solitary. In groups where two or more females breed, they may mate with one male in a unimale or multimale social system, or with several males in a multimale system.

Categories and subcategories of breeding systems

There are two main categories and several subcategories of primate breeding systems:

- **Singular breeders**
  - Semi-solitary breeders. Semi-solitary breeders occur in ecological situations where spatial and temporal separation between adults is advantageous to them. This develops when it is not efficient for males and females to forage together, share feeding sites and shelters, or where grouping may attract predators. Mouse lemurs that are spot feeders and insectivores benefit by not sharing gum-feeding sites that have limited productivity, and by limiting the foraging competition from conspecifics looking for insects to eat. Similarly, the solitariness of large, frugivorous orangutans minimizes within-species competition, given the irregular food supply and periods

Primate communities compared: ecology, morphology, and behavior 193

of scarcity in their mast-fruiting environment. Orangutan females live apart from one another in separate ranges but males' ranges can overlap with several females', and males may breed with different females.
- Pair-breeders. Pair-breeders like Asian gibbons and South American titi monkeys, live in monogamous social groups comprised of one male and one female with their young. Pair-breeding occurs in a variety of circumstances when both individuals benefit from cooperating to jointly defend territory where critical foods are located. In these monogamous breeding groups, the male also cooperates by providing offspring care.
- Singular cooperative breeders. Singular cooperative breeding is a system that occurs only in platyrrhines, in marmosets and tamarins (Figure 7.9). These New World monkeys always give birth to litters of twins, twice a year. Their social groups are composed of multiple males and females but only one dominant female is reproductively active at any one time. She may mate with one or more of the resident adult males. Infant care is contributed by the mother's partners, the non-breeding females in the group, and other group members. The high reproductive output – twins twice a year – creates the demand and selective pressure for all the group members to help with infant care.
- **Plural breeders**. In plural breeders, like howler monkeys, gorillas, spider monkeys, and chimpanzees, two or more females breed in each social group. Social groups are organized as either unimale-multifemale or multimale-multifemale groups, with multiple

Figure 7.9 An adult common marmoset carrying a set of twins on its back.

Photo credit: *Callithrix* family.jpg, by Paulo H. Alves (CC-BY-SA 4.0).

females either mating with only one male, or both males and females having multiple mating partners in each breeding season.

Convergent evolution of primate societies

The various forms of primate social organizations and breeding systems are found in all the major regions where primates live. Convergent evolution of these systems is a phenomenon found among strepsirhines, New World monkeys, Old World monkeys, and the great and lesser apes (Figure 7.10; Table 7.3). Unimale-multifemale groups are found in South American howler monkeys and African gorillas. Multimale-multifemale fission-fusion groups are found in neotropical spider monkeys, African red colobus monkeys, and chimpanzees. Pair-bonded social groups occur in South American titi monkeys and Asian gibbons. Semi-solitary groups occur in Madagascar's mouse lemurs and Asia's orangutans.

A different social organization, breeding system, and mating pattern can be effective in the same environment, and similar configurations can be found in different environments. Unimale and multimale groups are found in arboreal and terrestrial species, in species inhabiting forests and open country, in nocturnal and diurnal species, and among frugivores and folivores. With few exceptions, systems with the smallest group sizes, where the safety-in-numbers principal does not apply, semi-solitary and pair-bonded arrangements, are not found in terrestrial or in diurnal species. The pair-bonded, diurnal indris, which are lemurs, and the semi-solitary orangutan, a great ape, are exceptions. The huge arboreal orangutans have few natural predators, and the vertical-clinging-and-leaping indris are elusive locomotors.

An example of convergent social systems is found in spider monkeys and chimpanzees, two distantly related primates. They share the fission-fusion pattern – smaller foraging subgroups form temporarily – that occurs in some species that have multimale-multifemale social-groups. Spider monkeys and chimpanzees have several ecological and societal characteristics that are consistent with the formation of small foraging subgroups:

- Both are ripe-fruit specialists.
- The size of foraging groups correlates with the availability of ripe fruits in their habitats.
- Males and females tend to forage in separate subgroups.
- Males share strong bonds, are aggressive toward females, patrol territorial boundaries, and cooperatively defend their territories against neighboring troops.
- Females disperse from the natal group.

In both spider monkeys and chimpanzees, small fission-fusion subgroups enable females to concentrate on searching for productive patches of ripe fruit, while males range more widely to feed and to consolidate their control of feeding sites. Spider monkeys and chimpanzees are long-lived, slow-growing species. Dividing into female and male foraging subgroups means that females and their dependent young have good feeding prospects and undeterred socializing opportunities to learn about the home range, territorial boundaries and, possibly, neighboring groups. That knowledge is helpful when a female is old enough to leave her natal group and travel outside of its territorial boundaries to find her own breeding opportunities.

Primate communities compared: ecology, morphology, and behavior 195

Figure 7.10 Primates belonging to different regions and taxonomic groups that have similar social organizations. Taxa on the right are all catarrhines. Taxa on the left, except for the strepsirhine at the bottom, are platyrrhines. (a) howler monkey; (b) gorilla; (c) spider monkey; (d) chimpanzee; (e) titi monkey; (f) gibbon; (g) mouse lemur; (h) orangutan.

Photo credits: (a), *Alouatta palliata* 5 CR.JPG, by Cephas (CC-BY-SA 4.0); (b), Mountain gorilla (*Gorilla beringei beringei*) female, Volcanoes National Park, Rwanda, by Charles J. Sharp (CC-BY-SA-4.0); (c), Black-faced Black Spider Monkey (*Ateles chamek*), by Floro Ortiz Contreras (CC-BY-A 4.0); (d), Chimpanzee, Kibale, Uganda (15059241309).jpg, by Rod Waddington (CC-BY-AS 2.0); (e) Red Titi Monkey.jpg, by Steve Wilson (CC-BY- A 2.0); (f), Female yellow cheeked crested gibbon.jpg, by Carine06 (CC-BY-SA 2.0); (g), Nosy Be mouse lemur (*Microcebus mamiratra*) head.jpg, Charles J. Sharp (CC-BY-AS 4.0); (h), Frontal view of *Pongo tapanuliensis*, by Tim Laman (CC-BY-A 4.0).

Table 7.3 *A comparison of behavioral, anatomical, and ecological features associated with social organization and breeding systems in primates exhibiting convergent evolution of social organization*

	Howler monkeys	Gorillas	Spider monkeys	Chimpanzees	Titi monkeys	Gibbons	Mouse lemurs	Orangutans
Social organization								
Semi-solitary							✓	✓
Pair-bonded					✓	✓		
Unimale-multifemale	✓	✓						
Multimale-multifemale			✓	✓				
Breeding system	Plural	Plural	Plural	Plural	Pairs	Pairs	Semi-solitary	Semi-solitary
Domain								
Arboreal	✓		✓		✓	✓	✓	✓
Terrestrial		✓		✓				
Semi-terrestrial		✓						
Diet								
Frugivorous	✓		✓	✓	✓	✓	✓	✓
Folivorous	✓	✓						
Insectivorous							✓	
Morphology								
Sexual dimorphism	✓	✓		✓			✓	✓
Sexual monomorphism			✓		✓	✓		
Social factors								
Cohesive daily foraging	✓	✓			✓	✓		
Fission-fusion			✓	✓				
Male-dominant	✓	✓		✓				
Co-dominant			✓		✓	✓	✓	
Social dispersal	M, F	M, F	F	F	M, F	M, F		✓
Infanticidal males	✓	✓		✓				?

Abbreviations: M, male; F, female.

Social dispersal

Social dispersal in primate social systems is a method to avoid inbreeding. In various species, either males or females, or both, leave their natal group to establish breeding opportunities by joining a nearby group or forming one of their own. It tends to happen when an individual reaches puberty or adult body size. The precise timing of a dispersal may occur in response to aggression by other resident animals or if an individual is attracted to another social group. In some species, like pair-bonded gibbons, male dispersal occurs several years after they are grown, while in other pair-bonded species, like owl monkeys, the age when a male or female departs is variable. In the majority of primate species, it is the male that leaves. Closely related females generally form the core of a natal – and breeding – group.

Male dispersal is common in unimale-multifemale groups where there is a high degree of competition among males to monopolize females as mates. This is a widespread pattern among Old World monkeys and in strepsirhines. In multi-level societies comprised of very large groups and subgroups, like hamadryas baboons, a single male monopolizes more than half-a-dozen females as the breeding group. Males born into that group disperse to form their own groups, either a breeding group or a company of bachelor males, which may consist of related males. The females of a hamadryas breeding group are usually comprised of mothers, daughters, and aunts that are strongly bonded and interactive socially, and they do not disperse. They stay together and remain in the same area. To avoid inbreeding, in hamadryas and in the social groups of other primate species, males that have dispersed do not return to their natal kin-group.

Similar patterns occur in smaller social groups, both unimale-multifemale and multimale-multifemale, in some New World monkeys and strepsirhines, like the ring-tailed lemur. It also occurs in semi-solitary species, including orangutans and mouse lemurs. When female mouse lemurs mature, they stay close to the territory they occupied with their mothers, and form their own territory. This means that a certain area of the forest is occupied by a cluster of related females. That is why the dispersal distance of males born to any of those females is greater than the distance traveled by dispersing females. By moving farther away, males have a higher likelihood of encountering unrelated females.

Both males and females disperse in some species, such as gorillas and neotropical howler monkeys, both of which live in unimale-multifemale groups. Subordinate gorilla males disperse, and they may meet other unattached males in the forest. Females, who leave the natal group before they are of breeding age, may either join such a bachelor group or become part of a different, established breeding group.

Males and females disperse in pair-bonded groups, like titi monkeys and gibbons. A dispersing male gibbon tends to join with a female from a nearby social group. She may also be dispersing or be part of another monogamous group. A female gibbon travels farther than a male when dispersing, which decreases the likelihood of meeting a genetically related potential mate.

Female dispersal in fission-fusion groups, such as muriqui and chimpanzees, is connected to the composition of consistent foraging subgroups, which tend to be made up of closely related females. They are freer to range more widely than males. In chimpanzees, the hostile competitiveness of adjacent social groups that is largely carried out by males mitigates against male dispersal. Chimpanzees' social organization is also structured around strong male-male bonds that develop over many years, so the benefits of remaining in the natal group are many and the risks of transferring to a nearby social group are high.

Key concepts

Convergent evolution and parallel evolution
Coevolution
Phylogenetic constraint
Sexual selection
Social dispersal

Quizlet

1 What is meant by seed dispersal?
2 Why don't all primates eat leaves? Give two of the reasons.
3 What makes South America a primate fruit-eater's paradise?
4 Why do primates cooperate?
5 Give an example of body size correlating with diet.
6 The poor fruit supply in Madagascar's rainforest ecosystem has resulted in what consequences for lemurs?
7 Give two examples of primate societies that have evolved convergently.

Bibliography

Clutton-Brock, T. (2021). Social evolution in mammals. *Science*, 373(6561), eabc9699. https://doi.org/10.1126/science.abc9699

Corlett, R., & Primack, R. (2011). *Tropical Rain Forests: An Ecological and Biogeographical Comparison* (2nd edition). Oxford, Wiley-Blackwell.

Di Fiore, A, Link, A., & Campbell, C. J. (2010). The atelines: Behavioral and socioecological diversity in a New World radiation. In C. J. Campbell, A. Fuentes, K. C. MacKinnon, S. K. Bearder, & R. M. Stumpf (Eds.), *Primates in Perspective* (pp. 155–188). New York, Oxford University Press.

Ganzhorn, J. U., Arrigo-Nelson, S., Boinski, S., Bollen, A., Carrai, V., Derby, A., Donati, G., Koenig, A., Kowalewski, M., Lahann, P., Norscia, I., Polowinsky, S. Y., Schwitzer, C., Stevenson, P. R., Talebi, M. G., Tan, C., Vogel, E. R., & Wright, P. C. (2009). Possible fruit protein effects on primate communities in Madagascar and the neotropics. *PLOS ONE*, 4(12), e8253. https://doi.org/10.1371/journal.pone.0008253

Grabowski, M., Kopperud, B. T., Tsuboi, M., & Hansen, T. F. (2022). Both diet and sociality affect primate brain-size evolution. *Systematic Biology*, syac075. https://doi.org/10.1093/sysbio/syac075

Julliot, C. (1996). Seed dispersal by red howling monkeys (*Alouatta seniculus*) in the tropical rain forest of French Guiana. *International Journal of Primatology*, 17(2), 239–258. https://doi.org/10.1007/BF02735451

Norford, A. (2021). Seed dispersal is just as important as pollination. *Mongabay Environmental News*. https://news.mongabay.com/2021/08/seed-dispersal-is-just-as-important-as-pollination-commentary/

Swedell, L. (2012). Primate sociality and social systems. *Nature Education Knowledge*, 3, 84–91. https://www.nature.com/scitable/knowledge/library/primate-sociality-and-social-systems-58068905/

8 The primate fossil record
Highlights

Chapter Contents

The story of primate evolution is told through the teeth	201
The Paleocene Epoch: proto-primates	202
The Eocene Epoch: euprimates	203
Adapiforms and fossil tarsiiforms: early strepsirhines and haplorhines	205
The Eocene rise of toothcombed strepsirhines in Africa	206
Eocene strepsirhines and tarsiiforms from Asia	207
The Oligocene Epoch: proto-anthropoids and crown anthropoids	208
Late Eocene or early Oligocene: early platyrrhines	208
Africa's stem catarrhines: proto-apes	210
The Miocene Epoch: the early rise of modern primates	210
Platyrrhines	211
Catarrhines	213
Africa's great apes	214
Africa's Old World monkeys	214
Asia's Old World monkeys	215
Asia's lesser apes	215
Asia's great apes	216
Fossil apes from Europe	216
The Pliocene and Pleistocene Epochs: primates enter the modern world	216
The Holocene Epoch: subfossils from Madagascar and the Caribbean	217
The Anthropocene Epoch	219
Key concepts	219
Quizlet	219
Bibliography	220

The primate fossil record consists of remains of animals that have been found from geologic periods dating as far back as the Paleocene Epoch, between 66 and 56 million years ago, and throughout the seven geologic periods, up to the present day, that make up the Cenozoic Era (Table 8.1).

    Fossils continue to be discovered in various locations, from different time periods, adding more information to what we know about the primate order's evolution. Fossils reveal extinct species, new species, genera, and families, and present new issues for consideration.

DOI: 10.4324/9781003257257-8

The primate fossil record: highlights

Table 8.1 Major events of primate history shown in the fossil record

	Geological period	Age	Major events
Cenozoic Era	Holocene	12,000 Y–present	Final extinctions of continental and island-living large primates.
	Pleistocene	2.5 M–12,000 Y	Many large primates become extinct in Madagascar, South America, Africa, and Asia.
	Pliocene	5–2.5 M	Old World monkeys surpass apes in abundance.
	Miocene	23–5 M	Lineages of today's living anthropoids are found in South America, Africa, and Asia.
	Oligocene	34–23 M	African anthropoids are abundant. New World monkeys exist.
	Eocene	56–34 M	Strepsirhines and haplorhines appear in relatively modern form.
	Paleocene	66–56 M	Proto-primates called plesiadapiforms exist.

Abbreviation: M, Millions of years; Y, Standard years.

In this chapter we will look at a number of highlights from the fossil record in each of the Cenozoic's seven geologic periods that shed light on what we know about the seven major groups of primates extant in the world today, and the four regions they inhabit.

Primate fossils are not evenly distributed across geography and time. The fewest number of fossils have been found in Madagascar and South America – remains from Madagascar are actually subfossils, no more than a few thousand years old. Africa is the richest area in terms of material remains and taxa that are closely linked phylogenetically with living species. There are also abundant remains from Asia. Interestingly, there are primate fossils that have been found in North America and Europe, regions where no primates live today.

What we see in the fossil record:

- The primate order's commitment to the arboreal frugivorous adaptive zone
- Ecological and behavioral changes that coincide with the evolution of new lineages or clades
- Primate communities of the past and present diversify in similar ways despite phylogenetic and ecological differences
- Stability of today's primate communities reaching back at least 20 million years
- Mosaic evolution over time and within clades is a major feature of adaptive radiations

Table 8.1 shows the major milestones in primate history that occurred during the seven epochs of the Age of Mammals, geologically known collectively as the Cenozoic Era. During the long course of primate evolution, and during the preceding phase when the ancestors of primates surely existed, changes on a planetary scale, of climate, biomes, sea levels, and the configuration of landmasses, reshaped the geography of ecosystems and the distribution of flora and fauna. These environmental changes influenced primate evolution, as did the adaptive innovations that developed in the major taxonomic groups.

The story of primate evolution is told through the teeth

The study of teeth, of fossil and living species, has contributed greatly to understanding the lives of all mammals and their evolutionary history. Fossilized teeth are the one indispensable source of information about the origins and evolution of primates and all other mammals for a variety of reasons. Teeth are incredibly durable. The enamel cap that forms the crown is the hardest tissue produced by the body, measurably as hard as steel but more brittle, making teeth by far the most abundant fossilized anatomical parts. Teeth are what remain most frequently, while other parts of an animal more easily disintegrate. Through evolutionary adaptation, dentitions, and molar teeth in particular, become highly sensitive to diet. Specific dental patterns (e.g., the toothcomb complex in lemurs, lorises, and galagos), and especially the complexity of molar tooth structure, keenly reflect phylogeny (Figure 8.1).

The information offered by fossil teeth includes:

- Diet. In adapting to the physical characteristics of foods (e.g., soft, hard, squishy, pliable, protected by thick skin or shells), the masticatory system as a whole (teeth, jaws, and musculature) reflects how food is acquired, ingested, and chewed. The sizes and shapes of molar teeth are especially responsive to selective pressure and they relate to the physical properties of the foods eaten – fruits, leaves, insects, and more – how the material needs to be broken down for swallowing and digestion. The thickness of the

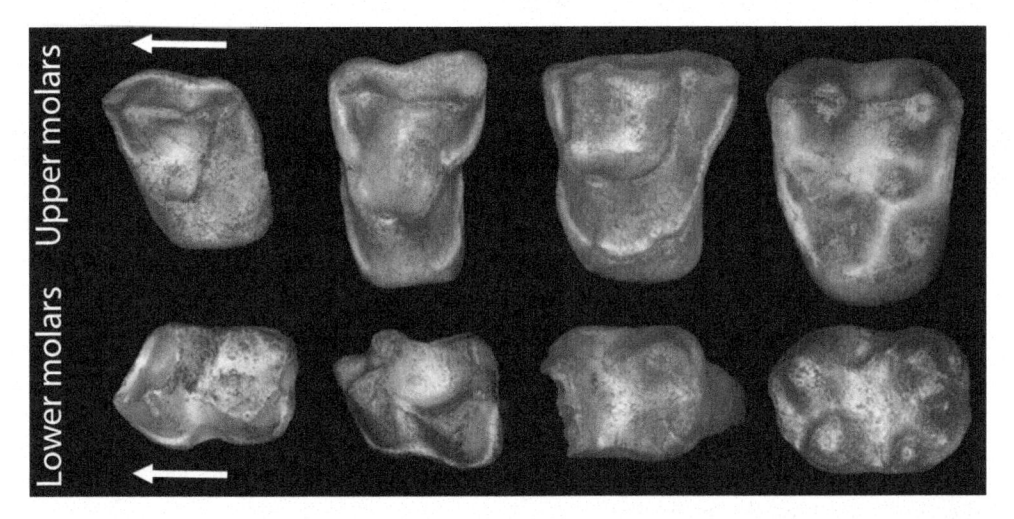

Figure 8.1 A sample of isolated fossil primate molars from four genera discovered in Libya at an Eocene site that is roughly 37–39 million years old. Upper teeth on top; lower teeth on bottom. All teeth are depicted at the same scale, and the largest tooth is about one-tenth of an inch (2.7 mm) long. Arrows are located on the outer sides of the crowns pointing to the front of the mouth. From left to right: *Karanisia*, a strepsirhine; *Afotarsius*, a tarsiiform; *Talahpithecus*, possibly an anthropoid; and *Biretia*, an anthropoid. Upper teeth are (left to right: M3, M2, M1, and M1. Lower teeth are: m2, m1 or m2, m1 or m2 (a broken partial tooth), and m2.

Adapted from Jaeger, J.-J., Beard, K. C., Chaimanee, Y., Salem, M., Benammi, M., Hlal, O., Coster, P., Bilal, A., Duringer, P., Schuster, M., Valentin, X., Bernard, M., Métais, E., Hammuda, O., & Brunet, M. (2010). Late middle Eocene epoch of Libya yields earliest known radiation of African anthropoids. *Nature*, 467, 1095–1098. https://doi.org/10.1038/nature09425

enamel cap that forms the crown of a tooth relates to food hardness. The activity of chewing also leaves marks as teeth wear – scratches, pits, and larger perforations of the enamel. These marks can indicate if hard foods like nuts in thick shells have been crushed while chewing, or if molars show a pattern of wear that maintains the sharpness of crests used to shear leaves.
- **Sociality.** The sizes and shapes of the canine teeth are often secondary sexual traits expressing dimorphism and monomorphism. For example, in baboons, males have very large dagger-shaped upper canines; females do not. So, even a single large dagger-like monkey canine fossil found in Africa leads us to infer that it is probably from a male. Further, we can infer that it probably came from a species with a social organization in which male dominance was an organizing factor, like many living primate species that are sexually dimorphic. On the other hand, finding a large sample of uniformly large dagger-like canines and no small ones in a fossil site in Asia would suggest that they belong to a hylobatid, following the monomorphic pattern of today's gibbons and siamangs.
- **Health and chronological age.** As a tooth crown forms, the enamel is layered down at a fixed rate in growth lines that resemble tree rings. The lines can be counted to estimate the age of an individual at death. If it was nutritionally challenged while the crown formed, enamel defects may be visible.
- **Environment.** About a third of the material that forms teeth consists of organic matter which absorbs carbon and other elements from the foods that are eaten. Since different types of carbon are found in different types of plants, the chemical signature of teeth can indicate if the plants eaten were fruits and leaves from a forest or grasses and grass-like plants from open country. Tooth chemistry also provides information on the degree of seasonality and rainfall intensity that an individual experienced as its teeth developed, its gestation length, and age at weaning.
- **Phylogeny.** When ancient DNA and proteins can be extracted from fossil and subfossil teeth, they provide phylogenetic information, as in the case of a two-million-year-old *Gigantopithecus* molar found in China. The DNA extracted from its enamel showed that this huge ape is closely related to orangutans. Continued improvements in the technology involved are making these methods applicable for older and older fossils.

The Paleocene Epoch: proto-primates

The beginning of primates starts with an event that fundamentally changed the biology of the entire planet and set the stage for the rise of primates and many other groups of mammals that are still extant today – Earth's collision with a large asteroid called Chicxulub 66 million years ago. It is named for the small town in Mexico near the site of the crash, marking the start of the Paleocene Epoch. The impact caused a chain reaction of destruction, a **mass extinction** that occurred over a brief period of time but had enduring consequences.

About 70% of the world's forests were wiped out and the food chains on land and in the oceans were devastated. Ultimately the dinosaurs were killed off as ecosystems collapsed. These animals had been dominant for 135 million years. The ecological breakdown was followed by a biological renewal, which included the adaptive radiations of the forerunners of the Order Primates, and the tropical rainforest ecosystems these animals inhabit. Within a million years, the biological rebound was underway. Ten million years after the Chicxulub asteroid crash, there is fossil evidence that early strepsirhine and haplorhine primates were spread across the warm, wet rainforests that had emerged on the planet, particularly in the northern continents – North America, Europe, and Asia were essentially one continuous mega-continent interrupted in places by regional seaways.

Plesiadapiforms existed from at least 65 million years ago – a million years after the Chicxulub impact – and for more than 25 million years thereafter. They lived in North America, Europe, and Asia, and probably in Africa as well. We can call them **proto-primates**, suggesting a primate predecessor or earlier version, because plesiadapiforms are the best available model of what the fossil forerunners of the strepsirhine-haplorhine common ancestor may have looked like adaptively. Their fossil remains provide a physical basis for how an early, non-primate placental mammal may have transitioned to evolve the characteristics of a genuine primate. Plesiadapiforms have also been called stem primates, which emphasizes the idea that they are the extinct mammals most closely related to today's primates and their extinct relatives.

The plesiadapiforms were abundant during the Paleocene. They are represented by thousands of fossils, mostly tiny teeth, attributed to more than a dozen families, about 50 genera, and more than 150 species – a very large and morphologically diverse order of mammals. Their skeletal anatomy is becoming better known because of new collecting methods that focus on blocks of rock that contain jumbles of well-preserved fossil bones that can be extracted, often as parts that can be reassembled as relatively complete skeletons. The use of three-dimensional CT-scanning and other visualization techniques that allow researchers to develop virtual fossils, and even reconstruct internal anatomy that can be studied, has contributed to major advances in the study of plesiadapiforms.

The oldest plesiadapiform fossil, *Purgatorius*, about 65 million years old, has features that are more primitive than the derived mammalian patterns found among primates in the skull, dentition, postcranial skeleton, and brain. The orbits were small and not forward-facing, there was no postorbital bar, the brain was smaller, their limbs were not indicative of leaping, and their digits were generally tipped with claws. The majority of plesiadapiform genera exhibit anatomical specializations of their own, too, like the elimination of several types teeth from the toothrow that are present in strepsirhines and haplorhines. Some have long, pointed, jutting lower incisors, or tall, wedge-like premolars that are not compatible with the morphology of any known primates. Traits such as these isolate most plesiadapiform genera as divergent taxa that cannot be directly ancestral to primates.

The plesiadapiforms were arboreal and some were omnivorous or insectivorous frugivores, reflecting an ecological combination that primates exploited with great success – the primate adaptive zone. *Purgatorius* had ankles that facilitated an inward rotation, or inversion, of the foot, and at least one other plesiadapiform, *Carpolestes*, had a nailed, opposable hallux, features facilitating locomotion in trees. But these animals did not move with the arboreal fluidity of a primate. Some had short, thick limb bones and digits tipped with large grappling claws, rather than long, slender limbs and sensitive fingertips tipped with nails. These differences highlight the adaptive diversity and phylogenetic complexity of plesiadapiforms. Some genera may have been more closely related to primates than others – plesiadapiforms are not a monophyletic group – which is why they are best considered as proto-primates, a taxonomic ancestral stock rather than a phylogenetic stem primate group.

The Eocene Epoch: euprimates

The oldest fossils that are clearly recognizable as primates belong to the strepsirhine and haplorhine clades. They first appeared in the fossil record about 56 million years ago during the early Eocene in North America, Europe, and Asia. They lived in northern tropical and subtropical ecosystems that were widespread, well away from the equator because the Eocene was an intensely hot period in Earth's history. It was over 100°F (38°C) every day of the year. The average global temperature at the start of the epoch was 55°F (13°C) warmer than the average world temperature of the early 2000s.

During the early Eocene, contiguity and connections of continents enabled the same families of early primates to attain enormous intercontinental distributions outside of today's regional patterns. One or two genera of strepsirhines are known to have lived on two continents that are now separated, Europe and North America, and one tarsiiform genus occurred in three, Asia, Europe, and North America. Newly discovered primate fossils from southern Africa have begun to show more taxonomic and morphological similarities among African, European, eastern Asian, and western North American forms. The Eocene world was productive for primates across the multiple continents that were physically interconnected – Afro-Eurasia is an example – to the extent that plants and animals were able to cross between them. The planet was warm, the poles were ice-free, grasslands were absent, and tropical and subtropical forests and animals predominated, even at low latitudes inside the Arctic Circle.

A crucial feature of the Eocene rainforest ecosystems in which primates were able to flourish was the dominance of angiosperm plants, particularly their physical structure and reproductive system. The abundance of angiosperm trees produced the first closed-canopy rainforests. This was conducive to the development of the primates' wide-ranging, tree-to-tree arboreal foraging patterns, the agile locomotor system, and the good eyesight which makes it possible. Angiosperms produce the food source that is central to primate diets and effectively explains their commitment to arboreal living: fruit. The seeds of angiosperms typically grow inside fleshy fruit that is nutritious. This is an adaptation that angiosperms evolved in order to enhance the survival of seeds by encouraging animals to feed on fruits and disperse their seeds via defecation. Primates are one of the chief angiosperm seed dispersers.

The flourishing of closed-canopy, angiosperm-dominant forests and early primates at the start of the Eocene Epoch are related phenomena. As major seed dispersers, primates promoted the growth and spatial distribution of new trees. As a result of this interaction, the primates were encouraging the proliferation of their own food sources and the development of a rich and varied forest habitat.

The early fossil strepsirhines and haplorhines were modern in morphology and adaptation, which is why they have been called **euprimates**, meaning real primates. As discussed further below, they had evolved a nimble arboreal skeleton. They had long, slender limbs, a muscular opposable hallux, and fully nailed digits (apart from grooming claws), evidence that they were able to leap about and grip branches with efficient grasping feet and sensitive finger tips. They were grasp-leapers, as discussed in Chapter 2. More derived than plesiadapiforms, the euprimates had smaller faces, eyes and brains that were relatively large, orbits that were rimmed laterally by the postorbital bar and positioned so that the eyeballs can look straight ahead, with a line of sight that was not blocked by a massive snout.

The appearance of euprimates in the Early Eocene, about ten million years after Chicxulub, suggests a prior history when the primate order first originated, which was followed by a period when the strepsirhine and haplorhine adaptive radiations were evolving their defining features, flourishing, and spreading across a vast intercontinental expanse. But physical evidence for these earliest phases in the form of fossils has proven elusive. No authentic primates have been identified among the relatively rich fossil collections from the earlier Paleocene Epoch, or during the close of the Age of Dinosaurs before that. The plesiadapiforms only show a small number of primate-like features. Also, the temporal gap in the fossil record of primates prior to the start of the Eocene at 56 million years may be acute, reaching further into deep time than the 65-million-year age of the oldest known plesiadapiform. Some molecular studies of living species estimate that primates originated long before that, as much as 85 million years ago. More modest estimates are near 75 million years.

The existence of modern-looking primates during the Eocene is part of a mammal-wide pattern. Many of today's orders of placental mammals also appear first as Eocene fossils, influenced by a favorable global climate and a favorable world geography. Like the Order Primates, other mammalian orders, like hooved mammals and bats, are recognizable at this early stage because their features demonstrate adaptations to life in the adaptive zones inhabited by their living descendants.

Adapiforms and fossil tarsiiforms: early strepsirhines and haplorhines

The fossil skulls and skeletons of the early strepsirhines are remarkably similar to Madagascar's lemurs. There is one major distinction between early Eocene fossils and living strepsirhines. The fossils do not have a toothcomb, the uniquely derived, universal trait shared by the lemuriforms, lemurs, lorises, and galagos, as discussed in Chapter 3. The fossils represent the ancestral stock from which toothcombed primates evolved, and are classified in a separate strepsirhine group called adapiforms.

In contrast to the adapiforms, the Eocene haplorhine fossils consist of species and specimens that are much smaller and more fragile, so less bony anatomy is preserved in the fossil record. They are related to tarsiers, their ancestral stock. Together with living tarsiers, the Eocene forms are classified in a distinct haplorhine group, the tarsiiforms, which separates them taxonomically from the other haplorhines, the anthropoids.

A substantial number of the early fossil tarsiiform genera exhibit crucial, characteristic adaptations found in extant tarsiers. Huge orbits in many of the fossil crania show they were nocturnal. Large, pointed premolars and molars show they were faunivorous. Very long legs and feet show they were vertical-clingers-and-leapers. The enormity of the fossils' orbits indicates more than nocturnality. The large-eyed living primates that come closest to the reconstructed eyeball proportions of some fossil tarsiiforms are extant tarsiers and several lorises, predaceous nocturnal hunters.

In the early Eocene, the adapiform and tarsiiform fossils seem to appear all at once and well established: phylogenetically diverse, taxonomically abundant, and modern in appearance. Adapiforms lived in a variety of ecological niches. The majority of genera were probably diurnal, although it is difficult to be certain. They varied in mass from small to medium and large body sizes, focused on eating fruits, leaves, or insects, and locomoted by quadrupedalism and leaping. The tarsiiforms were different. Judging from orbit size, probably all were nocturnal. Most of them were very small-bodied insectivores and frugivorous faunivores generally weighing from less than 3.5 oz. (100 g) to 10–15 oz. (300–400 g), and they were fully committed vertical-clingers-and-leapers. The adaptive radiation of fossil tarsiiforms consisted of many genera that were ecological specialists living at the edge of the primates' arboreal zone like their only surviving descendant, the nocturnal predaceous tarsiers.

The history of primates is not always a story of change. The similarity of features in fossil adapiforms and living lemuriforms, and in fossil tarsiiforms and living tarsiers, is an important finding. A North American Eocene adapiform like *Notharctus*, if not for the lack of a toothcomb, known from many specimens that include several nearly complete skulls, dentitions, and skeletons that are 50–55 million years old, could otherwise be mistaken for a living lemur if it was found on Madagascar. For fossil tarsiiforms, the similarities can be equally striking.

Over time, millions of years, taxa come and go, adaptive radiations arise, species become differentiated; many become extinct in the natural course of evolution. Yet, even as taxonomic assemblages change, morphologies can persist within monophyletic groups. Some of the critical adaptative morphologies and behaviors of extant strepsirhines and haplorhines have persisted since the early Eocene. In a sense, this means that some living lemurs and living tarsiers are **living fossils**, taxa whose characteristics have remained relatively unchanged for very long periods of geological time.

As noted, the adapiforms and tarsiiforms are called euprimates because of the clear phylogenetic and adaptive connections between them and the living strepsirhine and haplorhine radiations. They are also called **crown primates** because they and their descendants – especially the living species, also called crown primates – form the top of the primate phylogenetic tree, as opposed to being part of the more primitive base from which all primate lineages arose. The base is referred at as its stem. This is why some scientists refer to the proto-primates, the plesiadapiforms that preceded adapiforms and tarsiiforms in time, as stem primates, based on the supposition that primates evolved from a species that is classified in the plesiadapiform order. The stem vs. crown concept applies to other groups as well. The oldest fossils interpreted as anthropoids, for example, are called stem anthropoids to acknowledge their primitiveness and to clarify that later fossils and living species form a separate clade of crown anthropoids.

The Eocene rise of toothcombed strepsirhines in Africa

Fossil primates are well represented in mainland Africa. There are rich sites in the north from roughly 45–30 million years ago that extend into the Arabian Peninsula. They have produced strepsirhines, tarsiiforms, and many early anthropoids. A handful of older teeth from Morocco are potentially the oldest African primates, from the end of the Paleocene, about 57 million years ago, but their phylogenetic status is unclear; they may be proto-primates.

There are more than two dozen genera of primates that existed in Egypt, northwest Africa, and the Arabian Peninsula during a roughly eight-million-year interval spanning the late Eocene and early Oligocene. An exceptional, and still increasing, fossil trove was found in Egypt's desert south of Cairo, an area known geologically as the Fayum Depression, named after the city that is located nearby. It provides critical evidence bearing on lemuriform origins. At least two groups of strepsirhines lived there, including the adapiforms that lacked a toothcomb and tooth-combed lemuriforms, during the Eocene when this area was wet, warm and forested.

The Fayum fossils provide critical evidence regarding the early evolution of strepsirhines in mainland Africa and the origins of Madagascar's lemurs. For example, the adapiform *Afradapis*, about 37 million years old, known from teeth, jaws, and the ankle, is part of the ancestral stock from which toothcombed lemuriforms evolved. It was a medium-sized, slow-climbing quadrupedal folivore that in its dentition strongly resembled a neotropical howler monkey. *Afradapis* was closely related to adapiforms from the north, including the North American genus *Notharctus*. Their similarity highlights the extensive multi-continental dispersal experienced by primates and other mammals during the Eocene. Fayum fossils also demonstrate that lemuriforms, the extant branch of the strepsirhine radiation, were diversified in northern Africa.

Karanisia and *Saharagalago*, also about 37 million years old, are represented by dentitions. They demonstrate the antiquity of the two lorisoid families in Africa. *Karanisia* is related to lorisids; *Saharagalago* is related to galagids. Since the lorisoids and lemuroids share a common ancestor (Figure 1.10), these genera suggest that there might have been a broad radiation of toothcombed primates flourishing on the African mainland living alongside the more primitive adapiforms, prior to the colonization of Madagascar by lemurs.

Plesiopithecus, based on craniodental remains found in the Fayum, dated to 34 million years, and a younger genus from Kenya, *Propotto*, known from fossil teeth, provide evidence supporting the hypothesis that ancestors of Madagascar's primates had differentiated on the mainland before dispersing to the island. This is because *Plesiopithecus* and *Propotto* are most closely related to Madagascar's aye-aye, *Daubentonia*, and not to the lorisoids or any of the other strepsirhines found on the mainland. The unique dentition of *Plesiopithecus* demonstrates that the aye-aye's highly unusual gnawing anterior teeth, described in Chapter 3, began evolving from a toothcombed morphology in a predecessor that existed in Africa. That means the ancestral stock of the other Madagascar lemurs – which had a basic toothcomb, like the Egyptian *Karanisia* – split off from the daubentoniid family while it lived on the mainland. Madagascar's endemic lemurs are the survivors of a more widespread group that was rooted on the continent.

Lemurs probably dispersed from the African mainland to the island of Madagascar by way of a temporary landbridge that uplifted between the landmasses and connected them when sea level fell and the shorelines were closer. Geological evidence of the bridge exists on the seafloor of the Mozambique Channel that divides the mainland from the island. The landbridge, which may have been broken into a series of islands that breached the water's surface at times during the Cenozoic, was able to support forested vegetation that primates naturally use. When the bridge arose to connect the mainland with Madagascar, the geographical ranges of the flora and fauna living at water's edge on both sides merged.

Eocene strepsirhines and tarsiiforms from Asia

A significant adaptive radiation of non-toothcombed, early strepsirhines was once present in Asia, where today only a few species of nocturnal, arboreal lorises remain. About a dozen fossil genera, consisting of dental and postcranial remains widely regarded as Eocene and Oligocene strepsirhines, have been found. The Asian genera differ distinctly from the strepsirhine fossils discovered in Egypt that lived during the Eocene. By the middle Eocene, roughly 45–40 million years ago, the Asian and African adapiforms had become two regional clades with taxonomically distinct communities.

Tarsius is the only living tarsiiform, and it is now endemic to the islands of Southeast Asia. The fossil record of tarsiiforms, a taxonomic group of several dozen genera that includes the tarsiid family, is sparse in Asia, but this does not reflect the diversity of the tarsiids that had existed elsewhere. During the Eocene, these primates were widespread in the northern continents in Asia, Europe, and North America, and a few specimens have been found in northern Africa. Their fossil dentitions are relatively common at

Eocene sites. As noted above, all the fossil tarsiiforms were small primates. Many were insectivorous, or frugivorous insectivores, and they were nocturnal vertical-clingers-and-leapers. In addition to identifiable strepsirhines and tarsiiforms, a small Eocene collection of fossil primate teeth from East and South Asia, classified as eosimiids, have been interpreted as early anthropoids, but they are not morphologically distinct from various tarsiiforms. There is little reason to align them with anthropoids. It is very difficult to identify primitive anthropoids from isolated teeth and there are no cranial remains of eosimiids. Cranial anatomy provides very reliable evidence about anthropoid phylogeny.

The Oligocene Epoch: proto-anthropoids and crown anthropoids

From the Oligocene through the Pleistocene, from about 34 million years ago to 12,000 years ago, anthropoids are most prominent in the fossil record. By late Eocene and early Oligocene times, 37–34 million years ago, early anthropoids had become well established in Africa. Many of them are classified as parapithecoids. Several of them resemble living platyrrhines more than living catarrhines. Anthropoid primates were also present in South America near the end of the Eocene or the early Oligocene. There is no solid evidence that anthropoids were present in Asia during the Eocene, as noted above, nor in the Oligocene. The status of primates in Madagascar during these epochs remains unknown because there are no fossil land mammals that have been discovered on the island prior to the Holocene, the last 12,000 years.

A fossil called *Rooneyia*, discovered in Texas, in the United States, comes from a geological period that spans the late Eocene and early Oligocene, 37–35 million years ago. It is a proto-anthropoid that has relevance to the geographical origins of anthropoids. *Rooneyia* consists of a well-preserved cranium with an excellent dentition. Only one specimen of the genus has ever been found (Figure 8.2), and it bears little resemblance to any other fossil primates discovered in North America.

Among the dozens of fossil primate crania – many relatively complete – that have been discovered in Eocene sites from North America, Europe, Asia, and Africa, *Rooneyia* is the only one that shows incipient orbital closure (Figure 4.3). Postorbital closure is present in all anthropoids and is a critically important shared derived feature, as discussed in Chapter 4. The orbit size suggests *Rooneyia* was diurnal, and the position of the orbits below the forehead or frontal bone indicates that the eyes were deeply recessed in the socket. This morphology is uniquely anthropoid as well. The fossil's dentition is typically frugivorous. The phylogenetic interpretation favored here is that *Rooneyia* is more closely related to anthropoids than to any other primates. This means that the origin of the anthropoid lineage, when it separated from an ancestral haplorhine stock, involved primates that were present in northern continents.

Late Eocene or early Oligocene: early platyrrhines

The oldest South American fossil primates are estimated to be 41–29 million years old, between the late Eocene and early Oligocene. They consist of a handful of very small isolated teeth, mostly molars, discovered in Peru, and classified as two distinct genera, *Perupithecus* and *Ucayalipithecus*. These teeth are not classifiable in any of the three platyrrhine family-level clades that exist today. They belong to another, earlier branch of South America's primate radiation.

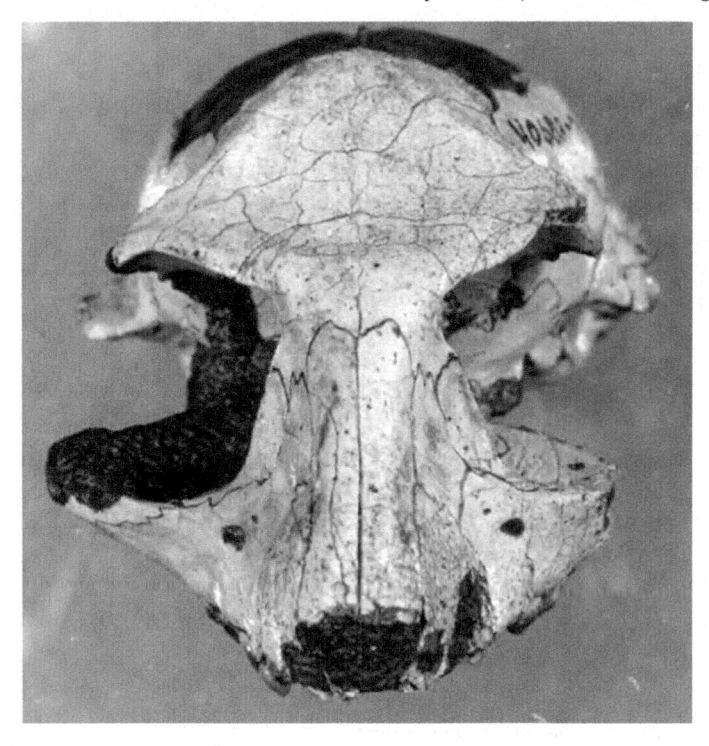

Figure 8.2 The fossil cranium of *Rooneyia*, a proto-anthropoid from the Late Eocene found in Texas, U.S.A.

Courtesy of T. Rowe, Vertebrate Paleontology Laboratory, Texas Memorial Museum, University of Texas at Austin.

It has been suggested that *Perupithecus* and *Ucayalipithecus* closely resemble early African anthropoids like those found in Egypt, including the parapithecoids, and that they the indicate the ancestors of platyrrhines rafted on a mat of vegetation across the South Atlantic Ocean – several hundred miles narrower during the Eocene than it is today. This far-fetched view, known as the Transatlantic Scenario, is obviously highly improbable. A more reasonable explanation for platyrrhine geographical origins is what is called the America's Scenario. In this reconstruction of events the dispersal of ancestral platyrrhines from the Old World would have happened gradually over a long period of time through natural range expansion, by way of an overland route that is known to have existed during the Eocene. This route enabled genera of fossil strepsirhines, tarsiiforms, and other mammals to cross between Europe and North America by way of Iceland and Canada. It is an example of the well-known method of **geodispersal**, geographical expansion of a species, or a set of coexisting animals and plants, by crossing a geological barrier that had been eliminated. The America's Scenario offers a reasoned explanation of how ancestral New World monkeys coming from the Old World could have gotten close to what may still have been the island continent of South America. But it is still no more than a speculative idea. In this case, it is possible that some fossils do have the power to tell us *where* an ancestral platyrrhine lineage may have come from geographically, but no fossils can tell us *how* they accomplished the move and managed to set foot in South America.

Africa's stem catarrhines: proto-apes

In very general terms, the stem catarrhine fossils are ape-like. They are proto-apes known from skulls, teeth, and postcranial remains. They show that Old World monkeys and their adaptations evolved from ape-like anthropoids.

In addition to the Fayum genus *Propliopithecus* (also called *Aegyptopithecus*), Africa's stem catarrhines consist of roughly two dozen fossil genera, mostly from Kenya and Ethiopia, that more closely resemble living apes – not living cercopithecoid monkeys – but still retain characteristics that are more primitive than living catarrhines. This large and complex fossil record spans the late Oligocene and Miocene Epochs, 27–5 million years ago.

The two or three families of African stem catarrhines are diverse anatomically, yet modern in important respects. As a group they are characterized by:

- A very large range of body sizes, 8 lbs. (3.5 kg) to 100 lbs. (45.3 kg), from strepsirhine-sized to chimpanzee-sized
- Dentitions that are generally frugivorous and resemble gibbons, chimpanzees, and gorillas but not Old World monkeys
- Crania resembling gorillas and gibbons, at the extremes
- Skeletons that were adapted to arboreality and resembled the morphology of the Fayum taxa and platyrrhines

The oldest stem catarrhine, dated to 30 million years, *Propliopithecus*, is the largest Fayum genus at 15 lbs. (6.7 kg). It is represented by several nearly complete and undistorted fossil crania, many dentitions, and postcranial remains. *Propliopithecus* had a relatively long snout compared to most anthropoids, but the face was not as long or large as a strepsirhine's (Figure 8.3). *Propliopithecus* was sexually dimorphic in canine size and skull shape, and at least some males developed a sagittal crest. Its cheek teeth had large, blunt cusps, a frugivorous pattern. The limb bones were robust and indicative of a quadrupedal locomotor style.

One of the most important features of *Propliopithecus* is its dental formula, 2.1.2.3, which is the same, derived tooth count found in Old World monkeys and apes. It is one of the reasons why *Propliopithecus* is recognized as a stem catarrhine. The ear region's ectotympanic bone, however, was platyrrhine-like, with a flattened, ring-like shape (Figure 4.4). That is one of the reasons why *Propliopithecus* is not a bona fide catarrhine. The ectotympanic later evolved into the tube-like structure found in all modern catarrhines. It is a feature that enables us to identify fossils belonging to the clade of crown catarrhines, and to separate them from stem catarrhines. The combination of a derived dental formula and primitive ear region in the same fossil species is an example of **mosaic evolution**, the evolution of different traits at different rates or times within a clade.

The Miocene Epoch: the early rise of modern primates

The ancestral stock, and potential direct ancestors, of today's living primates in South America, Africa, and Asia first appear in the Miocene, along with fossils that are less closely related to them but belong in the same clades. In the absence of a fossil record, as noted, no comparable information is available for Madagascar's lemurs.

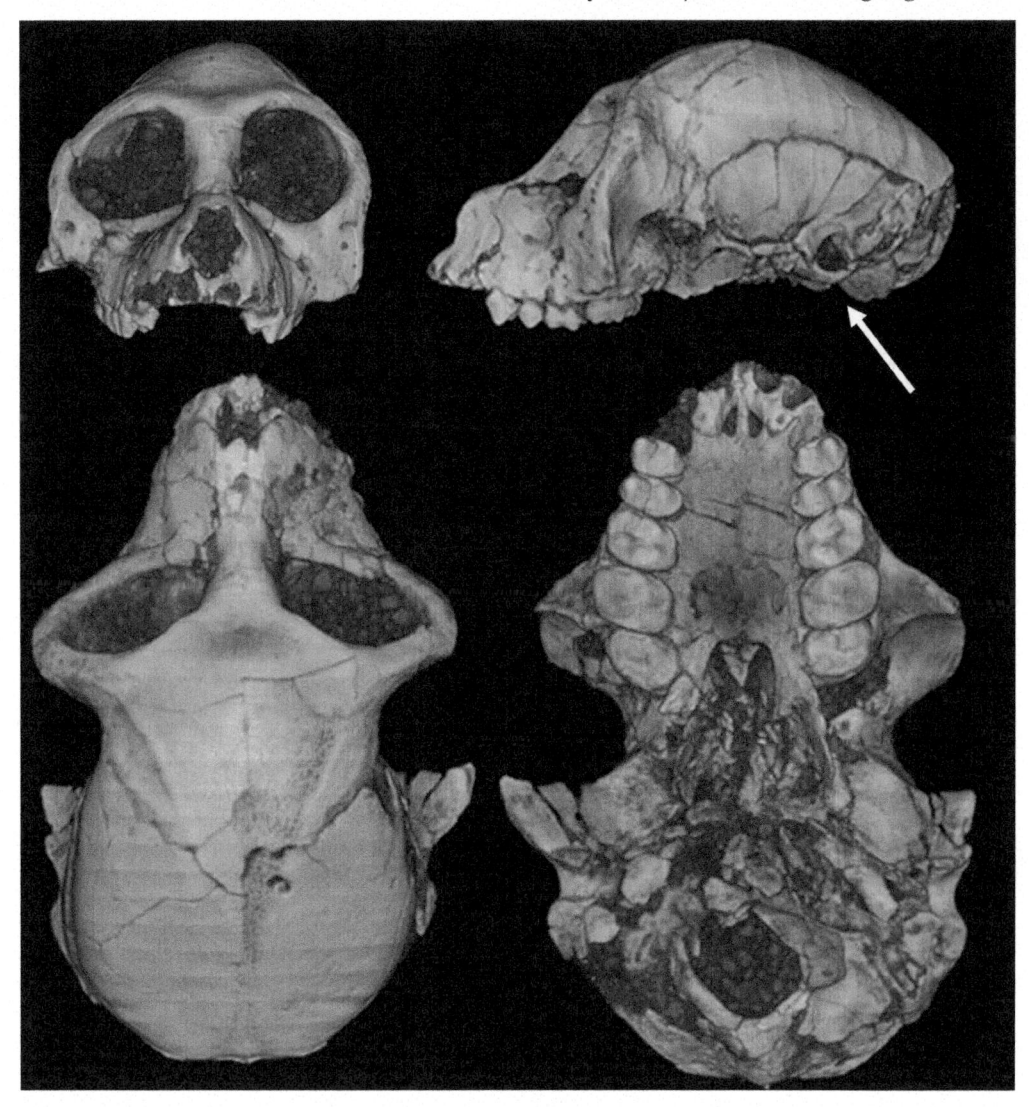

Figure 8.3 Four views of the cranium of a female fossil *Propliopithecus* (also called *Aegyptopithecus*), the oldest stem catarrhine, from the Oligocene, dated at 30 million years. The large circular opening for the middle ear (arrow), where the primitive ring-like ectotympanic bone for the eardrum attaches, is visible in the top-right image. The shapes of the cranium and dentition conform to the patterns found among fossil and living apes.

Photo credit: Simons, E.L., Seiffert, E.R., Ryan, T. & Attia, Y (2007). A remarkable female cranium of the early Oligocene anthropoid *Aegyptopithecus zeuxis* (Catarrihini, Propliopithecidae). *Proceedings of the National Academy of Sciences*, *104*(21), 8731–8736. www.pnas.orgcgidoi10.1073pnas.0703129104. Copyright (2007) National Academy of Sciences, U.S.A.

Platyrrhines

The fossil record of platyrrhines is very poor in comparison to the record of Old World anthropoids. There are very few individual specimens, fossil localities, and taxa. However, the living platyrrhines are so diverse and well differentiated morphologically, that

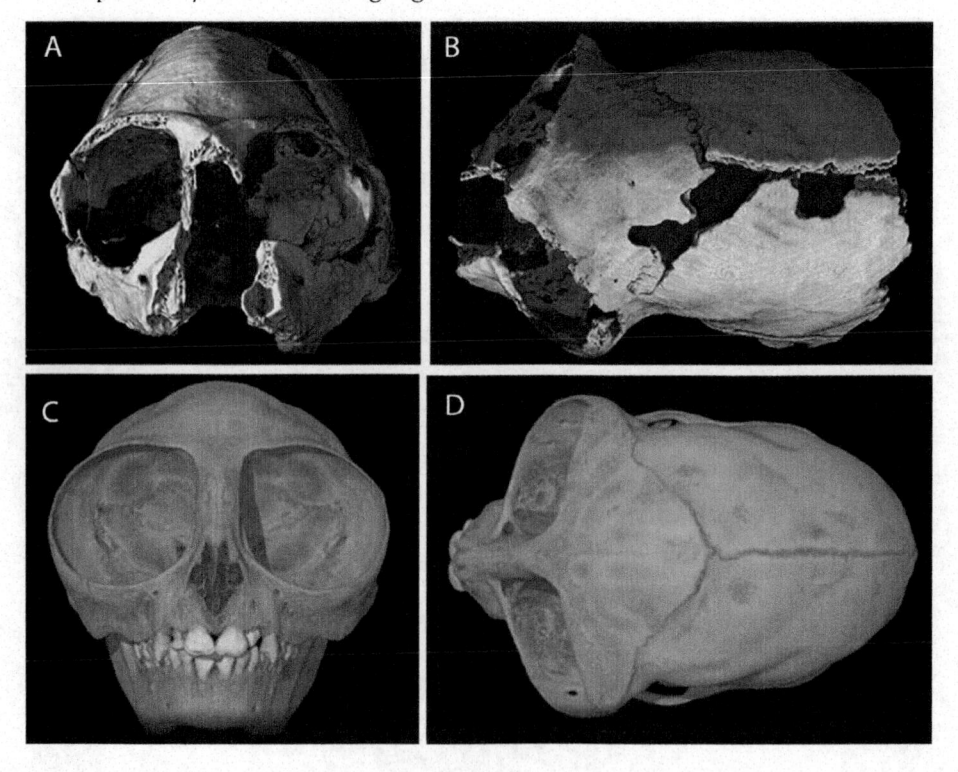

Figure 8.4 Frontal (a, c) and top views (b, d) of the owl monkey fossil, *Tremacebus* (above), and a skull of a living owl monkey (below) showing that both have very large orbits.

Credit: derived from Digimorph.org. https://digimorph.org/

their extinct relatives can often be identified on the basis of shared derived features, which often correspond with unique adaptations.

From the early Miocene, about 20 million years ago, we have two fossil genera from Argentina, *Dolichocebus* and *Tremacebus*, known from skulls and teeth, that are phylogenetically and adaptively linked with living forms of platyrrhines. *Dolichocebus* is related to squirrel monkeys. Its body size, skull, and teeth are similar to squirrel monkeys, suggesting it, too, was a frugivore that also ate insects. *Tremacebus* is related to owl monkeys (Figure 8.4). The skull's large orbits are like an owl monkey, indicating that it was also nocturnal.

These fossils demonstrate that lineages leading directly to living New World monkeys go back as far as 20 million years. Squirrel monkeys and owl monkeys are living fossils. The fact that *Dolichocebus* and *Tremacebus* come from the same time period and were found not very far from one another geographically, suggests that they could have belonged to the same primate community. Today's squirrel monkeys and owl monkeys live in primate communities together. These fossils are among the indicators that today's communities of neotropical primates are old as well.

The Long Lineage Hypothesis

The **Long Lineage Hypothesis** suggests that the majority of modern platyrrhine genera, the major family and subfamily clades, and the predecessors of the current Amazonian communities have a long, millions-of-years-old history. This hypothesis aligns with genetic evidence developed by a method called the **molecular clock,** which is used to estimate the time when a lineage originated. It is based on the observation that mutations accumulate in DNA at a constant rate over geological time, so the genetic differences between sets of living species can be used to calculate how long ago they diverged.

The Long Lineage Hypothesis is supported by fossil evidence indicating that many contemporary genera belong to ancient lineages. In taxonomic terms this means that if we trace the immediate ancestors of extant genera back in time, each of the fossils we identify would be classified in the same genus as those living today, or in a very closely related genus.

In fact, this is what we find. There is a paleocommunity of nearly a dozen fossil platyrrhine genera found at a middle Miocene site in Colombia, 12–14 million years old. More than a half-dozen fossil genera from the site, representing all the major platyrrhine clades, are directly aligned with extant New World monkey genera, phylogenetically and adaptively. This means that the structure of fossil platyrrhine communities in a region close to today's Amazon basin closely resembled modern Amazonian communities. Dental remains provide evidence that there were fossil squirrel monkeys, owl monkeys, howler monkeys, titi monkeys, Goeldi's monkeys, and sakis or uakaris in the middle Miocene, as there are today.

An important instance of genus-level continuity between the remote past and the present is the owl monkey lineage, as mentioned. It is represented by a 12–14-million-year old, large-eyed species, *Aotus dindensis*, from Colombia that is classified in the same genus as the living owl monkey, *Aotus*. Additionally, there is the 20-million-year-old fossil owl monkey from southern Argentina, *Tremacebus*, that is even older. A relative of modern squirrel monkeys, the 20-million-year-old fossil genus *Dolichocebus*, appears at the Colombian site as well, showing a continuity of six to eight million years between the Oligocene and Miocene fossils.

The pattern of long lineages demonstrated by the platyrrhines encompasses all the major ecological niches and more than half of today's extant genera. It shows that the core phylogenetic and ecological structure of platyrrhine communities has also been maintained over a long period of evolutionary time. The history of platyrrhines during the past 20 million years, roughly, is an example of a phenomenon called **evolutionary stasis,** the lack of evolutionary change.

Although we usually think of evolution as a process that produces biological change, adaptive stability has obvious advantages and it, too, is a product of natural selection. In fact, the lack of substantial change in a lineage over the course of evolutionary time is common. Natural selection is likely to preserve fundamental adaptive traits that enable a species to maintain its ecological roles so long as they are effective and others do not arise via genetic mechanisms to alter or replace them. The large eyes of the owl monkey lineage provide an example. The evolution of large eyes opened up an entirely new ecological niche that did not overlap with any other neotropical monkeys, a niche whose defining environmental context, nocturnality, would remain stable. By the same token, taxa may also remain stable over geological time.

Catarrhines

In the Oligocene, we have seen the existence of stem catarrhines, like *Propliopithecus* that was ape-like craniodentally but lacked diagnostic features found in all crown catarrhines.

Fossil discoveries from Miocene sites have shown that a diversity of abundant ape-like crown catarrhines existed in Africa, Asia, and Europe. One of the most important genera is *Proconsul*. *Proconsul* is a proto-ape from the early Miocene, 23–18 million years ago, that is known from nine or more partial skeletons found at one site in Kenya. Males, weighing about 44 lbs. (20 kg), were twice the weight of females. The face of *Proconsul* was relatively short and its cranial capacity was comparable to other anthropoids of similar body size. The ear region exhibits an important trait shared with all living catarrhines, a tube-shaped ectotympanic bone, rather than the more primitive ring-like structure seen in *Propliopithecus*. *Proconsul* incisors and cheek teeth indicate a frugivorous diet, and its skeleton, overall, is similar to arboreal, quadrupedal colobine monkeys and large platyrrhines. The postcranial remains show that it had not yet evolved the form of climbing quadrupedalism and suspensory behaviors that characterize today's apes. But how far along toward modern apes was it? Did *Proconsul* have a tail?

Some of the critical fossil specimens of *Proconsul* are damaged to the degree that a definitive answer is not possible. But the remains of another contemporaneous proto-ape, *Nacholapithecus*, is similar to *Proconsul* and includes an identifiable vertebra that belongs to the coccyx, the structure at the base of the vertebral column of living apes that is a remnant of the external tail. *Nacholapithecus* evidently lacked a tail. Did *Proconsul* also lack a tail? We can't be sure. Consequently, scientists differ in their views of *Proconsul* phylogeny. Some consider the genus a stem catarrhine while others maintain it presents a mosaic of primitive catarrhine features and derived ape features.

The rise of proto-apes, and then Old World monkeys, marks an important transformation in the primate fauna of Africa, a faunal turnover, when one set of species replaces another that lived in an ecosystem. The archaic anthropoids, stem catarrhines, and adapiforms that were prominent in the Fayum disappear from the community. The only forms that remained on the mainland were the small, nocturnal, frugivorous-insectivorous lorisoids that occupied niches that were not utilized by anthropoids. What drove the transition is not clear. How much was the result of climate and habitat changes, and how much was due to competition? In any case, the changeover in the composition of African primates from the Oligocene is apparent in the Miocene period when catarrhine lineages are ultimately formed, leading to the communities that exist today.

Africa's great apes

Fossil evidence of chimpanzees and gorillas is meager. It consists of a handful of isolated teeth, including several specimens that are poorly preserved, and the upper end of a single damaged femur. There are teeth from Kenya that have been classified as *Pan*, the same genus as living chimpanzees and bonobos. They are about half-a-million years old, while teeth attributed to a gorilla ancestor, from Ethiopia, are about ten million years old. Molecular estimates date the origins of the *Pan* lineage to about seven million years and the *Gorilla* lineage to roughly ten million years.

Africa's Old World monkeys

Old World monkeys originated in Africa in the early Miocene nearly 30 million years ago according to molecular estimates, when apes dominated the primate fauna. The date overlaps with the ages of the latest fossil primates discovered in the Fayum, particularly *Propliopithecus*. This is important because *Propliopithecus* has a very generalized dental and cranial morphology that serves as a "morphological model" from which modern hominoids, as well as cercopithecoid monkeys, can be derived.

The oldest fossil genera that can be identified as Old World monkeys are between 22 and 15 million years old. The older genus, *Alophe*, is represented by a few partial mandibles with teeth, and the 15–20 million-year-old *Victoriapithecus* is known by teeth, cranial, and postcranial remains. It was a medium-sized nine pound (4 kg), sexually dimorphic, quadrupedal monkey that would have been adept in either arboreal or terrestrial settings. *Victoriapithecus* is more primitive cranially than other cercopithecids because it lacks the derived specializations that define either of the two cercopithecid subfamilies. But, like *Alophe*, it is recognizable as a cercopithecoid by the bilophodont-like shapes of its molars. But in both genera the parallel ridges of the true bilophodont pattern were not yet fully developed.

Skulls, dentitions and skeletal remains of about two dozen fossil cercopithecid genera have been discovered in Africa. Nearly half of them belong to extant genera, including a macaque, *Macaca*, and a guenon, *Cercopithecus*, that are 5.6–6 million years old. Both cercopithecid subfamilies, the cheek-pouched monkeys and leaf-eating monkeys, are represented by fossils that are roughly the same age. Both groups occupied forested microhabitats, exploiting arboreal, terrestrial, and semi-terrestrial niches. Fossil geladas, *Theropithecus*, the highly terrestrial, open-country grass-eaters, were also present then. Today's most widespread open-country genus in Africa, *Papio*, is represented by abundant fossils. The earliest specimens are about 2.5 million years old.

Asia's Old World monkeys

Mesopithecus is a 7–8 million-year-old Miocene colobine, and the oldest classifiable Asian cercopithecid. It is a relatively primitive colobine, the remains of which have been found in China, Iran, Afghanistan, and Pakistan. It is also known from many sites in central and western Europe. The anatomy of *Mesopithecus* is well known from many fossils. It was a folivorous, extra-large monkey weighing 24 lbs. (11 kg), that occupied a variety of habitats. It may have been semi-terrestrial and somewhat omnivorous in parts of its range.

One of the most important points about fossil *Mesopithecus* and other Asian cercopithecids is their geographic connections with taxa that were widely distributed in Europe and Africa: they are Afro-Eurasian rather than Asian endemics. For example, dental remains of a one million-year-old Pleistocene species of the gelada, *Theropithecus*, has been discovered in central India. Geladas now are found only in a relatively small area of northeast Africa but during the Pleistocene they ranged widely across northern, eastern, and southern Africa, and into southern Europe and South Asia.

Asia's lesser apes

A variety of subfossil gibbons are known from China and eastern Asia. They are classified as extinct species of living genera, going as far back as two million years. The oldest hylobatid consists of 7–8 million-year-old specimens of *Yuanmoupithecus* that preserves diagnostic, derived hylobatid features of the face and dentition. This fossil is much younger than the time period when gibbons are thought to have split off as a lineage from the common ancestor that hylobatids shared with the great apes. The molecular clock estimates that happened 20–17 million years ago.

Asia's great apes

More is known about the fossil record of orangutans than any other extant great ape. Fossil orangutans classified in the living orangutan genus, *Pongo*, were once widely distributed in the Indonesian archipelago and on the mainland, living as far north as southern China during the Pleistocene, 2–1 million years ago. Other fossil relatives that belong to the orangutan clade come from earlier sites in Turkey and southern Asia. Two noteworthy examples are very large genera, one that existed in the Miocene and the other in the Pleistocene, which is discussed in the next section below.

Sivapithecus is a super-large ape (88–198 lbs.; 40–90 kg) whose 12–8 million-year-old Miocene fossils have been found in the Himalayan foothills of India, Pakistan, and Nepal. Its facial structure is very similar to the unique, derived morphology of orangutans. Also, like orangutans, *Sivapithecus* incisors are quite large and its cheek teeth are thickly enameled, adaptations for eating large, hard fruits. However, skeletal remains indicate that *Sivapithecus* had not evolved the characteristic quadrumanous locomotor adaptations exhibited by living orangutans. Their locomotor pattern was more comparable to the climbing quadrupedalism of African great apes.

Fossil apes from Europe

In addition to cercopithecids like macaques, geladas and the extinct *Mesopithecus*, there was a distinctive adaptive radiation of about ten genera of ape-like stem catarrhines called pliopithecids that also lived in Europe during the Miocene, between 18 and 8 million years ago. Several genera of the family ranged into Asia as well. One of the distinguishing features that sets apart the pliopithecids from more modern catarrhines is the lack of a tube-like ectotympanic bone, which was morphologically more similar to the structure found in *Propliopithecus* from the Fayum, Egypt. Cranially and dentally, they strongly resembled gibbons, but they did not have a gibbon-like skeleton suited for brachiation. They were quadrupedal climbers capable of some degree of suspensory locomotion.

More than a dozen Eurasian fossil apes belonging to the family in which modern apes are placed, the hominids, were also present in Europe in the Miocene between roughly 14 and 7 million years ago. They are represented by skulls, dentitions, and postcranial remains. They come from many parts of western Europe and western Asia, and from several circum-Mediterranean sites, a proximity which helps explain biogeographic connections to African apes. Many have a combination of skeletal morphologies consistent with the climbing and suspensory adaptations of modern apes but they were more varied in locomotor style, contrasting specializations like knuckle-walking in the semi-terrestrial chimpanzees and gorillas, and quadrumanous climbing in the orangutans.

The Pliocene and Pleistocene Epochs: primates enter the modern world

The Pliocene and Pleistocene Epochs were short time periods, covering the last 5.3 million years (Table 8.1). Few primate fossils are known from the Pliocene. More primate remains have been found in the Pleistocene, about a dozen genera of Old World monkeys and apes in Afro-Eurasia and several in the neotropics. Living genera, including macaques and geladas are found as fossils in Africa and Asia.

In the Pliocene and Pleistocene, about 5 million to 11,700 years ago, as part of a worldwide mammalian pattern that affected primates as well, the fossil record shows that many large-bodied primates began to become extinct in South America, Africa, Asia, and in Europe. This was also probably occurring in Madagascar but, as mentioned, we lack fossils from this region. At least four large-bodied genera of Old World monkeys vanished in Africa during the Pleistocene. Some ape species were gigantic, such as the Asian *Gigantopithecus blacki*.

The largest primates that ever existed were species of *Gigantopithecus*, close relatives of the orangutan. One was gorilla-sized and the other, the younger species, was even larger, 440–600 lbs. (200–272 kg). It appears to have gone extinct about 300,000 years ago. Though extraordinarily large by modern standards, *Gigantopithecus* was not ecologically anomalous. Sites that have produced *Gigantopithecus* fossils also preserve fossil remains of about nine living primate genera, including orangutans, gibbons, cheek-pouched monkeys and leaf monkeys, a large community of primates that lived in a forested, warm and humid habitat.

The massive size of *Gigantopithecus* means it must have been terrestrial. It is doubtful there are enough trees with branches strong enough to be able to support regular use by an animal weighing so much. The largest living arboreal mammal is the orangutan, which weighs far less at about 80–175 lbs. (36–80 kg). *Gigantopithecus* had powerful, enormous jaws and huge cheek teeth, with blunt cusps covered by thick enamel. Relative to molar size, the incisors were small and blocky, and the bluntly pointed, low-crowned canines tended to wear down heavily along with the incisors, the result of intensive biting. Dietarily, *Gigantopithecus* probably relied on an abrasive, varied diet of grasses, herbs, hard fruit, and roots and other vegetables dug out of the ground. *Gigantopithecus* required a huge volume of food each day, so it is likely that anything that seemed edible upon inspection would be tried.

A significant factor leading to the gradual extinction of the large primates during the 2.8 million years of the Pleistocene is attributable to global environmental changes that altered local habitats. In fact, it was their very large size that made many species particularly vulnerable because they were unable to withstand and/or adapt fast enough to radical changes that transformed their ranges and diminished the sizes of their populations to the point of unviability. They required more space and food, were slow-growing and slow to reproduce. Smaller primates that coexisted in some of the same communities were the ones most likely to survive.

The Holocene Epoch: subfossils from Madagascar and the Caribbean

During the Holocene, roughly the last 12,000 years, a rash of extinctions claimed both giant primates and medium-sized species confined to islands, including eight subfossil lemurs on Madagascar and an entire radiation of New World monkeys.

There is no primate fossil record on Madagascar other than the subfossils of the Holocene Epoch. However, as described in Chapter 3, there is an extraordinary series of subfossil lemurs, many that were much larger than any living lemurs, from dozens of sites that are several thousands of years old, and some that are only hundreds of years old (Figure 8.5). Many specimens are very well preserved. There are entire skeletons of species that no longer exist today. They belong to paleocommunities of lemurs that also include specimens of lemur species that are now alive.

Primates first arrived in the Caribbean from mainland South America during the Miocene and dispersed to Cuba, Hispaniola, and Jamaica (Figure 8.6). All of them are now extinct. Based on dates from several sites in Jamaica where the subfossil *Xenothrix* has been found, scientists estimate that this New World monkey was alive 900 years ago.

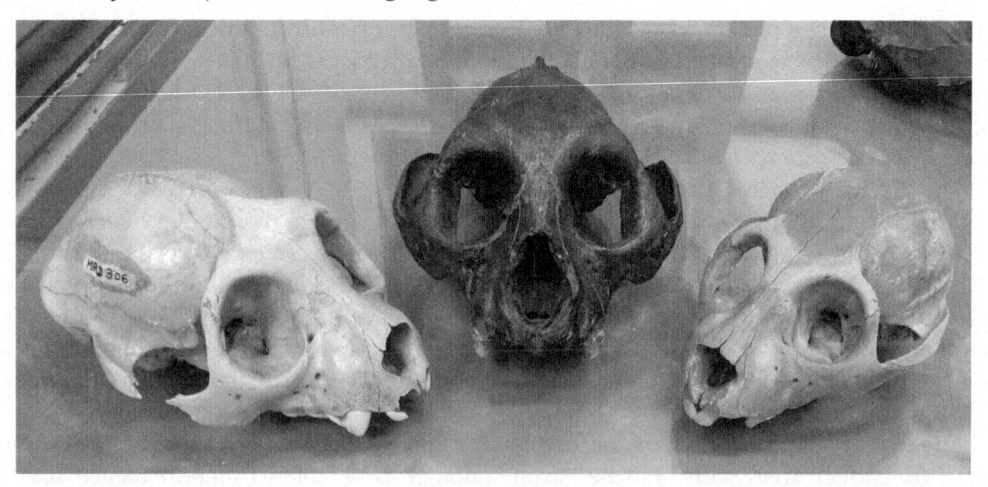

Figure 8.5 Three cranial specimens of the subfossil lemur *Archaeolemur* from Madagascar.
Photo credit: *Archaeolemur majori* skulls.jpg, by Ghedoghedo, (CC-BY-SA 3.0).

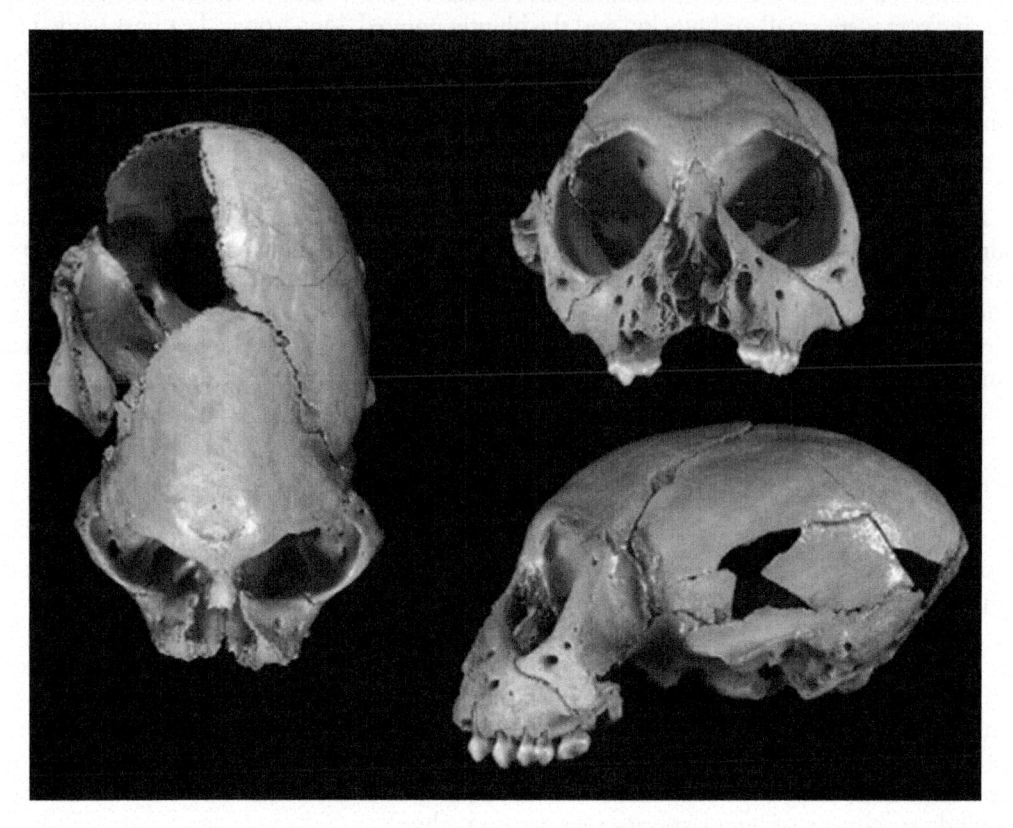

Figure 8.6 A subadult, subfossil cranium of the platyrrhine *Antillothrix* found in a submerged cave in the Dominican Republic.

Adapted from Rosenberger, A. L., Cooke, S. B., Rímoli, R., Ni, X., & Cardoso, L. (2011). First skull of *Antillothrix bernensis*, an extinct relict monkey from the Dominican Republic. *Proceedings of the Royal Society B: Biological Sciences*, 278(1702), 67–74. https://doi.org/10.1098/rspb.2010.1249

The Anthropocene Epoch

Geologists, whose wide-ranging research forms the basis of Earth history, are now in the process of adding a new geological epoch to Earth's timescale in order to recognize and study what has changed by the impact of humans on the planet itself. The proposed name for this new epoch, which will follow the Holocene, is the "Anthropocene." The root-word, *anthropos*, is from the Greek. It means "man" or "human."

To define an epoch there must be a starting point that has worldwide consequences which can be traced geologically in rocks and fossils, just as the beginning of the Paleocene, 66 million years ago, is marked by the effects of the Chicxulub asteroid collision, by the end of the dinosaurs, and the beginning of modern mammals.

When does the Anthropocene start? This has not yet been decided. One idea that is popular among geologists places the start of the Anthropocene at 1945, when the first atomic bomb was detonated. That was the beginning of an age in which mass destruction became possible by human design, on a scale unimaginable until then, rivaled only by the impact of an interstellar asteroid like Chicxulub. The Earth began changing at a rate that was never before seen in history, in new ways that were driven not by nature but by people. It is a rate of change that is not sustainable for many biological systems and processes that have been evolving for many millions of years without human interference. Unlike the natural, non-catastrophic extinctions of the past that occurred gradually over millennia and millions of years, the timescale of today's multifaceted threats to primate survival are measured in decades.

Many of the primate species that we have discussed in preceding chapters have been around for millions of years – they are living fossils that are surviving during the Anthropocene. The threats to their existence during the Anthropocene in each of the four main regions where primates are found are the same threats facing all of planet Earth, including altering the climate by raising the temperature beyond what species can tolerate, wholesale habitat destruction, and excessive pollution that contaminates the atmosphere and the oceans and changes the balance of nature. Yet, there is cause for optimism: primate living fossils have a future. From the perspective of the world's leading primate conservationists, Chapter 9 looks at the risks faced by primates and the efforts to mitigate them around the world.

Key concepts

Cenozoic Era
Mosaic evolution
Proto-primates
Asteroid collision
The Long Lineage Hypothesis

Quizlet

1. What can we learn from fossil teeth?
2. What are euprimates?
3. What is meant by "living fossil?" Give an example.
4. What is the likely origin of New World monkeys?
5. What is the concept of evolutionary stasis?

6 What is *Gigantopithecus*?
7 What is the Anthropocene Epoch?

Bibliography

Alba, D. M., Delson, E., Carnevale, G., Colombero, S., Delfino, M., Giuntelli, P., Pavia, M., & Pavia, G. (2014). First joint record of *Mesopithecus* and cf. *Macaca* in the Miocene of Europe. *Journal of Human Evolution*, 67, 1–18. https://doi.org/10.1016/j.jhevol.2013.11.001

Chester, S. G. B., Bloch, J. I., Boyer, D. M., & Clemens, W. A. (2015). Oldest known euarchontan tarsals and affinities of Paleocene *Purgatorius* to primates. *Proceedings of the National Academy of Sciences*, 112(5), 1487–1492. https://doi.org/10.1073/pnas.1421707112

Dunsworth, H. (2015). How to become a primate fossil. *Nature Education Knowledge*, 6(7), 1.

Génin, F., Mazza, P. P., Pellen, R., Rabineau, M., Aslanian, D., & Masters, J. C. (2022). Co-evolution assists geographic dispersal: The case of Madagascar. *Biological Journal of the Linnean Society*, 137(2), 163–182. https://doi.org/10.1093/biolinnean/blac090

Godfrey, L. R., Jungers, W. L., & Burney, D. A. (2010). Subfossil lemurs of Madagascar. In L. Werdelin (Ed.), *Cenozoic Mammals of Africa* (pp. 351–367). Berkely, University of California Press.

Harrison, T., Jin, C., Zhang, Y., Wang, Y., & Zhu, M. (2014). Fossil *Pongo* from the early Pleistocene *Gigantopithecus* fauna of Chongzuo, Guangxi, southern China. *Quaternary International*, 354, 59–67. https://doi.org/10.1016/j.quaint.2014.01.013

Kring, D. A. (2007). The Chicxulub impact event and its environmental consequences at the Cretaceous–Tertiary boundary. *Palaeogeography, Palaeoclimatology, Palaeoecology*, 255(1), 4–21. https://doi.org/10.1016/j.palaeo.2007.02.037

Rosenberger, A. L., Hogg, R., & Wong, S. M. (2008). *Rooneyia*, postorbital closure, and the beginnings of the Age of Anthropoidea. In E. J. Sargis & M. Dagosto (Eds.), *Mammalian Evolutionary Morphology: A Tribute to Frederick S. Szalay* (pp. 325–346). The Netherlands, Springer.

Seiffert, E. R., Simons, E. L., & Attia, Y. (2003). Fossil evidence for an ancient divergence of lorises and galagos. *Nature*, 422(6930), 421–424. https://doi.org/10.1038/nature01489

Silcox, M. T., Bloch, J. I., Boyer, D. M., Chester, S. G. B., & López-Torres, S. (2017). The evolutionary radiation of plesiadapiforms. *Evolutionary Anthropology*, 26(2), 74–94. https://doi.org/10.1002/evan.21526

9 Primates in crisis

Working under the aegis of the International Union for the Conservation of Nature and the Primate Specialist Group, 31 of the world's experts on primate biology and conservation published an extensive, data-driven analysis of the conservation status of the world's primates in 2017. The article is entitled "Impending extinction crisis of the world's primates: Why primates matter." It was published in *Science Advances*.

The authors are: Alejandro Estrada, Paul A. Garber, Anthony B. Rylands, Christian Roos, Eduardo Fernandez-Duque, Anthony Di Fiore, K. Anne-Isola Nekaris, Vincent Nijman, Eckhard W. Heymann, Joanna E. Lambert, Francesco Rovero, Claudia Barelli, Joanna M. Setchell, Thomas R. Gillespie, Russell A. Mittermeier, Luis Verde Arregoitia, Miguel de Guinea, Sidney Gouveia, Ricardo Dobrovolski, Sam Shanee, Noga Shanee, Sarah A. Boyle, Agustin Fuentes, Katherine C. MacKinnon, Katherine R. Amato, Andreas L. S. Meyer, Serge Wich, Robert W. Sussman, Ruliang Pan, Inza Kone, and Baoguo Li.

The information the authors culled comes from more than 200 published articles and numerous data sets, but their combined expertise goes well beyond the literature they cite. Together, this multi-national, multi-generational group of scientist-authors, academicians, administrators, and policy makers, represents many decades of experience on the ground, in the field, studying primates in their habitats. The experts' conclusions are inescapable and shocking. There is a worldwide crisis leading toward the extinction of hundreds of primate species. Still, they assert, there is cause for optimism if the work is done to mitigate the looming threats.

The following is excerpted from the article (p. 11):

"Despite the impending extinction facing many of the world's primates, we remain adamant that primate conservation is not yet a lost cause, and we are optimistic that the environmental and anthropogenic pressures leading to population declines can still be reversed. However, this is contingent on implementing effective scientific, political, and management decisions immediately. Unless we act, human-induced environmental threats in primate range regions will result in a continued and accelerated reduction in primate biodiversity. Primate taxa will be lost through a combination of habitat loss and degradation, population isolation in fragmented landscapes, population extirpation by hunting and trapping, and rapid population decline due to human and domestic animal-borne diseases, increasing human encroachment, and climate change. Perhaps the starkest conclusion of this review is that collectively – as researchers, educators, administrators, and politicians – we are failing to preserve primate species and their habitats. We face a formidable challenge moving forward, as success requires that sustainable solutions address

DOI: 10.4324/9781003257257-9

222 Primates in crisis

Table 9.1 Threats to primate survival and strategies to mitigate them

Some major causes of threats to primates	Consequences for primates	Mitigation strategies
Expansion of industrial agriculture and large-scale cattle ranching	Habitat loss, degradation, and fragmentation	Conservation research, education programs, and action plans
Industrial logging	Population decline	Continued asessment of species conservation status
Oil and gas drilling, mining, and fossil fuel extraction	Local species extinctions	Promote primate conservation while addressing human needs
Hunting and illegal pet trade	Loss of ecosystem health	Reduce human ecological footprint in primate habitats
Illegal trade in primate body parts	Impending extinction of 60% of the world's primates	Reforest cleared areas, reconnect forest fagments, and stabilize protected areas by enhancing law enforcement
Climate change	Continued and accelerated reduction of biodiversity	Reintroduce captive and rescued animals into the wild

Adapted from Estrada (2017), as cited.

the social, cultural, economic, and ecological interdependencies that are the basis of primate conservation.

By refocusing and publicizing our efforts to academics, government agencies, NGOs, businesses, and the public at large, we can build a comprehensive understanding of the consequences of primate population declines and encourage urgent and effective conservation policies. These policies will differ among countries, regions, habitats, and primate species based on the site-specific nature of each problem. We have one last opportunity to greatly reduce or even eliminate the human threats to primates and their habitats, to guide conservation efforts, and to raise worldwide awareness of their predicament. Primates are critically important to humanity."

Sci. adv.3,e1600946(2017).DOI:10.1126/sciadv.1600946

In June 2023, in a special section of the journal *Science*, two international teams including more than 100 researchers from 20 countries, began publishing the results of a revolutionary project to sequence the complete genomes of more than 200 primate species from all over the world. The immediate impact of this genetic research will improve conservation efforts dramatically. This research has already shown that most primate populations, despite habitat loss and other threats that can dangerously increase inbreeding, retain sufficient genetic diversity to provide hope for their future viability if mitigation efforts are implemented successfully.

Figure 9.1 An effort to save the orangutan: orphaned juveniles living in a reintroduction shelter are cared for and prepared to return to the wild to live on their own.

Courtesy of International Animal Rescue/Lisa Burtenshaw.

Glossary

Activity budget The amount and daily distribution of time (or energy) given to various activities.
Adaptation 1. A feature that is prevalent in a population because it is selectively advantageous. 2. The process by which a feature becomes widespread in a population through natural selection because of improved function.
Adaptive hypothermia The capacity to lower body temperature as a way to minimize energy expenditure when food is scarce.
Adaptive radiation A set of species that have diverged from a single common ancestor and have adapted to live in many ecological niches.
Adaptive zone The set of ecological niches occupied by a taxonomic group that shares features specifically adapted to utilizing a special array of resources.
Agonistic Behaviors that involve aggressive and submissive interactions.
Allogrooming The grooming of one animal by another belonging to the same species.
America's Scenario The explanation of platyrrhine origins that reconstructs their dispersal from the Old World along an overland route used by other mammals that entered North America during the Eocene.
Analogy/analogous A trait in one species that is functionally comparable to a trait in another species but evolved independently.
Ancestor, species The progenitor species from which a younger species evolved.
Ancestral characteristic A trait that has been retained unchanged from a preexisting state or condition.
Ancestral-descendant relationship The lineal phylogenetic link between older and younger species that form a genetic, inter-generational continuum.
Ancestral stock The taxonomic group that potentially includes the species ancestor of a younger species or taxonomic group.
Angiosperms The flowering plants.
Anthropogenic Alterations to the environment caused by humans.
Anthropoid Informal taxonomic name for the group (Anthropoidea) composed of New and Old World monkeys, and apes.
Ape Common name for the taxonomic group (Hominoidea) composed of great apes (chimpanzees, bonobos, gorillas, orangutans) and lesser apes (gibbons and siamang).
Apomorphy/apomorphic A derived characteristic.
Arboreal/arboreality/semi-arboreal Refers to living in trees.
Arboreal frugivory The adaptive zone of primates that is predicated on a set of evolutionary adaptations for fruit-eating and tree-dwelling.

Auditory bulla The shell-like bony housing of the middle and inner ears on the underside of the cranium.
Basal metabolic rate The rate at which energy is used by an organism in a resting state.
Bilophodont/bilophodonty A molar crown pattern built around two transverse ridges or crests aligned to connect two outer and two inner cusps.
Bimaturism Sex-specific differences in growth and development within the same species that may result in differences in appearance at the same life stage.
Binomen The two-term set (genus and species) that is the formal name of a species.
Biodiversity hotspot A region or area consisting of animal and plant species that are rare or endemic, taxonomically rich, and under threat of extinction.
Biological species concept The hypothesis that species consist of one or more populations the members of which can interbreed with one another, occupy the same ecological niche, and are reproductively isolated from other such groups in nature.
Brachiation Hand-over-hand, forelimb-powered suspensory locomotion.
Caecum The pouch-like structure forming the beginning of the large intestine.
Carbohydrate Organic compounds that are composed mostly of sugars and starches.
Cathemeral Active at various times of the day or night rather than being restricted to either period.
Cenozoic Era The geological era of Earth history that began 66 million years ago and shaped today's flora and fauna.
Cheek pouch A pocket situated between muscles inside the cheek that is used to store food temporarily.
Chest-beating Behavior displayed mostly by adult male gorillas, as they rhythmically slap their hands on a field of naked skin on the chest.
Clade A group of species descended from a single common ancestor.
Cladistics/cladistic analysis Refers to the sorting of traits as either primitive or homologously derived in order to develop sister-group phylogenetic hypotheses.
Cladogram A branching diagram depicting the phylogenetic relationship of taxa as sister-groups.
Classification, biological The ordering and arrangement of taxa in the form of a hierarchic, Linnaean organizational scheme.
Coevolution The joint evolution of ecologically interacting species that impose mutual selective pressures resulting in adaptation to one another's characteristics.
Community, primate The set of primate species that coexists in space and time, and form an identifiable ecological unit.
Conifer/coniferous Refers to trees (usually evergreen) and bushes that have needle-like leaves and wind-dispersed seeds that develop inside a cone-like structure.
Conservation status A designation by the International Union for the Conservation of Nature that refers to the existence of a species and the likelihood of its continued existence in the near future based on several observable and measurable factors, such as its reproductive viability, growth or decline in population size, and presence of threats.
Contact call A vocalization performed to monitor the spatial location of troop members.
Contest competition Competition that occurs when access to a resource can be controlled by one individual.
Convergence/convergent evolution The independent evolution of traits that are functionally similar and have similar roles in distantly related taxa.

Core area The area where a social group concentrates its activities.
Cranium The part of the skull that includes the face and braincase.
Critically endangered species A species identified by the International Union for the Conservation of Nature that presents an extremely high risk of becoming extinct in their native habitat.
Crown group, phylogenetic A group composed of living species and fossils that are cladistically nested within that group. Also applied to a group that includes all the living and fossil descendants of a single ancestor and their sister-group.
Crown primates The living primates and their extinct relatives that are cladistically nested within the group, as opposed to proto-primates or plesiadapiforms.
Crown, tooth The enamel-covered part of a tooth that forms the occlusal surface.
Dawn call Loud vocalizations regularly performed near dawn.
Dental formula A short-hand numerical identification system specifying the number and types of teeth in the mouth of a mammal.
Derived characteristic A feature (trait) that is changed from its preexisting, primitive condition.
Dichromatic Able to see color through two channels (cell types), with the photoreceptors not able to distinguish between red and green (a third channel).
Differential reproduction The situation in which some individuals in a population produce more offspring than others.
Dipterocarp A family of rainforest trees that produces seeds uniquely structured to be easily dispersed by the wind.
Dispersal, biogeographic The permanent movement of a population from one geographic area to another, unlike seasonal migrations that involve a return.
Dispersal, seed An aspect of plant propagation that involves the natural movement of seeds away from the confines of a parent tree.
Dispersal, social The departure of an individual from its social group, usually a maturing individual that leaves its natal group in order to establish breeding opportunities.
Diurnal Active during the day.
Dominance hierarchy The ranking of individuals within a social group that is generally correlated with resource access and is a function of agonistic interactions and/or kinship.
Dwarfism The evolutionary adaptation that results in a significant reduction in body size from the ancestral condition.
Ecological niche The position and role of a species in an ecosystem that enables its existence and interactions within the system.
Ecology The science that studies organisms and how they live in their environment.
Ecophylogenetics An approach to studying biodiversity and evolution by synthesizing information from ecology, phylogeny, and adaptation in a community context.
Ecosystem A system that encompasses all the organisms living in an area and its environment.
Ectotympanic bone The bone that supports the eardrum.
Endangered species A species identified by the International Union for the Conservation of Nature that is very likely to soon become extinct in its native habitat.
Endemic A species with a restricted, exclusive geographical distribution.
Energy maximizers Species that expend relatively large amounts of energy to obtain a nutritious diet.

Energy minimizers Species that conserve energy while maintaining a diet that is not nutritionally rich.
Estrous The phase of the reproductive cycle when females are fertile and display a willingness to mate.
Euprimates "Real" primates, phylogenetically related to living primates and modern in their appearance, adaptations, and ecological situation.
Evolution Change in inherited characteristics of a population over generational time.
Evolutionary stasis Lack of change over evolutionary/geological time.
Extractive foraging A food-collection method that requires removing objects that are embedded in trees or the ground.
Fallback food Foods, often difficult to procure or process, to which a species turns when widely available, preferred foods are scarce.
Feeding-foraging system The evolved coordination of adaptations relating to locating, acquiring, and eating food.
Fission-fusion social system A system in which a relatively large aggregation of individuals forming a social group divide up into smaller, temporary foraging parties.
Fovea centralis A small indentation on the retina, present in all haplorhines, where vision is most acute because of a high density of photoreceptor cells.
Genus/genera In the Linnean hierarchy, the rank that is situated above the species level.
Geodispersal The geographical expansion of a species, or a set of coexisting animals and plants, by crossing a geological barrier that has been eliminated.
Gigantism The evolutionary adaptation that results in a significant increase in body size from the ancestral condition.
Grasp-leaping The ancestral locomotor style of all primates, based on the ability to initiate a leap from a branch that is being held by grasping feet.
Grooming Tending to the fur or skin by removing particles, including organisms, using the fingers or teeth, often serving an important social function.
Grooming claw One or more digital "scratch claws" of the foot used in self-grooming.
Gum, tree/gumivore Refers to a semi-liquid substance produced by trees that can be eaten.
Gut The gastrointestinal system.
Hallux The large, innermost toe.
Haplorhine Informal version of the taxonomic name of the group (Haplorhini) composed of the clade consisting of tarsiers, New World monkeys, Old World monkeys, and apes.
Hibernation A state of inactivity (resembling sleep) when an animal's metabolic rate is lowered in order to conserve energy.
Home range The total area that is used by an adult individual or a social group.
Homology/homologous A characteristic that is shared by two or more species because it was present in the genetic makeup of the last common ancestor they shared.
Hypothesis A proposed explanation for an observable phenomenon that, in the realm of science, must be testable and capable of being rejected as false.
Intermembral index The ratio of forelimb (arm and forearm) to hindlimb (thigh and calf) lengths, which is correlated with different styles of locomotion.
Ischial callosities A sitting pad of thickened or hardened skin on the rump attached to the ischium bones of the pelvis.
Life-history The species-specific growth and development stages of an individual's lifetime.
Lineage A line of evolution such as a clade.

Linnaean hierarchy The tiered taxonomic arrangement of biological classification attributed to the Swedish botanist Carolus Linnaeus.
Living fossil A living species that has remained largely unchanged from its geologically remote fossil relatives.
Long Lineage Hypothesis Applied to platyrrhines, the hypothesis that clades and their particular ecological niches have persisted for long periods of geological time.
Mandible The lower jaw.
Mass extinction A global extinction event that wipes out a large proportion (~75%) of existing biodiversity in a relatively short period of geological time (< 3 million years).
Mast fruiting The synchronous production of a heavy fruit crop among species in a forest that occurs at long intervals, greater than one year.
Mate-guarding The behavior of a male or a coalition of males to shield a female from harassment or sexual advances by other males.
Matrilineal Kinship within the female line.
Megafauna The collection of large mammals in a habitat or geological period.
Microbiota The set of microscopic organisms that inhabits an environment, like the gastrointestinal tract.
Microhabitat The structural section of a habitat that is heavily used by a species.
Molecular clock A method that uses DNA to reconstruct the timing of splits in the branching network of a cladogram.
Monkey A common term of no particular scientific significance that is applied to platyrrhines and non-ape catarrhines.
Monogamy A breeding system in which a pair-bonded male and female are each other's exclusive mates.
Monophyletic Descent from a single shared ancestral species.
Mosaic evolution The evolution of different traits at different rates or times within a clade.
Multi-level society A social system involving a great many individuals with several layers of sub-units that are composed differently and may temporarily separate and rejoin the larger group.
Natural selection The differential survival and reproduction of certain individuals because they benefit from inherited traits that are well adapted to the environment.
Neotropics Central and South America, the Caribbean, and southern Mexico.
Niche breadth All the environmental resources to which a species or clade is adapted to satisfy their needs.
Niche differentiation The adaptive separation of species living in a community that enables each to occupy its own ecological niche.
Olfaction-vision tradeoff The balance between the senses of smell and sight when one is emphasized in favor of the other.
Olfactory bulb, primary and secondary The parts of the brain specialized to receive and transmit general information about environmental smells (primary) or information that is specific to individuals (secondary).
Opposable hallux, opposable thumb The ability to rotate and flex the large toe or thumb toward the sole or palm to grasp or control an object.
Order, taxonomic A major rank or category of the Linnaean hierarchy which, in mammals, tends to encompass a diverse clade that has radiated into a particular adaptive zone, like the Order Primates.

Orthograde Tending towards an upright form of body posture and locomotion.
Pair-bond The attachment between a single male and female as the basis of an exclusive social system and mating pattern.
Paleocommunity An assemblage of remains including species that no longer exist and occupied the same habitat for a period of time
Parallel evolution The evolution of similarities among relatively closely related taxa, ostensibly due to underlying genetic similarities or tendencies that are selected to perform similar functions.
Pelage The hair or fur of a mammal.
Petrosal bone A cranial bone that contains the middle and inner ears and forms a shell-like chamber (bulla) on the underside of the cranium in primates.
Phenology The periodic life-cycle processes of plants that produce leaves, fruits, flowers, and ripening schedules, for example.
Pheromone Chemical substance that produces scent used in communication.
Phylogenetic constraint A limitation in the possible course of evolution due to the genetic makeup of a clade.
Phylogeny The genetic or genealogical relationships among taxa.
Piloerection Raising hairs or fur as an agonistic behavioral display.
Polyspecific association A situation that occurs when different species join to move or feed or rest together.
Postorbital bar The bony strut that frames the outer side of the open orbit in strepsirhines and some other mammals.
Postorbital mosaic The region of the cranium on the outer and posterior side of the orbit that completes the closed eye socket of anthropoids by connecting the face to the braincase and is formed by different sets of bones.
Preferred food Foods that are generally abundant, available, and accessible.
Prehensile Able to grasp.
Primary sexual characteristics The genitalia of males and females.
Primitive characteristic A feature in a taxon that is retained unchanged from its original condition in an ancestor.
Pronograde Tending toward a horizontal form of body posture and locomotion.
Proprioception The awareness of limb and body position, and movement.
Prosimian An informal term based on an outmoded classification of primates that encompasses strepsirhines and tarsiers.
Proto-primates A group of mammals, classified as plesiadapiforms, and also called stem primates, that exhibit some primate-like arboreal features and may include the ancestral stock of primates.
Retina The layer of tissue at the back of the eyeball made of light-catching photoreceptors that send signals to the brain that generates the sensation of sight.
Rhinarium The patch of textured, always-moist skin that forms the nostrils and is continuous with the upper lip in strepsirhines and other mammals.
Sagittal crest A midline, longitudinal boney crest that forms on the braincase in adult males of some primate species.
Scent gland/scent mark Skin glands that exude pheromonal substances that are deposited and used in scent-based communication (marking).
Secondary compound Chemicals that protect plants from being eaten because they are distasteful, toxic, or hard to digest.

Secondary sexual characteristics Anatomical features that usually begin developing at puberty that distinguish the sexes (apart from the genitalia).
Selective pressure The evolutionary force that favors one trait over another because it is better suited to the environment.
Selfish herd The hypothesis that for an individual there is "safety in numbers" in the presence of a predator.
Semi-solitary A social system in which males and females generally meet only for mating or to temporarily share an overnight shelter.
Sexual monomorphism/dimorphism Adult males and females are similar (monomorphic) or different (dimorphic) in their secondary sexual characteristics.
Sexual selection Differential reproduction among individuals resulting from their ability to attract mates.
Sexual swelling Enlargement and often reddening of the skin around the genitals and perineum in females that accompanies estrus and attracts male attention.
Silverback male An adult male gorilla in his prime, as signified by a silver-furred lower back, that typically controls a group of females and sires their young.
Sister-group/sister taxa Taxa that are each other's nearest relatives because they share a unique common ancestor.
Skull Skeletal parts of the head including the cranium and mandible.
Speciation The processes by which new species are formed.
Species A biological unit defined by the ability of its members to exclusively reproduce with one another in nature.
Spot foods Foods that are concentrated in a relatively small location that provide limited access for feeding.
Stem group, phylogenetic A group composed of extinct species forming the ancestral stock of a clade.
Strepsirhine Informal version of the taxonomic name of the group (Strepsirhini) composed of the clade consisting of lemurs, lorises, and galagos, and characterized by the twisted-nostril and wet-nosed morphology.
Subfossil Paleontological remains that have not mineralized, and come from geologically recent periods, 10,000–12,000 years ago to the present.
Sympatric/sympatry Refers to populations occupying the same area and belonging to the same community.
Symplesiomorphy A primitive trait that is shared by taxa.
Synapomorphy A derived trait that is shared by taxa.
Tail-twining Twisting tails together.
Tapetum lucidum A membrane behind the retina in many nocturnal animals that amplifies dim light by reflecting it back toward the retina's photoreceptor cells to stimulate them a second time.
Tap-scanning The percussive method used by aye-ayes to echolocate the tunnels in trees where grubs live.
Taxon/taxa A group (singular and plural) that is formally named and placed in a biological classification.
Taxonomic inflation The dubious practice of greatly expanding the number of named taxa.
Taxonomy 1. A biological classification. 2. The rules and principles for developing a classification.

Terrestrial/semi-terrestrial Ground-dwelling, at all times or in tandem with arboreal activities as well.

Territory/territorial An area that is exclusively used and defended against other conspecific troops.

Theory A higher order "natural law" that explains natural phenomena and is supported by many related hypotheses that have been successfully tested.

Toothcomb A comb-like, horizontally oriented configuration of the lower incisors and canines found in strepsirhines that is used in grooming.

Torpor A short-term lowering of the basal metabolic rate that conserves energy and is accompanied by inactivity.

Transatlantic Scenario Applied to platyrrhines, the highly improbable proposition that suggests anthropoid primates colonized South America by rafting across the Atlantic Ocean from Africa on a mat of floating vegetation.

Trichromatic/trichromacy Able to see color through three channels to produce three- or full-color vision, made possible by the ability of photoreceptors to distinguish red and green.

Understory The area of trees and shrubs situated below the closed canopy of the forest.

Vomeronasal (Jacobson's) organ An olfactory organ that opens in the palate and is stimulated by the pheromones of conspecifics and predators.

Y-5 cusp pattern An arrangement of the five cusps on the lower molar crowns of apes that are separated by a set of indentations shaped like the letter Y.

About the author

Photo credit: Rivka Rosenberger

Alfred L. Rosenberger, PhD, is Professor Emeritus at Brooklyn College and the City University of New York Graduate Center, New York, USA, where he taught in the anthropology departments for more than a decade. He received the PhD from the CUNY Graduate Center, and a BA from the City College of New York. Dr. Rosenberger is a Fulbright Fellow and an elected Fellow of the American Association for the Advancement of Science. His research programs have taken him into the field to study the behavior and ecology of living monkeys and to sites in South America and Madagascar to find and study fossils. A leading figure in the evolutionary biology of New World monkeys, Rosenberger's broader research interests integrate primate morphology, adaptation, paleontology, behavior, and ecology in a phylogenetic and functional context. He has authored more than 100 articles and co-edited several volumes on living and fossil primates. He has written about the behavior of monkeys in the wild, feeding and locomotor adaptations, primate and anthropoid origins, the evolution of tarsiers, Eocene fossils, species concepts, and taxonomic inflation. An early proponent of the ecophylogenetic approach to primate evolution, Rosenberger's model of New World monkey evolutionary history, the Long Lineage Hypothesis, emphasizes the multi-million year continuity behind today's communities and their biodiversity. Rosenberger's definitive book *New World Monkeys: The Evolutionary Odyssey* was published by Princeton University Press in 2020.

Index

Note: **Bold** page numbers refer to tables; *italic* page numbers refer to figures.

activity budget 108
activity cycle: apes 119; lorisoids **52, 83,** 108, **123, 156**
activity rhythm 10–11; apes 19; cercopithecoid 115; cheirogaleids 53–54; galagos 15; lemurs 14, 49; lorises 14; New World monkeys 17; Old World monkeys 18; pitheciid 88; platyrrhine 76; tarsiers 16, 151
adapiform fossil 205–206
adaptations 5; African primates 122–133; Asian primates 155–163; aye-aye 64; convergent evolution of 178–180, 179; lemurs 51–65; phylogeny and 24; platyrrhines 81–99
adaptive hypothermia 53, 107
adaptive radiation 2–5, 44, 87, 168, 178, 202, 205–207
adaptive zone 6, 28–29
Aegyptopithecus 210, *211*
Afradapis 206
African primates: apes (*see* apes); fossil record 140–141, 214–215; galagos (*see* galagos/galagonids/galagids); geography and climate 104–105, *105*; living primates 2–5, *3*, **4**; lorises (*see* lorises/lorids/lorisids); Old World monkeys (*see* Old World monkeys); primate radiation 105–106; stem catarrhines 210; terrestrial baboons 28; toothcombed strepsirhines 206–207; tropical rainforests 175, *175*
agonistic behavior 87, 96
air plants *see* epiphytes
alarm calls 80
allogrooming 189–190; *see also* grooming
Alophe 215
Alouatta see howler monkey (*Alouatta*)
Amato, K. R. 221
Amazonia 70; tropical rainforests 172–174
America's Scenario 209

analogy/analogous characteristics 37
ancestor 21, 29, 72; bonobos 134; derived homologous characteristic 82; fossil record 39; gorilla 214; phylogeny reconstruction 37–38; platyrrhines 209; strepsirhine-haplorhine 203; tree-dwelling 29
ancestral characteristic 38
ancestral-descendant hypotheses 39
ancestral stock 39
Andasibe-Mantadia National Park 51
angiosperms 171, *171*, 173, 177, 204
angwantibos 106–107, *107*
Ankarafantsika National Park 51
Anthropocene Epoch 219
anthropogenic 26
anthropoids (Anthropoidea) 23, 24; crown 208; platyrrhines and 72–75; proto-anthropoids 208
Antillothrix 218
Aotus dindensis 99, 213
apes 4, 105–106, 109–110, 117–119; in Asia 146–155; bilophodont molar teeth 112; cheek pouches 112–114, *114*; color vision 111–112; communication 111–112; cranial and postcranial characteristics 112; distinguishing factors 19; ecological dominance of 109; families of 23; feeding strategy in 31; fossil record 214–216; geographical distribution 3, 3–4, **4**; guts 112–114; olfaction 111–112; profile 119–121; proto-apes 210; taxonomic groups 21; teeth 111
apomorphy 38
arboreal frugivores 6, 28–40, 44; adaptive zone 28–40; brain 32; diet and dentition 29–30, *30*; in ecological context 168–170; epicenter of primate life 177–178; feeding 37; food-handling 30–31; fossil record

39; functional-adaptive patterns and attributes 29–33; grasping feet and hands 35, 36; hard-anatomy features 35; lemurs (*see* lemurs); life-history strategies 33; locomotion 30–31, 36–37; paleocommunities 39–40; phylogeny reconstruction 37–38; posture 30–31; reproductive systems and attributes 35; sensory systems 31–32; social systems and group size 32; special senses 37; uniqueness 33–34; *see also* frugivores
arboreality 2, 28, 33–34
Archaeolemur 218
Arregoitia, L. V. 221
Asian primates: adaptation 155–163; apes (*see* apes); communities 155–163; conservation status 164, 164–165; diversity 155–163; ecology 155–163; fossil record 215–216; geography and climate 145, 145–146; Hanuman langurs 159–160; living primates 2–5, *3*, *4*; lorises (*see* lorises/lorids/lorisids); macaques 159–160; odd-nosed monkeys 160–161, *161*; Old World monkeys (*see* Old World monkeys); orangutans 161–163, *162*; paleocommunities 163–164; radiation 146–155; tarsiers (*see* tarsiers); tarsiiforms 207–208; tropical rainforests 176–177, *177*
atelids 70, 82; diet and locomotion 84–85; mating 85–86; social systems 85–86
Atlantic Forest: 71, 82, **98**; muriqui 77, 86–87; New World monkeys (*see* New World monkeys)
auditory bulla 36
Avahi (wooly lemur) 61
aye-aye (*Daubentonia madagascariensis*) 62–65, *63*

baboon 2, *31*, 130, 202; Africa's terrestrial 28–29, 112; bleeding heart 131, *131*; craniodental morphology 115; evolutionary adaptations 66; frugivorous-omnivorous savanna 132; hamadryas 190, **197**; open-mouthed subadult male, tooth groups 30, *30*; sexual selection 190; skeletons *118*; social system 133; terrestrial (*see* terrestrial baboon)
Barelli, C. 221
basal metabolic rate 55; capuchins 92; lemurs 51, 55, 180
behavior: agonistic 87, 96; climbing 11, 62, 66, 85, 90, 92, 117, 119, 136, 176, 187, 216; primate community 168–198; social (*see* social behavior);

strepsirhines 46; submissive, gelada 116; tail-twining 89, 191; tool use 31, 135–136, 162
behavioral display 137, 154
bilophodont molars 183–184
bilophodonty 112
bimaturism 161
binomen 23
biodiversity hotspots 43, 67, 71, 146
biological classification 5, 23
biological diversity of primates 8–9
biological echolocation 65
biological species concept 4
black-and-white colobus 126
bleeding heart baboons 131, *131*
body mass/weight 10; apes 19, 119; cercopithecoid 115; galagos 15; lemurs 14, 49; lorises 14; lorisoids 107–108; New World monkeys 17; Old World monkeys 18; platyrrhine 76; primate communities 180–182; tarsiers 16, 151
Bohol tarsier 18; *see also* tarsiers
bonobo 4, 5, 106, 134; *vs*. chimpanzees **134–135**
Boyle, S. A. 221
brachiation 11, 77–78, 185–186; acrobatic form of 119; folivores 159; gibbon-like skeleton suited for 216; hand-over-hand 153
Brachyteles (muriqui) 77, 86–87
brain: arboreal frugivores 32; anthropoid 75; aye-aye 64; capuchin 92; and diet 85; encephalization 81, *179*; haplorhines 72–73; and olfaction 47; primate 32, 35–37
breeding system 12, 99, **196**; categories and subcategories of 192–194, *193*; convergent evolution 194–195, *195*, **196**

caecum 57
Callicebus see titi monkey (*Callicebus*)
callitrichines 93–96, *94*, 182
call signals, types of 79–80
Campbell's monkey 126
canine honing 111
capuchin (*Cebus*) 77; dry season niche partitioning 96–97; facial expressions 72, *73*; hard-anatomy features 73–75, *74*; South American 6
carbohydrate 161
Caribbean Sea 70
carnivorans 106
Carpolestes 203
catarrhines (Catarrhini) 23, 109–110, *110*, **111**; bilophodont molar teeth 112; cercopithecoids 112; cheek pouches 112–114, *114*; color vision 111–112;

communication 111–112; cranial and postcranial characteristics 112; fossil record 213–214; guts 112–114; ischial callosities 115; olfaction 111–112; platyrrhines and 75, 75–76; stem, in Africa 210; teeth 111
cathemerality 2; lemurs 58–61; owl monkeys 89, *89*; primate communities 182–183
cebids 90–91; diet 77; dry season niche partitioning 96–97; non-Amazonian 97–99
Cebus (capuchin) 77, 91–93; dry season niche partitioning 96–97; facial expressions 72, *73*; hard-anatomy features 73–75, *74*
Cebus albifrons 97
Cebus apella 97
Cenozoic Era, fossil record 199–200, **200**; Eocene Epoch 203–208; Holocene Epoch 208, 217, *218*, 219; Miocene Epoch 210–216; Oligocene 206, 208–210, 213, 214; Paleocene Epoch 199, 202–203, 205, 219; Pleistocene Epoch 208, 215–217; Pliocene Epoch 216–217; study of teeth 201–202
cercopithecids 109, 112, 122; in Asia 157–159; and colobine specializations 112–114; profile 115–117, *116*; size, diet, and locomotion 126–129; social systems and mating 129–130; in Tai Forest 126–130, *127*, *128*; *see also* Old World monkeys
cheek-pouched monkeys 109, 112–114, *114*
cheirogaleids: activity rhythm 53–54; diet 54–55; locomotion 55–56; mating 56; postural behavior of 55; social systems 56
Cheirogaleus see fat-tailed dwarf lemur (*Cheirogaleus medius*)
chest-beating 121, 138, 139
Chicxulub 202, 204, 219
chimpanzees 4, 5, 9, 23, 106, 133; bonobo *vs.* **134–135**; breeding systems 194–195, *195*, **196**; diet and locomotion 135–137; fossil evidence 214; parenting styles 121; social organizations 194–195, *195*, **196**, 197; social systems and mating 137–138
clade 38, 53, 55, 58, 61, 90, 106, 182, 187, 206, 210
cladistics/cladistic analysis 21
cladogram 21, *22*; branching patterns of 21; taxonomic terms 21
clambering 11, 62, 85, 119, 187
classification, biological 5, 22–23
climate: Africa 104–105, *105*; Asia 145, 145–146; Madagascar 41–43, *42*; South American 69–71, *70*

climbing behaviors 11, 62, 66, 85, 90, 92, 117, 119, 136, 176, 187, 216
closed canopy 29, 34, 124, 145, 171, 204
coevolution 168, 169
color vision, catarrhines 111–112
common ancestor *see* ancestor
common chimpanzee *see* chimpanzees
communication 12; apes 19, 121; catarrhines 111–112; cercopithecoid 116, *116*; galagos 15; lemurs 14, 51; lorises 15, 108; New World monkeys 17, 79–80; Old World monkeys 18; owl monkeys 79; platyrrhines 79–80; tarsiers 16, 152
community, primate 6–7; Africa 122–133; Asia 155–163; behavior 168–198; ecology 168–198; lemurs 66; morphology 168–198; platyrrhines 81–99
conifer/coniferous 146, 160
conservation status 24, 26; African primates *141*, 141–142; lemurs 66–67, *67*; New World monkeys 99–101, *100*; ring-tailed lemur 66–67; Southeast Asian primates *164*, 164–165; spider monkeys 100
contact calls 79
contest competition 113
Convention on International Trade in Endangered Species of Wild Fauna and Flora (CITES) 26
convergent evolution 37; of adaptations 178–180, **179**; social organizations and breeding systems 194–195, *195*, **196**
core area 6
craniodental morphology: apes 120, *120*; baboon 115; cercopithecoid 115; howler monkey 78; lemurs 50; lorisoids 108; platyrrhine 78–79; tarsiers 152
cranium 36–37, 191; of adult female and male baboons *190*; adult male gorilla *120*; of *Antillothrix 218*; catarrhines 76, 112; female fossil *Propliopithecus* 211; lemurs 50; morphology 186–187; of *Rooneyia 209*; subfossil 217, *218*; tarsiers 152
crisis, primates in 221–223
critically endangered species 13–19, 24, 66, 199, 141, 164
crown anthropoids 208
crown primates 206
crown, tooth 202

Darwin, C. 6
Daubentonia (aye-aye) 62–65, *63*
dawn call 88; indri 61–62
de-encephalization, platyrrhines 81
De Guinea, M. 221

dental formula 61, 62, 76
dentition: arboreal frugivores 29–30, 30; howler monkey 86; primate communities 183–186; sportive lemur 55
derived characteristic 38
dichromatic 80, 115
diet 11; apes 19, 119; arboreal frugivores 29–30, 30; atelids 84–85; callitrichines 93–95; cathemeral lemurs 58–59; cebines 91–92; cercopithecids 115, 126–129, 158; cheirogaleids 54–55; chimpanzees 135–137; fork-marked lemur 57; galagos 15; gorillas 138–139; howler monkey 76, 84; hylobatids 159; lemurs 14, 49; long-faced papionins 132; lorisoids 14, 108, 124–125; New World monkeys 17; Old World monkeys 18; orangutans 162–163; pitheciid 88; platyrrhine 76–77; primate communities 180–181, 183–186; primate fossil record 201–202; slow loris 155–157; tarsiers 16, 151
differential reproduction 5
Di Fiore, A. 221
dipterocarps 172, 176–177; in Borneo 171; structurally distinctive 173
diurnality 2; primate communities 182–183
diversity: African primates 122–133; Asian primates 155–163; lemurs 66; platyrrhines 81–99
Dobrovolski, R. 221
Dolichocebus 212
dominance hierarchies 60
dominance rank 60
drills 130
drumming technique 65
dry-nosed primates *see* haplorhines (haplorhines)
dry-season niche partitioning 96–97
duikers 119
dwarfism 90
dwarfs 181, 181–182; hippopotamus 66; lemurs 53–56

echolocation 62
ecological domain: apes 119; cercopithecoids 115; lemurs 49; lorisoids 108; platyrrhines 76; tarsiers 151
ecological niches 2, 6
ecological turnover 164
ecology: African primates 122–133; Asian primates 155–163; lemurs 51–65; opportunities 7–8, 8; platyrrhines 81–99; primate communities 168–198
ecophylogenetics 2
ecosystems 7; rainforests 28, 145, 172, 177, 202, 204

ectotympanic bone 76, 210, 211
enamel cap 183–184, 201–202
encephalization 81; howler monkey 81; New World monkeys 81
endangered species 13–19, 24, 66, 199, 141, 164
Endangered Species Act 26
endemics 43, 71, 82, 98, 101, 104, 146–147
energy maximizer 85
energy minimizer 84
environment 1, 2, 5, 6, 9, 12, 24, 28, 32, 33, 34, 35, 43, 45, 51, 56, 59, 70, 71, 109, 111, 115, 122, 132, 133, 164, 168, 171, 174, 175, 178, 186, 193, 194, 202
Eocene Epoch 203–208
epiphytes 98
Estrada, A. 221
estrous 60, 93, 129–132, 137
Eulemur (brown lemur) 58–59
euprimates 203, 204; adapiforms and fossil tarsiiforms 205–206; strepsirhines and tarsiiforms from Asia 207–208; toothcombed strepsirhines in Africa 206–207
Europe, fossil apes from 216
evolution 2–3; of primate traits 37–38
evolutionary principles: natural selection 5–6; selective pressure 5–6
evolutionary stasis 213
extinction: causes of 13; crisis 221–222; gradual 217; mass 202
extractive foraging 62, 64

fallback foods 96, 97, 161
fat storage 184
fat-tailed dwarf lemur (*Cheirogaleus medius*) 43, 44, 45, 54
fatted-male syndrome 93, 191
Fayum 140, 206–207, 210, 214
feeding-and-foraging system 20, 132
feeding calls 80
Fernandez-Duque, E. 221
fission-fusion groups 12, 59; female dispersal in 197; social organizations 195
flanged males 161
folivores 20, 44, 82, 84; brachiating 159; cercopithecids 126–130; nocturnal gumivorous 155–159; semi-terrestrial 133–138; *see also* frugivores
food-handling 184; arboreal frugivores 30–31
food sharing and defense 188–189
foregut 114
fork-marked lemur (*Phaner*) 56–58, 57
Fossey, D. 139–140, 142
fossil record 39, 199–200; African primates 140–141, 214–215; Anthropocene Epoch 219; apes from Europe 216; Asian primates 215–216; catarrhines

213–214; cranium of *Rooneyia* 208, *209*; Eocene Epoch 203–208; gibbons 216; great apes 216; haplorhines 204–206; Holocene Epoch 217, *218*; lesser apes 215; owl monkeys 212, *212*, 213; Paleocene Epoch 202–203, 205, 219; platyrrhines 211–213; Pliocene and Pleistocene Epoch 216–217; *Propliopithecus*, cranium *211*; squirrel monkeys 212, 213; teeth evolution *201*, 201–202
fovea, retinal 72, 150
frugivores 20, 44, 82, 84, 212; brachiating 159; cercopithecids 126–130; gibbons 153; nocturnal gumivorous 155–159; omnivorous 76, 90–91; semi-terrestrial 133–138; *see also arboreal frugivores*; *folivores*
fruit-huskers 87–90
Fuentes, A. 221

galagos/galagonids/galagids 105–107, *107*; distinguishing factors 15, *16*; family 23; geographical distribution *3*, 3–4, *4*
Garber, P. A. 221
gelada (*Theropithecus*) 130, 215; male *19*; submissive behavior *116*
gender-specific dominance hierarchies 60
genus (genera) 2, 4–6; living primates 7, *8*
geodispersal 209
geography 10; African primates 104–105, *105*; apes *3*, 3–4, *4*, 19; Asian primates 145, 145–146; galagos/galagonids/ galagids *3*, 3–4, *4*, 15; lemurs *3*, 3–4, *4*, 14; living primates 10; lorises 14; Madagascar 41–43, *42*; New World monkeys *3*, 3–4, *4*, 17; Old World monkeys *3*, 3–4, *4*, 18; South American 69–71, *70*; tarsiers *3*, 3–4, *4*, 16
giants **181**, 181–182; lemur 66
gibbons *118*, *185*; diet and locomotion 159; fossil 216; genera 153; pair-bonded 189–190; social dispersal 197; social organizations and breeding systems 194–195, *195*, **196**
gigantism 90
Gigantopithecus 138, 182, 202, 217
Gillespie, T. R. 221
Goeldi's monkey 93–96; skulls of 73, *74*
golden lion tamarin 100
Goodall, J. 136, 142
gorillas 5, 9; diet and locomotion 138–139; fossil evidence 214; mountain (*see* mountain gorilla); size and diet 180; social organizations and breeding systems 194–195, *195*, **196**; social systems and mating 139–140

Gouveia, S. 221
grasping tails 187–188
grasp-leaping 36
Gray's bald-faced saki monkey *18*
great apes *see* apes
grooming 48, 50; between adult males and females 139; toothcomb 48, 51
grooming claw 48, 186
gumivorous 11
Gunung Palung Orangutan Project 165
gut 29, 55, 84, 87, 183–186; catarrhines 112–114; in primate communities 183–186

habitat generalist 84, 91, 146
habitats 10; lemurs 44–45; Madagascar, geographical location 41, *42*; platyrrhine 99; primate in Southeast Asia (*see* Asian primates); *see also* microhabitat
habitat specialists 95, 160, 182
hallux 30
hand-over-hand brachiation 153
hanging 185–186
Hanuman langur (*Semnopithecus entellus*) 160
Hapalemur (bamboo lemur) 59
haplorhines (haplorrhines) 21–24, 45; adapiforms and fossil tarsiiforms 205–206; dry nose 150; families of 23; features of 71–72; fossil 204–206; nocturnal lineages 147; phylogeny 151; retina 183
hard-fruit eaters 183–184
health and chronological age, primate fossil record 202
Heymann, E. W. 221
hibernation 43, 53, 54; lemur *44* (*see also* lemurs)
Holocene Epoch, fossil record 208, 217, *218*, 219
home ranges 56, 92
hominids 117–119; profile 119–121; semi-terrestrial frugivore-folivores 133–140
homology/homologous characteristics 37; ancestral 38; primitive and derived 38
howler monkey (*Alouatta*) 34, *34*, 77, 173, 181; convergent evolution 194, **196**; craniodental morphology 78; dentition 86; diet 76, 84; ecological context 169; encephalization 81; locomotion 84; neotropical 207; prehensile-tailed 185; slow quadrupedal locomotion 82; social dispersal 197; social organizations and breeding systems 194–195, *195*, **196**; vocal communication 79, 191, *192*
hylobatids 153–155; in Asia 153–155
hypothermia, adaptive 53, 107

indriids 61–62
infanticide 93, 121, 133, 138, 160
insectivore 53, 140, 192, 205, 208
intermembral index 49; lemurs 49–50; galagos 124; hylobatids 153, 185; tarsiers 151
ischial callosities 115
IUCN SSC Primate Specialist Group 2, 24, 100, 141, 221

Jacobson's organ *see* vomeronasal organ
Japanese macaques *see* snow monkey (*Macaca fuscata*)

Karanisia 207
king colobus 126
knuckle-walking 120, 135, 136, 138, *175*, 216
Kone, I. 221
Kuala Lompat 155–160, **156**, *157*

Lambert, J. E. 221
langurs 155
La Venta 99
leaf monkeys 109, 114, *177*
leaping adaptations 184–185, *185*
Lemur catta (ring-tailed lemur) 60; breeding season 60–61; conservation status of 66–67; dietary and locomotor adaptations 59; ecological domain 49
Lemurids 58–61
lemurs *14*, 44–45, 48; activity rhythm 49; adaptation 51–65; aye-aye 62–65; body mass 49; cathemeral 58–61; cheirogaleids 53–56; communication 51; community ecology 66; conservation status of 66–67, *67*; craniodental morphology 50; crowned, skull of 74; diet 49; distinguishing factors 14, *14*; diversity 51–66; dwarf and mouse 53–56; ecological domain 49; ecology 51–65, *52*; evolutionary adaptations 66; families 44, *45*; features of 45–48; geographical distribution *3*, 3–4, **4**; habitats 44–45; indriids 61–62; intermembral index 49–50; locomotion 49–50; low basal metabolic rate 51; mating 51; paleocommunities 65–66; parenting styles 51; pelage 50; phylogeny 49; profile 48–51; reproductive pattern 51; sexual monomorphism/dimorphism 50–51; social systems 51; subfossils 217, *218*; woolly 61–62; *see also* Madame Berthe's mouse lemur
Leontocebus weddelli 96
Leontopithecus see lion marmoset (*Leontopithecus*)
Lepilemur see sportive lemur (*Lepilemur*)

lesser apes *see* hylobatids
Li, B. 221
life-history strategies, arboreal frugivores 33
lineage 10, 31, 39, 215; ancestral platyrrhine 209; anthropoid 208; fossils and 212; gigantism 182; nocturnal 147; owl monkey 180; terrestrial 109
lion marmoset (*Leontopithecus*) 82, 98; *see also* marmosets
lion tamarin 82
living fossils 206, 212, 219
living primates: activity rhythm 10–11; adaptive radiation of 2–5; body weight 10; communication 12; diet 11; genealogical relationships 21, 22; genera 7, 8; geography *3*, **4**, 10; habitat 10; locomotion and posture 11; parenting style 13; phylogenetic tree 21, 22; phylogeny 10; reproductive pattern 12–13; sensory modality 10–11; social systems 11–12; taxonomically organized 21–25; ten primary factors 9–13, 20; *see also specific primates*
locomotion 11; apes 19, 119; arboreal frugivores 30–31; atelids 84–85; callitrichines 93–95; cathemeral lemurs 58–59; cebines 91–92; cercopithecids 115, 126–129, 158; cheirogaleids 55–56; chimpanzees 135–137; galagos 15; gorillas 138–139; howler monkey 82, 84; hylobatids 159; lemurs 14, 49–50; lorisoids 15, 108, 124–125; New World monkeys 17; Old World monkeys 18; orangutans 162–163; pitheciids 88; platyrrhines 77–78; slow loris 155–157; tarsiers 16, 151–152; variations in 184
long-distance vocalizations 191, *192*
long-faced papionins *see* papionins
Long Lineage Hypothesis 212–213
lorises/lorids/lorisids *15*, 105–106, *107*, 147; in Asia 146–149, *149*; biological profile 107–109; diet and locomotion 124–125; distinguishing factors 14–15; family 23; general features **147**; social systems and mating 125
loud calls 80

Macaca fuscata (snow monkey) 9, 146
MacKinnon, K. C. 221
Madagascar 7; geography and climate 41–43, *42*; lemurs (*see* lemurs); living primates 2–5, *3*, **4**; tropical rainforests 172
Madame Berthe's mouse lemur 53, *54*, 182; activity rhythm 53; body mass 76; body sizes 34; *see also* lemurs
mainland Africa *see* African primates

male dominance hierarchy 60
mandible 73
mandrill (*Mandrillus*) 130; dual dominance hierarchies 190; feeding-foraging specializations 132; in Gabon 133; pelage 115; sexual monomorphism/dimorphism 116, 190; *see also* Old World monkeys
Manu National Park 81
marmosets 82, *94*, 98–99, 182, *193*; body mass 76; diet and locomotion 93–95; dry season niche partitioning 96–97; parenting styles 81; social systems and mating 95–96
marsupials 35
mass extinction 202
mast fruiting 162, 177
mate-guarding 88
mating: apes 121; atelids 85–86; callitrichines 95–96; cathemeral lemurs 59; cebines 92–93; cercopithecids 116–117, 129–130, 158–159; cheirogaleids 56; chimpanzees 137–138; gorillas 139–140; hylobatids 159; lemurs 51; long-faced papionins 133; lorisoids 108, 125; New World monkeys 81; opportunities 189; orangutans 163; pitheciid 88–89; slow loris 157; tarsiers 152
matrilines 160
mega-fauna 66
Mesopithecus 215, 216
Meyer, A. L. S. 221
microbiota 55
microhabitat 39, 51, 90, 92, 95, 97, 158, 215; chimpanzees 135; Kuala Lompat primates **156**, 157; lorises 153; slow loris 155–157; specialists 182; sub-canopy 176, 185; Tai Forest primates 123; *see also* habitats
mimicry 148
Miocene Epoch 163, 210; Africa's great apes 214; Africa's Old World monkeys 214–215; Asia's great apes 216; Asia's lesser apes 215; catarrhines 213–214; fossil apes from Europe 215; platyrrhines 211–213
mitigation strategy, primate populations 13, 26, 100, 197, 221
Mittermeier, R. A. 221
molar teeth 201, *201*; bilophodont 112
molecular clock 213
monogamy 59, 62, 81, 88, 193
monophyletic 21
monotremes 35
morphology: convergent evolution of 178; primate communities 168–198; skeletal 117, *118*; social behavior and 189–192

mosaic evolution 38
mountain gorilla *20*
mouse lemurs 53–56, *54*, 182; activity rhythm 53; body mass 76; body sizes 34; breeding systems 194–195, *195*, **196**; social organization 194–195, *195*, **196**; *see also* Madame Berthe's mouse lemur
moving calls 79
Mozambique Channel 43, 207
Müllerian mimicry 148
multi-level societies 12, 133, 197
multimale-multifemale groups 12, 194
muriqui *see Brachyteles* (muriqui)

Nacholapithecus 214
natural laws 1
natural selection 5–6
Nekaris, K. A. I. 221
neotropical monkeys 70, 173, 174, 213
neotropics 2–5, *3*, **4**
New World monkeys 4, 69; adaptation 81–99; anthropoids 72–75; atelids 82–86; *Brachyteles* 86–87; callitrichines 93–96, *94*; and catarrhines 75, 75–76; cebids 90–91; cebines 91–93; communities 81–99; conservation status of 99–101, *100*; distinguishing factors 17; diversity 81–99; dry season niche partitioning 96–97; ecology 81–99, *83*; families of 23; features of 71–72; feeding strategy in 31; geographical distribution *3*, 3–4, *4*; Goeldi's monkey 93–96; hard-anatomy features 73–75, *74*; marmosets 93–96, *94*; neotropical 23; non-Amazonian cebids 97–99; paleocommunities 99; pitheciid 87–89, *89*; platyrrhine (*see* platyrrhines (Platyrrhini)); profile characteristics 76–81; tamarins 93–96, *94*; taxonomic groups 21; *see also* platyrrhines
niche breadth 91
niche differentiation: in arboreal realm 178; cercopithecids 126; lemurs 51, 58, 66; New World monkeys 98; Tai Forest monkeys 126, *128*
niche partitioning, dry season 96–97
Nijman, V. 221
nocturnal frugivorous-gumivorous predators 124–125
nocturnal species 2, 10–12, 24, 45, 53, 182–183, 205, 213; ambush predators 149–153; aye-aye 62; gumivorous frugivore-faunivore 155, 157; haplorhines 72; lorisoids (*see* lorises/lorids/lorisids); owl monkeys (*see* owl monkeys); primate communities

182–183; stalking predators 147–149; strepsirhine eye 72; woolly lemur 61
non-Amazonian cebids 97–99
non-governmental organizations (NGOs) 67

odd-nosed monkeys 160–161, *161*
Old World monkeys 105–106, 109–110; in Asia 146–155, 215; bilophodont molar teeth 112; cheek pouches 112–114, *114*; color vision 111–112; communication 111–112; cranial and postcranial characteristics 112; distinguishing factors 18; ecological dominance of 109; families of 23; fossil record 214–215; geographical distribution 3, 3–4, *4*; guts 112–114; neotropical 23; olfaction 111–112; taxonomic groups 21; teeth 111; *see also cercopithecids*
olfaction, catarrhines 111–112
olfaction-vision tradeoff 24, 71
olfactory bulb 47, *48*
Oligocene 140, 206, 208–210, 213, 214
opposable hallux 30, 36, 203, 204
orangutans 121, 182, 217; diet 162; fossil 216; locomotion and posture 184; orphaned 222, *223*; paleocommunities 163; parenting styles 121; semi-solitary breeders 193; size and diet 180; social organizations and breeding systems 194–195, *195*, **196**
orbital closure 73
order 28
Order Primates 28, 33, 202, 205
orthograde 117, 119, 152, 186
owl monkeys 98, 182–183; activity rhythm 76; canine size 191; cathemeral 89, *89*; communication 79; diversity, ecology, and adaptation 82; fossil 212, *212*, 213; fruit-huskers and seedeaters 87; genus-level continuity 213; lineage 180, 213; paleocommunities 99; social dispersal 197; social systems and mating 88

pair-bonded systems 12; monogamous groups 59, 62, 81, 88, 193; social organizations *195*
pair-breeders 193
Paleocene Epoch 199, 202–203, 205, 219
paleocommunities 7, 30; African primates 140–141; arboreal frugivory 39–40; lemurs 65–66; Madagascar lemurs 65; New World monkeys 99; orangutans 163; owl monkeys 99; Southeast Asian primates 163–164

Pan 4, 5
Pan, R. 221
Pan paniscus (bonobo) *see* chimpanzees
Pan troglodytes (common chimpanzee) *see* chimpanzees
papionins 112, 122, 130; diet and feeding-foraging specializations 132; general features 131–132; social systems and mating 133
parallel evolution, in platyrrhines 187–188
parapithecoids 208, 209
parenting styles 13; apes 19, 121; cercopithecoids 117; galagos 15; lemurs 14, 51; lorisoids 15, 108–109; New World monkeys 17, 81; Old World monkeys 18; tarsiers 16, 152
pelage: apes 120; cercopithecoids 115–116; lemurs 50; lorisoids 108; New World monkeys 79, *80*; tarsiers 152
Perupithecus 208, 209
petrosal bone 36
Phaner (fork-marked lemur) 56–58, *57*
phenology 162, 165, 176
pheromones 51
phylogenetic constraint 178
phylogenetic hypotheses 39
phylogenetic stem primate group 203
phylogeny 5–6, 10; and adaptation 24; apes 19, 119; cercopithecoid 115; convergent evolution of adaptations 178–180, **179**; galagos 15; lemurs 14, 49; lorises (*see* lorises/lorids/lorisids); New World monkeys 17; Old World monkeys 18; platyrrhines 76; primate communities 178–180; primate fossil record 202; reconstruction 37–38; tarsiers 16, 151
Pithecia see saki monkey (*Pithecia*)
pitheciid 87–89, *89*
placentals 35
platyrrhines (Platyrrhini) 4, 23, 69; adaptation 81–99; as anthropoids 72–75; and catarrhines 75, 75–76; communication 79–80; communities 81–99; conservation status of 99–101, *100*; diversity 81–99; ecology 81–99, *83*; encephalization 81; families and subfamilies 77; features of 71–72; fossil record 211–213; late Eocene/early Oligocene 208–209; parallel evolution in 187–188; parenting styles 81; profile characteristics 76–81; reproductive pattern 81; sexual monomorphism and dimorphism 79; skulls of 73, *74*; social systems and mating 81; traits of **111**; vocal communication 79–80; *see also* New World monkeys

Pleistocene Epoch, fossil record 208, 215–217
plesiadapiforms 203, 205, 206
Plesiopithecus 207
Pliocene Epoch, fossil record 216–217
pliopithecids 216
plural breeders 193–194
polyspecific associations 122, 137
postcranial characteristics, catarrhines 112
postorbital bar 73, *74*
postorbital mosaic 76
posture 11; arboreal frugivores 30–31; of cheirogaleids 55; variations in 184
pottos 106–107, *107*
predation 188
predatory omnivorous frugivores 90–93
preferred foods 91, 185
prehensile 31
prehensile tail 77
primary olfactory bulb 47
primary sexual characteristics 50, 131, 190
primates: biological diversity of 8–9; conservation status of 24, 26; in crisis 221–223; early rise of modern 210–216; endangered species and subspecies of 25; grasping feet and hands of 35, 36; living (*see* living primates); mitigation strategies **222**; *vs.* other mammals 35–37; seven groups 2–5, *3*, **4**; species (*see* species of primate); threats to survival **222**; without tails 186
Primate Specialist Group of the IUCN Species Survival Commission 2, 24
primatologists 1, 4
primatology 1–2
primitive characteristic 38
primitive-to-derived transformation 38
Proconsul 214
Projeto Muriqui de Caratinga 100
pronograde 117, 186
Propliopithecus 210, *211*, 213–216
proprioception 37
proto-anthropoids 208
proto-apes 210
proto-primates 202–203, 206
Purgatorius 203
pygmy marmoset 82, 98–99, 182, *193*; body mass 76; diet and locomotion 93–95; dry season niche partitioning 96–97; parenting styles 81; social systems and mating 95–96

quadrumanous 108; arboreal locomotion 163; climbers 176; locomotion 119, 124, 184, 216; quadrupedalism 11, 55, 66, 205; arboreal 184; cercopithecoid profile 115; climbing 216; climbing-and-clambering style 85; knuckle-walking 136, 138; quadrumanous 157; woolly monkeys 82

rainforests: African primates 104, *105*; Asian primates 145, *145*; biome 170–172; closed-canopy 145, 204; ecosystem 28, 145, 172, 177, 202, 204; Madagascar lemurs 42, 51; South America primates 70, *70*, 97–98; *see also* tropical rainforests
reciprocal allogrooming 189; *see also* grooming
reintroduction 223
reproductive pattern 12–13; apes 19, 121; cercopithecoid 117; galagos 15; lemurs 14, 51; lorisoids 15, 108; mammals 35; New World monkeys 17, 81; Old World monkeys 18; tarsiers 16, 152
reserve populations 26
retina 46, 72; daytime 150; haplorhine 183; photoreceptor cells 182
rhinarium 47, *47*
ring-tailed lemur *see Lemur catta* (ring-tailed lemur)
Rooneyia 208
Roos, C. 221
Rovero, F. 221
ruffed lemur (*Varecia*) 59
Rusinga Island 140
Rylands, A. B. 221

sagittal crest 120
Saguinus sp. 97; *S. fuscicollis* 96
Saharagalago 207
Saimiri see squirrel monkeys (*Saimiri*)
saki monkey (*Pithecia*) 77, 87–89; Gray's bald-faced *18*; white-faced 79, *80*
Sapajus macrocephalus (tufted capuchin) 96
scent glands 50–51
scent-marking 61
secondary compounds 58
secondary sexual characteristics 87, 131, 133
sectorial tooth 111
seed dispersal 169, 204
seedeater 77, 87–90
selective pressure 5–6
self-grooming 189–190; *see also* grooming
selfish herd effect 130, 137
semi-solitary 12, 56, 188; breeders 192–194; social organizations *195*
semi-terrestrial 133–138; frugivore-folivores 133–140
Semnopithecus entellus (Hanuman langur) 160

sensory modality 10–11; apes 19; arboreal frugivores 31–32; galagos 15; lemurs 14; lorises 14; New World monkeys 17; Old World monkeys 18; tarsiers 16
serially monogamous 155
Setchell, J. C. 221
sexual monomorphism/dimorphism *190*, 190–191; apes 120–121; cercopithecoids 116; lemurs 50–51; lorisoids 108; New World monkeys 79; tarsiers 152
sexual selection 190
sexual swelling 129, 131
Shanee, N. 221
Shanee, S. 221
shape: apes 119; cercopithecoids 115; lorisoids 107–108; tarsiers 151
sifakas 50, 61, 62, 185, *185*
silverbacks 120, 138
singular breeders 192–193
singular cooperative breeders 193, *193*
sister-groups 21
Sivapithecus 216
size 10; arboreal frugivores 32; cercopithecids 126–129; primate communities 180–181
skulls: Goeldi's monkey and macaque 73, *74*; lemur and anthropoid *74*; male gorilla *120*; male howler monkey 191, *192*; male spider monkey *192*; platyrrhine and catarrhine 75, *75*
slow loris *15*; in Asia 155–157; mimicry 148, *149*
snow monkey (*Macaca fuscata*) 9, 146
snub-nosed monkey 160
social behavior 188–189; allogrooming 189–190; long-distance vocalizations 191, *192*; morphology 189–192; self-grooming 189–190; sexual dimorphism and monomorphism *190*, 190–191; tail-twining 191
social dispersal **196**, 197
social dominance 60, 85, 132
social grooming *see* allogrooming
socialization and learning 189
social organizations, convergent evolution 194–195, *195*, **196**
social systems 11–12; apes 19, 121; arboreal frugivores 32; atelids 85–86; callitrichines 95–96; cathemeral lemurs 59; cebines 92–93; cercopithecids 116–117, 129–130, 158–159; cheirogaleids 56; chimpanzees 137–138; convergent 194; galagos 15; gorillas 139–140; hylobatids 159; lemurs 14, 51; long-faced papionins 133; lorisoids 15, 108, 125; New World monkeys 17, 81; Old World monkeys 18; orangutans 163; pitheciid 88–89; slow loris 157; tarsiers 16, 152
song 154
South American primates: geography and climate 69–71, *70*; howler monkey 34, *34*; New World monkeys (*see* New World monkeys)
Southeast Asia *see* Asian primates
speciation 3
species of primate 2–5; as communities in nature 6–7; endangered 24, *25*; giants and dwarfs **181**, 181–182
spider monkeys 11, *78*, *82*, *87*, 186; conservation status of 100; convergent evolution of 194, **196**; diet and locomotion 85–86; locomotion 77–78; parallel evolution 187–188; social organizations and breeding systems 194–195, *195*, **196**
sportive lemur (*Lepilemur*) 53; dental formula 62; dentition and digestive tract 55; locomotion 55
spot foods 56
squirrel monkeys (*Saimiri*) 91–93; dry season niche partitioning 96–97; encephalization 81; fossils 212, *213*; parallel evolution 187–188; sexual dimorphism 191; social systems and mating 81, 91
stasis, evolutionary 213
stem group, phylogenetic 203
stem primates 203, 206; anthropoids 206; catarrhines, in Africa 210
strepsirhines (strepsirrhines) 21, 22–23, 24, 70–71; adapiforms and fossil tarsiiforms 205–206; from Asia 207–208; behavior 46; eye 46, *46*; features of 45–47; olfactory bulb 47, *48*; toothcombed 206–207
subfossils 39; gibbons 215; lemurs 65–66, 182, 217, *218*; platyrrhine primates 99, *218*
submissive behavior, gelada *116*
subspecies of primates 24, *25*
Sussman, R. W. 221
sympatry 6
symplesiomorphy 38
synapomorphy 38

Tai Forest primates 106, 124, *127*; cercopithecids 126–130; ecological characteristics of **123**, *128*; lorisoids 124–125
tails: grasping 187–188; prehensile 77; prehensile-tailed howler monkey 185; primates without 186; twining behavior 89, 191

tail-twining behavior 89, 191
talapoin monkey 115, 122
tamarins 77, 93–97, 94
tannins 176
tapetum lucidum 46, 46
tap-scanning 65
tarsiers 4, 17, 72, 182–183; activity rhythm 16, 151; in Asia 149–153, 151; body mass and shape 151; communication 16, 152; craniodental morphology 152; cranium 152; diet 16, 151; distinguishing factors 16; ecological domain 151; geographical distribution 3, 3–4, 4, 16; locomotion 16, 151–152; mating 152; nocturnal haplorhines 72; parenting styles 16, 152; pelage 152; phylogeny 16, 151; reproductive pattern 16, 152; sensory modality 16; sexual monomorphism 152; social systems 16, 152; vertical-clinging-and-leaping 150
tarsiiforms 205–208
Tarsius pumilus 182
taxonomy 5–7
taxon/taxa 5
teeth: dental formula 62; fossil 201–202
terrestrial baboon: African 28–29, 112; skeletons 118; see also baboon
terrestrial primates 2, 173, 178
territory 56
theory 1
Theropithecus (gelada) see gelada (*Theropithecus*)
titi monkey (*Callicebus*) 77, 87–89; pair-bonded 191, 197; pair-breeders 193; parallel evolution 188; parenting styles 81; social organizations and breeding systems 194–195, 195, 196; tail-twining behavior 191
tool use behaviors 31, 135–136, 162
toothcomb 47, 47, 48
tooth crown 202
torpor 53
trace fossils 39
Transatlantic Scenario 209
tree canopy 34, 54, 55, 174
tree-gougers 183
Tremacebus 212

trichromatic vision 84, 115
tropical rainforests: Africa 175, 175; Amazonia 172–174; Madagascar 172; Southeast Asia 176–177, 177 (see also Asian primates); structural profiles of 171; see also rainforests
Tropics of Cancer and Capricorn 7
tufted capuchin (*Sapajus macrocephalus*) 96

Ucayalipithecus 208, 209
understory 11, 34, 64, 88, 90, 93, 104, 153, 171, 172–174; Campbell's monkey 126; dim-lit 122, 176; dwarf galago forages 124; locomotor behavior 182; *Tarsius* 153
unflanged males 162
unimale-multifemale groups 12, 194, 195, 197
United Nations Educational, Scientific and Cultural Organization (UNESCO) Biosphere Reserve 82

Varecia (quadrupedal ruffed lemur) 59
vascular plants 170
vertical-clinging-and-leaping (VCL) 11, 88, 106, 124; indriids 61–62; locomotion 49–50, 55, 168, 180, 182; platyrrhines 78, 184; tarsiers 150
Victoriapithecus 215
vocal communication: howler monkey 79, 191, 192; long-distance 191, 192; New World monkeys 79–80; patterns 21
vomeronasal organ 47

well-being 189
wet-nosed primates see strepsirhines (strepsirrhines)
white-faced saki monkey 79, 80
Wich, S. 221
woolly lemurs 61–62
woolly monkeys: diet and locomotion 84–85; rapid quadrupedalism in 82; social systems and mating 86

Xenothrix 217

Y-5 crown pattern 120
Yuanmoupithecus 215